LAKE POLLUTION RESEARCH PROGRESS

Lake Pollution Research Progress

Franko R. Miranda
and
Luc M. Bernard
Editors

Nova Science Publishers, Inc.
New York

For permission to use material from this book please contact us:
Telephone 631-231-7269; Fax 631-231-8175
Web Site: http://www.novapublishers.com

LIBRARY OF CONGRESS CATALOGING-IN-PUBLICATION DATA

Lake pollution research progress / Franko R. Miranda, Luc M. Bernard (editor).
p. cm.
ISBN 978-1-60692-106-7 (hardcover)
1. Water--Pollution--Research. 2. Lakes--Environmental conditions--Research. I. Miranda, Franko R. II. Bernard, Luc M.
TD424.5.L35 2009
628.1'68091692--dc22
2008036568

Published by Nova Science Publishers, Inc. ✝ New York

CONTENTS

PREFACE

Water pollution is the contamination of water bodies such as lakes, rivers, oceans, and groundwater caused by human activities, which can be harmful to organisms and plants which live in these water bodies. Although natural phenomena such as volcanoes, algae blooms, storms, and earthquakes also cause major changes in water quality and the ecological status of water, water is typically referred to as polluted when it impaired by anthropogenic contaminants and either does not support a human use (like serving as drinking water) or undergoes a marked shift in its ability to support its constituent biotic communities. Water pollution has many causes and characteristics. This new book concentrates on lake pollution.

Expert Commentary - Albano, Bolsena and Bracciano are the most important lakes in Central Italy; the relevance and the potential vulnerability of these lakes is enhanced by their location in a populous area, with a high water demand for agriculture and other public uses. The waters of Lake Bracciano are already utilized for drinking supply to the city of Rome.

The aim of this paper is to update the information on the water chemistry of these lakes, on the basis of samplings carried out by the authors; moreover experimental data are compared with similar analyses available from the literature.

Besides the mass hydraulic balance of the lake system, the whole volcanic basin was considered and data related to the period 2000-2005 were also highlighted.

Chapter 1 - The main goal of this chapter is to present the role of the environmetric approaches in quality assessment of lake water as tools for classification, modeling and interpretation of large data sets from lake pollution monitoring procedures.

In the first part of the chapter the major multivariate statistical methods used for environmental data mining will be briefly described, namely cluster analysis, principal components analysis, principal components regression (as source apportioning approach), N-way principal components analysis, and self – organizing maps (SOM) of Kohonen.

The information power of the environmetrics as a tool for lake pollution research will be then demonstrated on three case studies:

Case study 1: Assessment of the pollution of high – mountain lakes in Bulgaria (Rila and Pirin mountain lakes) using intelligent data analysis

Case study 2: Lake water quality assessment of a lake system in Northern Greece with identification of pollution sources by exploratory data analysis

Case study 3: Evaluation of the quality of water sources for human consumption from the vicinity of Athens, Greece by various environmetric methods for data classification and data modeling.

The lake water data treatment by various multivariate statistical methods makes it possible to reveal similarities and dissimilarities between sampling objects and between chemical and physicochemical parameters describing the lake water quality, to find structures in the monitoring data sets, to identify sources of pollution and to quantitatively determine the contribution of each identified source to the formation of the total species concentration. Thus, environmental problems could be easier solved and the decisions are based on solid information. Additionally, the compressed information delivered by environmetric tools enables the optimization of the monitoring procedures.

Chapter 2 - Karst environments are typically characterized by scarce presence of water at the surface: after a generally short runoff, water infiltrates underground through swallow holes and discontinuities in the rock mass to develop the subterranean complex karst systems made of variable size conduits and caves. However, locally the presence of thick residual deposits, prevailingly consisting of clays, may determine stagnancy of water at the surface, and formation of karst lakes. In Apulia, that is one of the most important regions in the Mediterranean as regards karst because of the extensive outcropping of soluble rocks, several areas show such peculiarities, that led to development of karst lakes. These landforms had a remarkable role in man's history in Apulia, since many ancient settlements were established nearby the lakes, due to availability of water. Even during more recent times, exploitation of the hydric resources contained therein resulted of extreme importance, as testified by the many dry-stone wells built within the lakes. The present article is intended to describe the geological and morphological reasons which controlled the lakes formation, the historical and social aspects related to these karst landforms, and the degradation they have been experiencing in recent years. As a consequence of the latter, at the present the lakes do not have great importance as water supplies but, nevertheless, represent habitats of great naturalistic value that are still able to support the ecological functionality and the wet environments with self-vegetation.

Chapter 3 - *Cylindrospermopsis raciborskii* is a commonly encountered cyanobacterium (blue-green algae) found in water bodies worldwide, particularly tropical and subtropical lakes and reservoirs. The changing global climate, increasing eutrophication from agricultural runoff and other factors are contributing to an increased occurrence of this alga in lakes and reservoirs worldwide, with blooms now also reported in temperate areas. Blooms pose a serious concern for water managers, since *C. raciborskii* growth is often accompanied by production of the toxin, cylindrospermopsin (CYN). This toxin clearly represents a human health concern but also targets a range of aquatic plants and animals. In recent years, research into the potential risks associated with CYN contamination of human drinking water supplies has grown considerably. However, by comparison, few studies have examined the potential for, and the influences on, ecological (environmental) effects associated with toxin-producing blooms. Consequently, whilst the World Health Organisation is working towards a provisional guideline for CYN in drinking water, the environmental management of toxic blooms remains underdeveloped: ecological risks appear to be poorly recognised, evaluated or minimised. This chapter will review the current research available on the toxicity and bioaccumulation of toxin from *C. raciborskii*, with a focus on managing the environmental toxicity of blooms. Aspects including algal growth dynamics, toxin production, modes of exposure in the aquatic environment, influences on uptake routes and potential for toxic metabolism and bioaccumulation in target organisms are discussed. Furthermore, to address the gap in environmental research concerning CYN, a summary of recent, laboratory-based ecotoxicity studies is given. Finally, a model to calculate the ecological risks associated with *Cylindrospermopsis* blooms is provided: this involves a consideration of cell concentrations, toxin concentrations and the relative proportions of cell-bound and aqueous toxin. This chapter is designed to raise awareness of the potential ecological effects associated with *C. raciborskii* in lakes and reservoirs worldwide, and to stimulate further work towards the development of ecologically relevant guidelines for toxic blooms in affected water bodies.

Chapter 4 - Fresh surface water accounts for just 1/10,000 of the total water available on the planet, and lakes contain almost all of the fresh surface water. Once contaminated, it is difficult to restore fresh-water quality: lakes therefore require special protection from contamination. This chapter aims to (i) understand the basic relation between lakes and pollution and (ii) answer a fundamental question: "How can we manage lakes effectively and properly?"

Models allow administrators to predict changes within the natural lake system resulting from management actions; however, extensive data input and a large number of parameters are required to simulate complex lake ecosystems. Heat loss from Arctic/subsractic lakes tends to be rapid in late summer and results in complete mixing of the water. Based on such subarctic shallow lakes, a case study (68°02'N and 33°11E) was carried out to investigate water quality, sediment profile and bulk deposition. Although there are various sources of heavy metals - weathering, industrial processes and so on, the presented study shows that there are prime (cautious) factors responsible for changing the lake/sediment quality. This finding suggests that the combination of some key factors with a simple model may be practical in daily lake management.

Chapter 5 - Aquatic pollutant testing using biological assays is useful for ranking the toxicity of different chemicals and other stressors, for determining acceptable concentrations in receiving systems and for elucidating cause and effect relationships in the environment. This 'ecotoxicological' testing approach supplants previous approaches that indirectly estimated toxicity using chemical and physical surrogate measurements alone.

Nevertheless, many published aquatic pollution studies are restricted to examining the effects of a single toxicant on only a single species. Moreover, laboratory-based ecotoxicity tests often intrinsically suffer from a number of limitations due to their small-scale. For example, a major criticism of single-species bioassays is their failure to integrate and link toxicants (and other associated abiotic components) with higher scales of biological and ecological complexity (predation, competition, etc.). Many researchers have suggested that single-species toxicant testing has become so widely entrenched that it has hindered the development and greater use of testing at more ecologically-relevant scales. An improvement to single-species laboratory tests are microcosm and mesocosm studies using more complex and relevant measures to aquatic biotic communities.

Nevertheless, mesocosms still do not entirely simulate the ecosystem they come from, rather they mirror its general properties. As a result, there is increasing interest in correlating pollution measures from field surveys with measures of aquatic biotic community structure to determine a toxicant's scale of effect. However, field assessments, although extremely useful in determining site-specific impacts, may be limited by lack of experimental controls, too few or poorly-positioned regional reference sites and by confounding effects from impacts unrelated to the disturbance of concern.

Chapter 6 - The Barra Bonita reservoir (S=310 km^2; $\bar{z}$=10,1 m) is an ecosystem subjected to strong anthropogenic stress. It is located in the central part of the Tietê river basin - São

Paulo State, characterized by the widespread urbanization, industrial development and intensive agricultural use of the soil. The Barra Bonita reservoir has high social-economic importance due to hydropower generation, navigation and fisheries. From a limnological point of view, it is classified as polymictic and eutrophic. The purpose of this study was to characterize the actual fish assemblage in the reservoir and, by means of the species-abundance relationship, value its organization in relation to the disturbances caused by the anthropogenic impact. For the fish sampling we chose 24 sites located in three different habitats: reservoir shoreline, mouth of tributary and centre. Samples were taken in two periods of the year: dry season (winter, August-September 2003), and rainy season (summer-February 2004). Fish sampling was standardized by using 10 gillnets with mesh sizes ranging from 3 to 12 cm between opposite knots and funnel traps. At each point we also measured some morphological, physical-chemical and environmental variables. Multivariate analyses (three way-anova and ancova) were employed to, respectively, point out space and temporal variations in the Catch per Unit Effort in weight ($CPUE_W$) and to detect the importance of morphological, physical-chemical and environmental variables on Shannon species diversity (H' in number and weight). A total of 35 species, belonging to 14 families and 4 orders, were caught. Fish assemblage is composed of small body-sized species, with wide feed flexibility and high reproductive compensation. The superposition of the biological cycles of the fishes with the hydrological management of the reservoir, suggests that only those with multiple spawn and/or parental care have success. The $CPUE_W$ distribution shows significant statistical differences between seasons, zones and habitat, indicating the presence of transversal migration from the centre to the shoreline and mouth of the tributary habitats, especially in the rainy season. The variables that influenced the diversity of the fish assemblage were: depth, transparency and landscape for H'_N and H'_W; temperature (H'_N); conductibility and macrophytes (H'_W). The correlation of diversity with the depth was negative, showing that the tributary mouths and the shoreline habitats of the reservoir are the most explored by the ichthyofauna. The environmental variables selected (surrounding landscape and macrophytes beds) act respectively as external feeding support and shelter/nursery habitat, revealing the essential importance of the structural environmental complexity for the diversity maintenance in reservoirs.

Chapter 7 - Basing on the results of studies in the industrially developed Arctic region - Russian Kola - there are discussed the features anthropogenically-induced processes in lakes under pollution impacts: distribution and fate of metals, features of acidification under influence of local and trans-boundary acid deposition, eutrophication under domestic sewage pollution. Main attention is paid to pollution from copper-nickel smelters and mining industry. The estimation of metal accumulation in lake sediments in a historical retrospective is given. The influence of a combination of pollutants on water quality and aquatic ecosystems is discussed. Critical levels of an integrated toxicity index for Arctic waters, the excess of which creates a risk of fish pathology, are indicated.

Chapter 8 - The USA Environmental Protection Agency (EPA) was established in 1970 to consolidate in one agency a variety of federal research, monitoring, standard-setting and enforcement activities to ensure environmental protection. The EPA's mission is to protect human health and to safeguard the natural environment—air, water, and land—upon which life depends. Since the establishment of the EPA scientific advancements have been made in the determination of health risks from human exposure to pollution. Methodology has been developed that can provide an educated and quantitative guess of the possible effect of pollution on human health.

Chapter 9 - The validity of using deuterium excess (*d*) as a tracer of water mean residence times (MRTs) was tested for a watershed having thick soil and weathered gneiss layers (Tsukuba Experimental Watershed). The MRTs of stream and spring water were estimated to be approximately 1 to 3 years. Transit-time distributions estimated for stream and spring water indicated almost no contribution from rain that fell within the previous 90 days, which can be explained by the time needed for rainwater to permeate the thick soil and weathered gneiss layers and reach the groundwater table. A dispersion model and exponential piston flow model estimated the MRTs of stream water for the whole watershed to be 1.3 years and 2.3 years, respectively, corresponding to respective mobile water storage volumes of 839 and 1454 mm (i.e., $V_m = \tau Q$, where τ and Q are the MRT and annual discharge, respectively). These values are within the limits for the water storage capacities of brown forest soil and heavily weathered gneiss layers (2148 mm).

Chapter 10 - Internal loading can account for a significant percentage of phosphorus entering a lake, and may prevent the recovery of lake water quality even after external loads are reduced. Prior studies in two west Michigan coastal lakes showed that internal phosphorus loading can account for more than 80% of the total phosphorus load, depending on the time of year. However, these studies occurred on lakes with dense shoreline development and in rapidly urbanizing watersheds. To provide a contrast, the authors conducted an analysis of internal phosphorus loading in White Lake, MI, which has a less-developed shoreline and a watershed with much lower urban/developed land cover, and greater agricultural land use, than lakes they previously studied.

Sediment cores were removed from 4 sites in White Lake and incubated in the laboratory under aerobic (with oxygen) and anaerobic (without oxygen) conditions. Phosphorus flux from the sediments into the overlaying water column was measured over a 27-day period and compared to rates measured from sediment cores collected previously from Mona Lake and Spring Lake, MI. Internal total phosphorus (TP) loading from White Lake sediments ranged from 1.55 to 7.78 mg TP/m^2/d in anaerobic conditions and from -0.18 to 0.14 mg/m^2/d in aerobic conditions. The negative value suggests that the sediments in some areas of White Lake could act as a sink for TP during aerobic periods. The anaerobic phosphorus release rates were approximately ½ the rates measured during summer months in Mona Lake and ¼ those measured in Spring Lake in previous years.

Internal loading contributed 1.24 tons of total phosphorus based on our laboratory study, with about half coming from the eastern-most basin in White Lake. Compared to an estimated external total phosphorus load of 15.48 tons/yr, internal loading of TP accounted for ~7.4% of the total TP load entering White Lake. This percentage is much lower than what has been measured in Spring Lake (~55-67% of TP load) and Mona Lake (~9-82% of TP load), indicating internal phosphorus loading is less important in White Lake than in the other lakes. These data suggest that management strategies should be currently focused on reducing external phosphorus loads to White Lake.

Chapter 11 - Because of arsenic contamination in shallow groundwater, characterization of deeper aquifers and assess their hydraulic connectivity is now an important issue in Bangladesh. To determine the hydraulic characteristics of aquifers and development potential of deep aquifer for sustainable long-term use, study was undertaken by assessing water levels of different aquifers and conducting pumping test in deep aquifer under Meghna floodplain area of southeastern Bangladesh. Study shows that groundwater pumping for irrigation and other uses cause large seasonal water level fluctuations that is between 2 and 4.5m, 6.5 and 11m and 6.5m in the shallow, main and deep aquifer formations, respectively. The trend of groundwater level fluctuations supports the hydraulic connectivity of these aquifers. Aquitards separating aquifers are not continuous regionally. This implies that uncontrolled development of deeper aquifers may cause leakage of arsenic from contaminated shallow depths to aquifers below. Water levels dropping below sea level for over withdrawal may eventually cause saline water intrusion as well. However, during the 98.5 hours constant-discharge pumping test for deep aquifer, water levels in observation wells open to the shallow and main aquifers showed no noticeable effect from pumping in the deep aquifer i.e. under conditions of moderate groundwater use for public supply, arsenic-rich groundwater in the shallow aquifer are not likely to be drawn into the deep aquifer. The transmissivity values of the aquifer is generally favorable for groundwater development and ranged from about 1,070 m^2/day using Jacob's straight line method to 2,948 m^2/day using Chow's method at a distance of 44 m from the pumped well. Transmissivity ranged between 1,570 m^2/day using Jacob's straight line method and 2,956 m^2/day using Hantush-Jacob solution at a distance of 120m. Transmissivity was calculated as 2,385 m^2/day using recovery data. Estimated storage coefficient values ranged between 0.0000375 and 0.00268, indicates that the aquifer is confined to leaky-confined or semi-confined in nature.

Chapter 12 - This paper presents the approach used to determine the impacts of metal contamination from the Aldermac abandoned mine to Dasserat Lake. Rather than relying on a single risk quotient approach, the authors built a weight of evidence assessment of impacts at Dasserat Lake. As lines of evidence of impacts, they used regional surface water pH, water quality criteria, acute biotic ligand models, predicted no effect concentrations for three groups of species (i.e. algae, invertebrates and fish), transplanted bivalve experiments, indigenous bivalve population survey, invertebrate community surveys, fish condition indicators and finally, fish demographic information. pH levels were acidic in Lake and Bay Arnoux while close to neutrality in Dasserat Lake. Cadmium levels in most parts of Dasserat Lake were above the Canadian water quality guideline for the protection of aquatic life. Correspondingly, transplanted bivalves did not survive in Dasserat Lake 1 km downstream of Arnoux Bay and the natural population of bivalves was confined to the north-west and south east portion of Dasserat Lake. All these lines of evidence indicated that along with Arnoux Lake, who is currently completely acidified, almost 75% of Dasserat Lake was negatively impacted by drainage from the Aldermac abandoned mine site. These results were used to evaluate if the current risks at Dasserat Lake were unreasonable.

In: Lake Pollution Research Progress
Editors: F. R. Miranda and L. M. Bernard

ISBN: 978-1-60692-106-7

Short Communication

AN UPDATED REPORT ON THE WATER CHEMISTRY OF THE LAKES OF CENTRAL ITALY

Franco Medici and Gilberto Rinaldi
Department of Chemical Engineering Materials and Environment, "La Sapienza" University of Roma, Via Eudossiana – 00184 Roma (Italy)

ABSTRACT

Albano, Bolsena and Bracciano are the most important lakes in Central Italy; the relevance and the potential vulnerability of these lakes is enhanced by their location in a populous area, with a high water demand for agriculture and other public uses. The waters of Lake Bracciano are already utilized for drinking supply to the city of Rome.

The aim of this paper is to update the information on the water chemistry of these lakes, on the basis of samplings carried out by the authors; moreover experimental data are compared with similar analyses available from the literature.

Besides the mass hydraulic balance of the lake system, the whole volcanic basin was considered and data related to the period 2000-2005 were also highlighted.

INTRODUCTION

The lake system of Central Italy, composed of five medium-sized lakes (Trasimeno, Bolsena, Bracciano, Vico and Albano) with a total water volume of 15.6 km^3, is the second in importance in Italy after the Alpine Lake district in the region of Lombardia (Mosello, 2004). Moreover, in terms of water volume, it collects approximately 11% of the Italian hydro lake resource, (the total volume of collected water being 150 km^3). The larger lakes of Central Italy are reported in table 1 with their main characteristics; a number of smaller lakes (area < 4 km^2), also located in the same area, are reported in table 2.

Table 1. Morphological characteristics of the main lakes (tw = theoretical water renewal time)

	Level (m)	Watershed area (km^2)	Lake area (km^2)	Volume (10^6 m^3)	Mean depth (m)	Max depth (m)	tw (y)
Trasimeno	259	383	128.7	590	4	6	24.4
Bolsena	305	159	113.6	9200	81	151	121
Bracciano	164	90	57	5050	89	165	137
Vico	510	28	12.1	261	21.6	48.5	17
Albano	293	16	5.9	464	77	175	48

Table 2. Morphological characteristics of the smaller lakes

	Level (m)	Watershed area (km^2)	Lake area (km^2)	Volume (10^6 m^3)	Max depth (m)
Chiusi	251	146	3.87	12.9	6
Martignano	207	6.2	2.49	71	60
Montepulciano	249	119	1.88	2.2	3
Nemi	318	11	1.67	32.5	33
Ripasottile	372	46	1.05	4	8
Canterno	538	67	0.65	13.6	30
Monterosi	237	0.8	0.32	2.2	8

The most important lakes are: Bolsena, Bracciano and Albano, which represent 59%, 32% and 3% (9.2, 5.1 and 0.46 km^3) of the area's collected water, respectively. Lake Trasimeno and Lake Vico, although significant, can be nevertheless considered less important.

According to an approved national classification (Tonolli, 2001), these lakes are classified as "regional", having been created by natural events which shaped the region and gave it a distinctive and prominent character. Of the Lazio lakes, those of volcanic origin are of particular interest, especially Bracciano, Vico and Albano. These are crater lakes, in that their water lies inside a primordial volcanic crater; whereas Lake Bolsena, similar in origin to Crater Lake in Oregon (USA), is located in a volcanic depression caused by the sinking of the centre of the volcanic cone, and is thus classified as a caldera volcanic lake. Lake Trasimeno, in the region of Umbria, differs from the Lazio lakes, being of tectonic origin.

The main lakes, (Bolsena, Bracciano and Albano), apart from being a fundamental source of drinking water and of irrigation, are also considered important for nature and tourism. The Lakes of Albano and Bracciano are public sites of natural and historic interest, and together with Lake Vico, they are part of three regional natural parks.

From the geographical and hydrografic point of view, Lake Albano and Lake Bracciano (see figures 1 and 2), are part of the district of Rome, and their water constitutes the main reservoir basin for the city; whereas Lake Bolsena and Lake Vico (see figure 1), are separated from the hydrographic basin of the Tiber river. The area is of strategic importance and interest for the region of Lazio and for the city of Rome, hence the management of these water systems becomes a very complex matter; what is more, the problems are worsened by the fact that the Mediterranean area is undergoing a phase of climate aridity, with a subsequent negative influence on the hydraulic equilibrium of all lake systems. (Medici, 2005 and 2007).

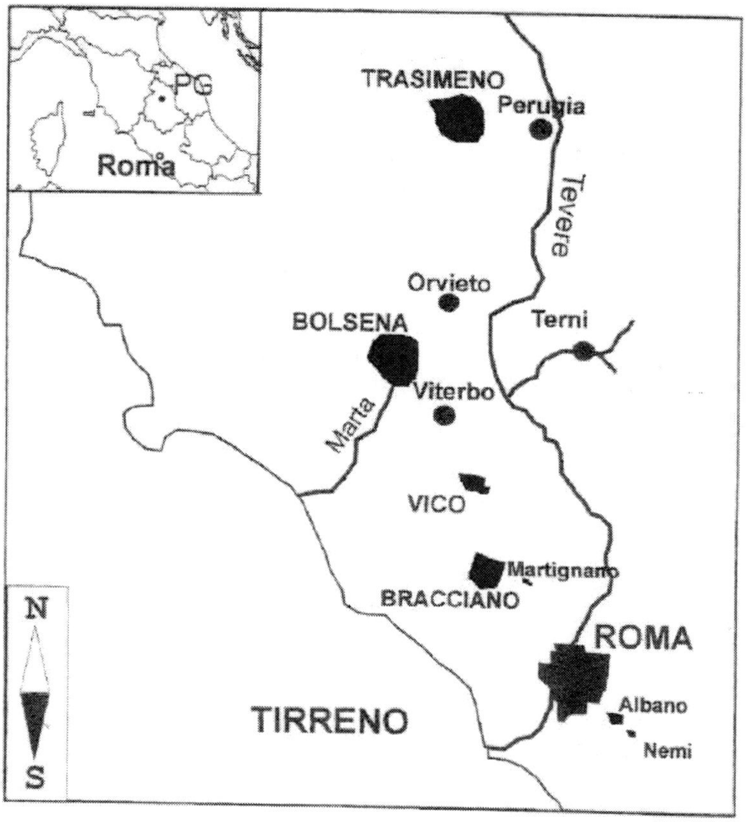

Figure 1. Main lakes of Central Italy.

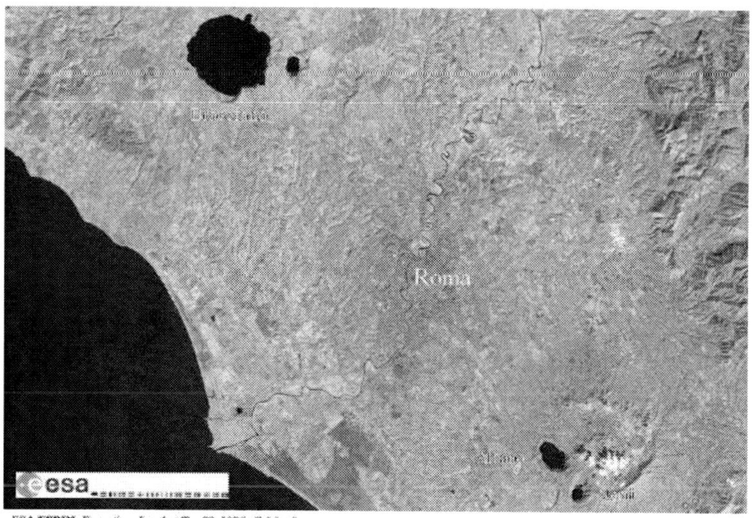

Figure 2. Lakes of the district of Rome.

History

Lake Albano is fed by underwater pollen and has an artificial outlet, excavated by the Romans in 398-397 b.C. However, since 1992 no surplus water flows from the lake.

Lake Bracciano, on the other hand, has various underground springs, for example the hot-springs of Vicarello, and it has two outlet distributaries: the Arrone River, which once carried the lake's drainage into the Tirrenian Sea, but no longer does so today, and the Paul Aqueduct (built by Pope Paul V in 1611), which still supplies some areas of Rome and the Vatican City gardens. It is important to note that of the thirteen aqueducts which supplied Ancient Rome only two collected lake water, supplying the Alsietina aqueduct (Augustus 2 b. C. with water taken from Lake Martignano), nevertheless already in that period considered undrinkable, and the Traiano aqueduct (Traiano 109. a. C.) which collected water from the foothills of the Sabatini Mountains in the area of Vicarello (Lake Bracciano), also regarded as low quality drinking water. In practice, both aqueducts supplied only water mills in the Gianicolo area of Rome, as well as the naumachia basin of Trastevere; any surplus water was used for irrigation purposes. The same waterspring sources of the Traiano aqueduct supplied the Paul aqueduct, which in fact used part of the Traiano arches to carry its low quality "Paola" water, the same which is still used today in some districts of Rome and the Vatican gardens, flowing along the same 16th century aqueduct structures. More recently, in order to ensure the supply of safe drinking water for the city of Rome, ACEA, Rome's Waterboard Authority, built an "emergency" pipeline which has a maximum carrying capacity of 8 m^3/s, with direct water withdrawal from the lake, at a depth of 50 m. An off take of this water pipeline also supplies the city of Civitavecchia. The total average removal is about 0.8 m^3/s = 25 (Mm^3/y).

Finally, Lake Bolsena is supplied by underground springs reduced by a third over recent years (Pagano, 2000). The lake's outlet is the River Marta, also reduced to a minimum vital ebb flow of 0.5 (m^3/s) = 15.768 (Mm^3/y).

The following sections take into account the most important lakes of the Lazio region (Bolsena, Bracciano and Albano). Their hydraulic balance and water quality is analyzed and commented on, followed by a section on the problem of increasing water pollution over the last two decades.

Water Balance

The analysis was carried out by applying the traditional methods of mass balance, using equations from the literature (Medici, 2007) and the meteorological data available for the region of Lazio in the period 2000 – 2005 (average monthly temperatures, humidity and precipitation). The aim is to evaluate the direct withdrawal of water in stationary conditions carried out to avoid a decrease in the hydrometric level, which makes it compatible with the mass balance. The frame of reference used is reported in figure 3

The lake's recharge consists of direct rainfall water over the lake surface (P), of surface run off coming from the drainage basin (R), of underground inflow springs (Se), and of inflow capacity (I_N). The water outflows are made up of evaporation components (Ev), of direct withdrawals from the lake area (Prel), as well as the system of underground outflows

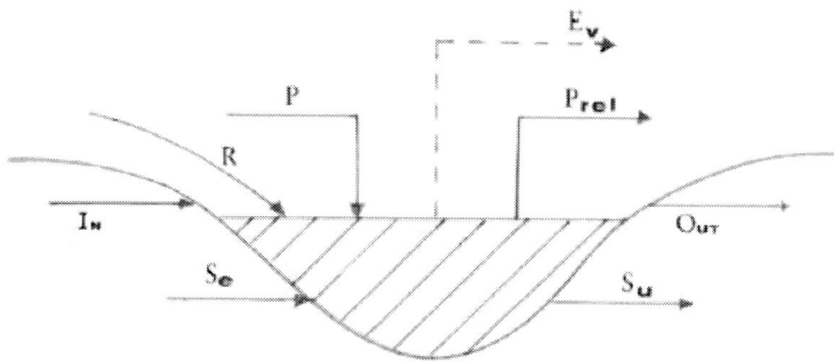

Figure 3. Conceptual scheme of a lake model.

(Su) and of the outlet flow capacity (Out). In stationary conditions, it is possible to write the following balance equation:

$$P + R + Se + I_N = Ev + Su + Prel + Out \quad (1)$$

In this case, the inflow capacity (I_N) is usually zero. In fact, the lakes of Lazio do not have inflows; hence, with the term ΔS = (Se - Su), referring to the water table recharge, the equation (1) can be written more simply as:

$$P + R + \Delta S = Ev + Prel + Out \quad (2)$$

All the parameters of the equation are deducible from annual meteorological data published by the region of Lazio or from the literature, in particular the ΔS values (water table recharge) have been taken from two specific studies (Gazzetti, 2005; Pagano, 2000). The evaporation component measurement (Ev) which is the total sum of the surface lake-water evaporation and the evapo-traspiration of the rain-collecting basin, is based on data from the literature (Medici, 2007, Dragoni, 2006). The global results of the processed data are reported in table 3.

Table 3. Hydraulic balance (data in Mm^3/y)

	Albano	Bracciano	Bolsena
P +R	7.044	86.013	128.620
ΔS	2.428	18.761	67.475
Ev	10.146	90.117	177.270
Out	-	-	15.768
Balance	- 0.674	+ 14.657	+ 3.057

The hydraulic balance results in the above table, report Lake Bracciano and Lake Bolsena as $[(P + R) + \Delta S > Ev]$: that is, the surface evaporation is lower than the total sum of the rain-water recharge and the water table inflows; whereas the equation for Lake Albano is $[(P+R) + \Delta S) < Ev]$: that is, the surface evaporation is higher than the total sum of the rain water recharge and the water table inflows.

The balance analysis, performed in stationary conditions, clearly identifies the maximum volume of water which could be drawn from the lake surface, in order to avoid the decrease of the reference levels. Water should not be taken from Lake Albano, whereas 14.7 and 3 (Mm^3/y) can be drawn from Lake Bracciano and Lake Bolsena, respectively.

In actual fact, the lakes undergo heavy water drawings for various uses (agriculture, drinking water and ornamental purposes). From Lake Bracciano in particular, ACEA draws approximately 25 (Mm^3/y) for drinking water for the city's supply (Musmeci, 2002); Lake Bolsena supplies 10 (Mm^3/y) for agricultural use (personal communication to the authors from the Province of Viterbo), and finally Lake Albano supplies approximately (1 Mm^3/y) for ornamental purposes (for the gardens of Villa Barberini, the Vatican City gardens and other private parks).

In conclusion, taking into account direct water removal, the hydraulic balance is highly negative for all three lakes; the calculated decrease, according to reference levels, is about 30, 18, 11 (cm/y) for Lake Albano, Bracciano and Bolsena, respectively. To have an idea of the decrease of the reference levels of the Lakes of Albano and Bracciano, see figures 4, 5 ,6 and 7.

Figure 4. Lake Albano (1970), circled: a pier.

Figure 5. Lake Albano (2005), circled: the same particular of the figure 5.

Figure 6. Lake Bracciano (2006): the pier of Trevignano.

Figure 7. Lake Bracciano (2006): lowering of lake' s water.

WATER QUALITY

The surface water of all three lakes under analysis, is characterized by an average ion concentration of between 5 to 6 (meq/l), with a prevalence of bicarbonate ion among the anions. Table 4 reports the average analytical results recorded for some chemical parameters, related to the samples of surface water taken from different points and concerning recent studies (2004-2007), together with recent data obtained for Lake Nemi.

Table 4. Analytical results (surface samples)

	pH (-)	Conductivity (μS)	Total alkalinity (meq/l)	Total hardness (meq/l)	Number of samples
Albano (Medici, 2004)	7.8	380	4.7	2.9	16
Bracciano (Catalani, 2006)	7.7	390	3.8	2.1	24
Bolsena (Bruni, 2007)	8.5	480	4.1	2.3	8
Nemi (Medici, 2004)	7.6	277	3.27	2.1	8

An important study (Mosello, 2004) made it possible to identify the ion balance for the waters of the five lakes in the region of Lazio, the results of which are reported in figure 8.

As regards anions, the ion balance suggests that $HCO_3^- > (SO_4^{2-} + Cl^-)$; regarding cations, on the basis of total hardness the following occurs: $(Ca^{+2} + Mg^{+2}) > Na^+ > K^+$, for all the examined lakes.

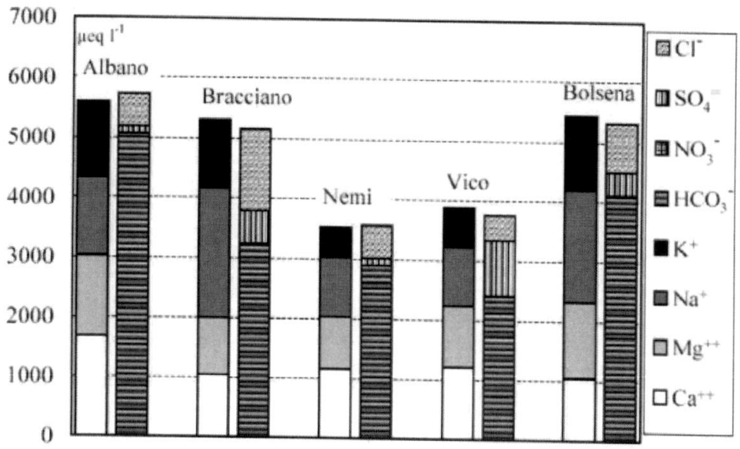

Figure 8. Ion balance of five lakes of Central Italy (Mosello, 2004).

Botrè and others (1975) in one of their studies, concluded that the water conditions at Lakes Albano and Bracciano were to be defined as discrete, in relation to the parameters set for nitrogen and phosphorus, and in relation to pollution levels from agricultural activities and urban waste. Further studies (1986-2007) analyzed Lakes Albano, Bracciano and Bolsena and a comparison of total nitrogen and phosphorus detected is reported in tables 5, 6 and 7.

Table 5. Lake Albano, total nitrogen (TN) and phosphorus (TP): a comparison

	TN (mg/l)	TP (μg/l)	Number of samples
Botrè, 1975	0.18	-	9
Pagnotta, 1986	0.21	68.4	not indicated
Pettine, 2001	0.31	24.5	4
Medici, 2004	0.87	31	32

Table 6. Lake Bracciano, total nitrogen (TN) and phosphorus (TP): a comparison

	TN (mg/l)	TP (μ/l)	Number of samples
Botrè, 1975	not detectable	not detectable	13
Pagnotta, 1986	0.08	11.5	not indicated
Ferrara, 2002	0.29	9	not indicated
Catalani, 2006	0.37	< 15	48

Table 7. Lake Bolsena, total nitrogen (TN) and phosphorus (TP): a comparison

	TN (mg/l)	TP (μ/l)	Number of samples
Pagnotta, 1986	0.11	11	not indicated
Mosello, 2004	0.21	8	2
Bruni, 2007	0.27	8	7

CONCLUSIONS

It is evident that the basins of Lakes Albano, Bracciano and Bolsena, which constitute a naturalistic and environmental heritage for Europe, are undergoing an intense exploitation of their waters.

Excluding direct removal from the lake surfaces, the state of natural hydrological equilibrium results negative for Lake Albano (- 0.674 Mm^3/y), but positive instead, for Lake Bolsena (+3.057 Mm^3/y) and Lake Bracciano (+14.657 Mm^3/y). However even if actual direct withdrawal of water is also taken into consideration, (Villa Barberini in the case of Lake Albano, ACEA at Lake Bracciano and private concerns at Lake Bolsena), it is easy to demonstrate, also theoretically, what has been amply seen in the last decade, that is, the constant lowering of the lakes' surface equal to 30-18-11 (cm/y) respectively for Lake Albano, Lake Bracciano and Lake Bolsena (see figures 4, 5, 6, and 7).

The concentration of nitrogen measured in different studies over the years demonstrates the alarming increase in the eutrophication of Lake Albano and Lake Bracciano, though less serious at Lake Bolsena. The concentration of nitrogen between 1986 and 2006 increased nearly four times at Lakes Albano and Bracciano, and about 2.5 times at Lake Bolsena.

The situation is particularly worrying at Lake Bracciano, whose water is used as drinking water in certain areas of Rome, and whose replacement time is in theory 137 years, which is longer compared to that of Lake Albano (48 years) and of Lake Bolsena (121 years).

Finally it must be remembered that the waters of Lake Bracciano, as has been noted since 1800, have always contained a certain amount of fluoride ion, recently measured at 1.41 mg/l (Catalani, 2006), very close to the limits of 1.50 mg/l set by the European Directive for Drinkable Water (98/83/EEC). In the event of "global warming", an increase in temperature could determine a growth in evaporative components which would exceed the above limits, with easily predictable consequences.

In the absence of urgent intervention, which are currently unforeseen, Central Italy could very soon find itself facing an environmental disaster. Lake Albano could be reduced to a polluted, stagnant pond while Lake Bracciano could very well dry up and in any case its "last" waters would be so fluoride enriched as to be rendered undrinkable without costly purifying treatment.

It must be remembered an emblematic intervention at the beginning of the last century (1905-1913): the construction of an aqueduct for the growing city of Los Angeles to supply water from a distance of 400 km caused intense desertification of the surrounding area and the salinization of Lake Owens (Sierra Nevada, USA).

Such scenarios, the scientific community agrees, irregardless of opinions on the causes of climate change and reasons for the intense exploitation of the water table recharge of lakes, must be taken into serious consideration by those responsible for lake management. Urgent intervention by the State is needed, given the technical and economic dimensions involved, that cannot be delegated to local public or private entities.

Acknowledgement

Authors thank Dr. Emanuele Loret (c/o Earth Observation Program Science Application Department of European Space Research Institute of Frascati – Italy) for the satellite image of figure 2.

References

Bruni P. (2007). I laghi vulcanici della provincia di Viterbo. www.bolsenaforum.it

Botrè C., Ielmini M., Sanna M., Bielli G. (1975). Contributo alla conoscenza dello stato di inquinamento dei laghi in provincia di Roma: Albano, Bracciano e Nemi. *Rassegna Chimica*, 2, 76-92.

Catalani A., Medici F., Rinaldi G. (2006).Bracciano lake waters: an experimental survey on the surface layer pollution. *Annali di Chimica,* 96, 743-749.

Dragoni W., Piscopo V., Di Matteo L., Gnucci L., Leone A., Lotti F., Melillo M., Petitta M. (2006). Risultati del progetto di ricerca PRIN "laghi 2003-05". *Giornale di Geologia Applicata*, 3, 39-46.

Ferrara O., Vagaggini D., Margaritora F. G. (2002). Zooplankton abundance and diversity in lake Bracciano, Latium, Italy. *Journal of Limnology*, 61(2), 169-175.

Gazzetti C., Loy A., Mazza R., Rossi S., Sarandrea P. (2005) Fattori morfologici, lito – pedologici e territoriali. In *Strumenti e strategie per la tutela e l' uso compatibile della risorsa idrica Lazio,* Pitagora Ed. Bologna-Italy, 95 – 103.

Medici F., Rinaldi G. (2004) Problemi di qualità delle acque dei laghi Albano e di Nemi. *Acqua e Aria*, 35(7), 32-36.

Medici F. (2005). Laghi Albano e di Nemi: carenza idrica e alterazione della qualità delle acque. *Geologia dell' Ambiente*, 13 (1), 8 -11.

Medici F. (2007). Laghi Albano e di Bracciano: bilancio idrico e valutazione dei prelievi. *Geologia dell' Ambiente,* 15 (2), 2-5.

Mosello R., Arisci S., Bruni P. (2004). Lake Bolsena (Central Italy): an updating study on its water chemistry. *Journal of Limnology*, 63(1), 1-12.

Musmeci F., Correnti A. (2002). Elementi per il bilancio idrico del lago di Bracciano. E.N.E.A. Bracciano-Italy, LIFE 02 ENV/IT/000111 Project.

Pagano G., Meneghini A., Floris S. (2000). Ground water budget of the Vulsini basin. *Geologia Tecnica and Ambientale*, 3, 24 - 35.

Pagnotta R., La Noce T., Pettine M., Puddu A. (1986). I laghi dell' Italia Centrale: classificazione trofica ed analisi dei fattori che la influenzano. In *Proceedings of Seventh A.I.O.L. Congress*, Trieste-Italy, June 1986, 385-396.

Pettine M., Tartari G. (2001). Studio sulla caratterizzazione delle acque superficiali del parco dei Castelli Romani. *Proceedings workshop Ricerca sulla qualità delle acque superficiali del parco regionale dei Castelli Romani*, Castel Gandolfo, March 2001.

Tonolli V. (2001). Introduzione allo studio della limnologia. *Edition of Istituto Italiano di Idrobiologia*, C.N.R. Pallanza-Italy, 8 - 24. 13th May 2008

In: Lake Pollution Research Progress
Editors: F. R. Miranda and L. M. Bernard
ISBN: 978-1-60692-106-7

Chapter 1

ENVIRONMETRICS AS A TOOL FOR LAKE POLLUTION ASSESSMENT

***Aleksander Astel**[*1] **and Vasil Simeonov**[2*]*

[1] Environmental Chemistry Research Unit, Biology and Environmental Protection Institute, Pomeranian Academy, 22a Arciszewskiego Str., 76-200 Słupsk, Poland;
[2] Chair of Analytical Chemistry, Faculty of Chemistry, University of Sofia "St. Kl. Okhridski", 1164 Sofia, J. Bourchier Blvd. 1, Bulgaria

ABSTRACT

The main goal of this chapter is to present the role of the environmetric approaches in quality assessment of lake water as tools for classification, modeling and interpretation of large data sets from lake pollution monitoring procedures.

In the first part of the chapter the major multivariate statistical methods used for environmental data mining will be briefly described, namely cluster analysis, principal

[*] Aleksander Astel: e–mail: astel@pap.edu.pl
[*] Corresponding author: Vasil Simeonov:e-mail:vsimeonov@chem.uni-sofia.bg

components analysis, principal components regression (as source apportioning approach), N-way principal components analysis, and self – organizing maps (SOM) of Kohonen.

The information power of the environmetrics as a tool for lake pollution research will be then demonstrated on three case studies:

- Case study 1: Assessment of the pollution of high – mountain lakes in Bulgaria (Rila and Pirin mountain lakes) using intelligent data analysis
- Case study 2: Lake water quality assessment of a lake system in Northern Greece with identification of pollution sources by exploratory data analysis
- Case study 3: Evaluation of the quality of water sources for human consumption from the vicinity of Athens, Greece by various environmetric methods for data classification and data modeling.

The lake water data treatment by various multivariate statistical methods makes it possible to reveal similarities and dissimilarities between sampling objects and between chemical and physicochemical parameters describing the lake water quality, to find structures in the monitoring data sets, to identify sources of pollution and to quantitatively determine the contribution of each identified source to the formation of the total species concentration. Thus, environmental problems could be easier solved and the decisions are based on solid information. Additionally, the compressed information delivered by environmetric tools enables the optimization of the monitoring procedures.

INTRODUCTION

The careful monitoring of natural water systems like river streams, lakes, wells and underground sources is a very responsible task. Usually, a set of chemical and physicochemical parameters reflecting the surface or underground water quality are carefully analysed and the results obtained are compared to certain threshold values in order to decide if the water quality fulfils the quality desired. The choice of quality parameters is normally standardized and described in various instructions and directives for individual countries or unions like EU [1-5]. Recently, a very specific attention requires a different type of water quality parameter called "toxicity" ("ecotoxicity") [6-8]. It is well known that biological test are already often applied to environmental samples, especially water for drinking purposes but in case of testing toxicity the aim is to determine the level of pollution in the systems of interest.

It is also important to note that lake waters are important element of the aquatic ecosystems. They constitute ecological niches (especially in combination with the lake bottom sediments) supporting not only fish but also benthic organisms, i.e. animals and plants living on the bottom of bodies of water and are a source of nutrients for aquatic organisms such as small invertebrates and protozoans. An assessment of the effect of pollution on life in lake water bodies requires also monitoring of bottom sediment samples both for chemical, physicochemical, and toxicity parameters. They are very useful material for various environmental studies because they act as sorption column and provide a clear image of all events taking place in the overlying water layer.

Very often, however, the monitoring data are considered in an "univariate" way – each parameter separately. In the reality the state of an ecosystem is depending simultaneously on many factors and parameters. Therefore, these systems are multivariate in nature. That is why the classification, modeling and interpretation of the monitoring data sets have to be performed by the use of the chemometrics and environmetrics [9-15], where the references given are only a tiny part of many environmetric studies. The specific point in the studies of lake waters is that there is lack of intelligent data analysis of the monitoring sets comprising of different water quality parameters simultaneously interpreted.

The aim of the present chapter is to demonstrate the role of environmetric classification, modeling and interpretation of monitoring data from the different lake systems in Bulgaria and Greece. The final assessment of the quality of the lake ecosystem and the possible pollution sources will make it possible to reveal various relationships between the quality parameters and the systems studied.

THEORETICAL DESCRIPTION

The modern chemometrics is a branch of chemistry (very often related to analytical chemistry) which deals with the application of mathematical and statistical methods in order to evaluate, classify, model and interpret chemical and analytical data, to optimize and model chemical and analytical processes and experiments and to extract a maximum of chemical and analytical information from experimental data. When the methods of chemometrics are applied to data sets obtained by monitoring of various environmental compartments (surface water, atmosphere, soil, sediments, biota, etc.) the term environmetrics is used to stress the information ability of the methods to gain specific information from samples of the total environment.

The most important methods of multivariate statistics employed in environmetrics are divided in unsupervised learning methods (like cluster analysis, non-linear mapping, minimal spanning tree, principal components analysis), supervised learning methods (like multivariate analysis of variance or MANOVA, multivariate discriminant analysis, K – nearest neighbor, neuron net classification etc), factorial methods (like principal components analysis, partial least squares modeling, etc.). In this section only several major environmetric methods for data classification, modeling and interpretation like cluster analysis, principal components analysis, principal components regression, N-way principal components analysis, self-organizing maps of Kohonen will be briefly presented as reliable tools for lake water pollution research.

Cluster Analysis

Cluster analysis (CA) is an exploratory data analysis tool for solving classification problems, based on unsupervised learning [16]. CA enables objects stepwise aggregation according to the similarity of their features. As a result hierarchically or non-hierarchically ordered clusters are formed. A single cluster describes a group of objects that are more similar to each other than to objects outside the group. Similarity understood in the term of CA measures how alike two cases are. While the term similarity has not unique definitions, it is common to refer to all similarity measures as "distance in multi-features space" measures since the same function is served. A similarity between two objects i and i' is a distance if:

$$(D_{i'i} = D_{ii'}) \leq 0 \ \ where \ \ D_{ii'} = 1 \ \ if \ \ x_i = x_{i'} \tag{1}$$

(where x_i and $x_{i'}$ are the row-vectors of the data table **X** with the features measurements describing objects i and i'). When two or more features are used to define their similarity, the one with the largest magnitude dominates. This is why primary standardization of features becomes necessary. There are a variety of different measures of inter-cases distances and inter-cluster similarities and distances to use as criteria when merging nearest clusters into broader groups or when considering the relation of an object to a cluster. A few most popular ways of determining how similar interval measured objects are to each other are as follows:

1. Euclidean distance – the distance between two objects x_i and $x_{i'}$ is defined by equation 2 where j presents repetition of measurements.

$$d_{x_i x_{i'}} = \sqrt{\sum_{j=1}^{J} (x_{ij} - x_{i'j})^2} \tag{2}$$

2. Squared Euclidean distance – removes the sign and places greater emphasis on objects further apart, thus increasing the effect of outliers (Eq. 3).

$$d_{(x_i, x_{i'})} = \sum_{j=1}^{J} (x_{ij} - x_{i'j})^2 \tag{3}$$

3. Manhattan distance (city-block distance, block distance) is the average absolute difference across the two or more dimensions which are used to define distance. The Manhattan distance is defined slightly differently to the Euclidean distance. Except for some specific cases when Manhattan distance is equal to Euclidean distance, it is always higher than Euclidean distance (Eq. 4).

$$d_{(x_i, x_{i'})} = \sum_{j=1}^{J} \left| x_{ij} - x_{i'j} \right| \tag{4}$$

4. Chebychev distance is the maximum absolute difference between a pair of cases on any one of the two or more dimensions (features) which are being used to define distance. Pairs will be defined as different according to their difference on a single dimension, ignoring their similarity on the remaining dimensions (Eq. 5).

$$d_{(x_i, x_{i'})} = \max \left| x_i - x_{i'} \right| \tag{5}$$

5. Mahalanobis distance takes into account that some features may be correlated and so defines roughly the same object's properties (Eq. 6) (C is the variance-covariance matrix of the features).

$$d_{ii'} = \sqrt{(x_i - x_{i'}) \cdot C^{-1} (x_i - x_{i'})'} \tag{6}$$

6. Minkowski distance should be applied if the object weight is increasing related to the dimensions in each compared objects and indicates the lowest similarity.

$$d_{(x_i, x_{i'})} = \sum_{j=1}^{J} \sqrt[r]{(x_i - x_{i'})^p} \tag{7}$$

7. Pearson correlation is based on correlation coefficient. Since for Pearson correlation, high negative as well as high positive values indicate similarity, the researchers usually select absolute values.

There are several other related distance measures (weighted Euclidean distance, standardized Euclidean distance, cosine, customized, etc.) but usually specific reasons are required if a very sophisticated distance measure is to be applied.

In case of CA one task is related with determination of similarity between measured objects, but equally important task is to define how objects or clusters are combined at each step of similarity assessment procedure. One possibility for clustering objects is their hierarchical aggregation. In this case the objects are combined according to their distances from or similarities to each other. Within hierarchical aggregation agglomerative and divisive methods can be distinguished. Divisive clustering is based on splitting the whole set of objects into individual clusters, while in case of more frequently used agglomerative clustering one starts with single objects and gradually merges them in broader groups. Usually some objects create one broader group, while rest of them creates the other. As in case of distance measure various algorithms (linkage techniques) are available to decide on the number of clusters. They result in slightly different clustering pattern. A few most popular linkage algorithms are:

1. Nearest neighbor (single linkage) – the distance between two clusters is the distance between their closest neighboring objects, in other words the similarity of the new group from all other groups is given by the highest similarity of either of the original objects to each other object (Eq. 8).

$$d_{mj} = \frac{d_{ij} + d_{i'j}}{2} - \frac{|d_{ij} - d_{i'j}|}{2} = \min(d_{ij}, d_{i'j}) \tag{8}$$

where:

m – new object or cluster,
i', i, j – clustered before objects.

This algorithm works well when the plotted clusters are elongated or chain-like, moreover the sizes of the clusters and their weight are assumed to be equal.

2. Furthest neighbor (complete linkage) – the distance between two clusters is the distance between their furthest member objects. Furthest neighbor algorithm of linkage refers only to

the calculation of similarity measures after new clusters are formed, and the two clusters (or objects) with highest similarity are always joined first (Eq. 9).

$$d_{mj} = \frac{d_{ij} + d_{i'j}}{2} - \frac{|d_{ij} - d_{i'j}|}{2} = \max(d_{ij}, d_{i'j}) \tag{9}$$

This algorithm works well when the plotted clusters form distinct clumps (not elongated chains). Application of the procedure presented above leads to well separated, small compact spherical clusters.

3. Average linkage - the distance between two clusters is the average distance between all inter-cluster pairs. There are two possible ways of calculating average linkage algorithm: non weighted (Eq. 10) and weighted (Eq. 11), according to the size of each group being compared (n). When the clusters' size is equal both algorithms give identical results.

$$d_{mj} = \frac{d_{ij} + d_{i'j}}{2} \tag{10}$$

$$d_{mj} = \frac{n_i}{n} d_{ij} + \frac{n_{i'}}{n} d_{i'j} \quad with \quad n = n_i + n_{i'} \tag{11}$$

Applying weighted average linkage algorithm no deformation of the clusters is observed. To some extent small clusters consisting of outliers might arise.

4. Ward's method is a minimum distance hierarchical method which calculates the sum of squared Euclidean distances from each case in a cluster to the mean of all variables (Eq. 12). The cluster to be merged is the one which will increase the sum the least. Thus, this method minimizes the sum of squares of any pair of clusters to be formed at a given step.

$$d_{mj} = \frac{n_i + n_j}{n + n_j} d_{ij} + \frac{n_{i'} + n_j}{n + n_j} d_{i'j} - \frac{n_j}{n + n_j} d_{ii'} \tag{12}$$

5. Centroid linkage is calculated as the average of a cluster is applied as the basis for aggregation without distorting the cluster space (Eq.13).

$$d_{mj} = \frac{n_{ij}}{n} d_{ij} + \frac{n_{i'j}}{n} d_{i'j} - \frac{n_i n_{i'}}{n} d_i d_{i'} \tag{13}$$

6. Median linkage is calculated as the median of a cluster is applied as the basis for aggregation without distorting the cluster space (Eq. 14).

$$d_{mj} = \frac{d_{ij}}{2} + \frac{d_{i'j}}{2} - \frac{d_{ii'}}{4} \tag{14}$$

An advantage of median linkage algorithm is that the importance of a small cluster is preserved after aggregation with a large one.

There are variety of additional linkage algorithms (correlation of items, binary matching, etc.), but it would be rare that a researcher needs to apply too many combination of distance and linkage measures, however comparing of many approaches may be a way of clustering pattern validation.

In hierarchical agglomerative clustering the graphical output of the analysis is usually a dendrogram – a tree-like graphics, which indicates the linkage between the clustered objects with respect to their similarity (distance measure). Decision about the number of statistically significant clusters could be made for different reasons. Often a fixed number of clusters is to be assumed. Sometimes a distance measure or an allowed difference between clusters (classes) is used for evaluating the number of significant clusters. For practical reasons the Sneath's index of cluster significance is widely used. It represents this significance on two levels of distance measure D/D_{max} relation: 1/3 D_{max} and 2/3 D_{max}. Only clusters remaining compact after breaking the linkage at these two distances are considered significant and are object of interpretation.

The algorithms for non – hierarchical clustering offer the division of the studied objects into a priori given number of clusters (determined by some practical or theoretical reasons).

In principle, the data set could be considered as a matrix consisting of rows (the objects) and columns (the features describing the objects). CA makes it possible to classify both the objects and variables. This is very important from practical point of view because in environmetric studies it is very interesting to get information on relationships between the sampling locations and between the monitoring parameters. The whole idea of risk assessment is based actually on estimation of relationship between monitoring features (variables).

Principal Components Analysis

Principal component analysis (PCA) seems to be the most widespread multivariate chemometric technique and is a typical display method (also known as eigenvector analysis, eigenvector decomposition or Karhunen-Loéve expansion). It enables revealing the "hidden" structure of the data set and helps to explain the influence of latent factors on the data distribution. PCA is done on covariance matrix when the data are centered or on correlation matrix when the data are standardized [17, 18]. PCA transforms the original data matrix into a product of two matrices, one of which contains the information about the objects (e.g. samples) and the other about the features (e.g. analyte's concentration). The matrix characterizing objects contains the scores (understood as projection) of objects on principal components (PCs). The other one, characterizing features is a square matrix and contains the set of eigenvectors (understood as weights, in PCA terminology called "loadings") of the original features in each PC. In matrix terms, this can be expressed as:

$$X = S \bullet L + E \qquad (15)$$

where:

X – is the original data matrix (features as columns, cases as rows),
S – is a scores matrix (has as many rows as the original data matrix),
L – is a loadings matrix (has as many columns as the original data matrix),
E – is an error matrix.

The number of columns in the matrix S equals the number of rows in the matrix L. It is possible to calculate scores and loadings matrices as large as desired, provided the "common" dimension is no larger than the smaller dimension of the original data matrix, and corresponds to the number of PCs that are calculated. Each scores matrix consists of a series of column vectors, and each loadings matrix a series of row vectors. Many authors use s_a and l_a notation to express these vectors, where a is the number of the PC. The matrices S and L are composed of several such vectors, one of each PC. The first scores vector and first loadings vector are often called the eigenvectors of the first PC. Each successive component is characterized by a pair of eigenvectors. Using f eigenvectors in one dimension, where f is smaller than, or equal to the rank of the data, f PCs can be obtained. Usually, a small number of PCs is needed to represent most of the information in the data. The minor PCs which explain little of the data structure can be eliminated, thus simplifying the analysis. Also, these minor PCs contain most of the random error, so eliminating them tends to remove extraneous variability from the analysis. In the ecosystems monitoring studies PCA and related multivariate techniques are often applied to determine the possible influence and contribution of natural and anthropogenic factors in data structuring.

Some important features of PCA could be summarized as follows. The principal components axes (the axes of the hidden variables) are orthogonal to each other. Most of the variance of the data is contained in the first principal component. In the second component there is more information than in the third one etc. For interpretation of the projected data both the score and the loading vectors are plotted. In the score plots, the grouping of objects can be recognized. A loading plot reveals the importance of the individual variables with respect to the principal component model.

A very important task in PCA is the estimating the number of principal components necessary for a particular PC model. Several criteria exist in determining the number of components in the PCA model:

percentage of explained variance,
eigenvalue – one criterion (Kaiser criterion),
Scree – test,
cross validation.

The percentage of explained variance is applied in sense of a heuristic criterion. It can be used if enough experience is gained by analyzing similar data sets. If all possible principal components are used in the model the variance can be explained by 100 %. Usually, a fixed percentage of explained variance is specified, e.g. 80 %. In environmental studies even 75 % of explained variance is a satisfactory measure for the adequateness of the PCA model chosen.

The eigenvalue – one criterion (Kaiser criterion) is based on the fact that the average eigenvalue of autoscaled data is just one. In this case only eigenvalues greater than 1 are considered important.

The Scree – test is based on the phenomena that the residual variance levels off when the proper number of principal components is reached. Visually the residuals or more often the eigenvalues are plotted against the number of latent factors in a Scree plot. The principal component number is then derived from the leveling-off in the plot.

The forth approach of deciding on the number of principal components uses the following idea. In the simplest case, every object of the input matrix X is removed (leave-one-out method) from the data set once and a model with the remaining data is computed. Then the removed data are predicted by the use of the PCA model and the sum of the square root of residuals over all removed objects is calculated. In case of large data sets, the leave-one-out method can be replaced by leaving out a whole group of objects.

Interpretation of the results of PCA is usually carried out by visualization of the component scores and loadings. In the score plot, the linear projection of objects is found, representing the main part of the total variance of the data (in the plot PC1 vs. PC2). Other projection plots are also available (e.g. PC1 vs. PC3 or PC2 vs. PC3) but they represent less percentage of explained total variance of the system in consideration. Correlation and importance of feature variables is to be decided from the factor loadings plots.

Principal Components Regression

In case of many studies related to natural ecosystems PCA and other multivariate statistical techniques are used to determine possible natural or anthropogenic influences in the formation of the determinants total mass. However, PCA does not provide a direct balancing and apportionment. After the pollution sources identification by the application of PCA, the next calculation step in modeling and balancing of pollution impacts is the apportioning itself. It is performed mostly by absolute principal components analysis (APCA). The procedure introduced by Thurston and Spengler [19] is well developed and often applied for apportionment purposes. As the factor scores obtained from PCA are normalized, with mean zero and standard deviation equal to unity, the true zero for each factor score is calculated by introducing an artificial sample with concentration equal to zero for all variables (Eq. 16).

$$(Z_0)_i = \frac{(0 - C_i)}{s_i} = -\frac{C_i}{s_i} \qquad (16)$$

where:

C_i – arithmetic mean concentration of analyte *i* (understood as feature),
s_i – standard deviation of variable *i*.

Finally, the source composition profiles and source contributions are estimated by multiple linear regression of the total mass concentration against the APCS (absolute principal component scores) values. According to this procedure regressing the analytes concentration data on APC gives estimation of the coefficients which convert the APCS into pollutant source mass contributions from each source for each sample. The source contributions to C_i

can be calculated by mentioned above linear regression procedure according to the following Eq. (17):

$$C_i = (b_0)_i + \sum APCS_p \cdot b_{pi}, \quad p = 1,2,...,n \tag{17}$$

where:

$(b_0)_i$ – constant term of multiple regression for variable i,
b_{pi} – the coefficient of multiple regression of the source p for variable i,
$APCS_p$ – scaled value of the rotated factor p for the considered sample,
$APCS_p \cdot b_{pi}$ represents the contribution of source p to C_i.

The mean of the product $APCS_p \cdot b_{pi}$ on all samples represents the average contribution of the sources. The method estimates source profiles and contributions but its serious disadvantage is error propagation in centering and uncentering of data. This balancing approach accepts that all sources have been identified by the principal components analysis and all of them participate in the source contribution procedure.

N-Way Principal Components Analysis

Most common chemometric techniques are performed using two-way data sets, often presented in the form of matrices. However, due to increasing interest of handling of three-way data matrices three-dimensional analogies to two-dimensional techniques are required. For example, a two-dimensional PCA has its analogue in the form of Tucker3 or PARAFAC (parallel factor analysis) models. The generality of the Tucker3 model, and the fact that it covers the PARAFAC model as a special case, had made it an often used model for decomposition, compression and interpretation in many applications. The use of multiway models provides a better insight into the data structure, reduces the noise, and shows which of the original variables are correlated and which of them are most significant for a certain environmental problem description. It should be emphasized at this point that the most typical environmental data sets have the following form of a three-dimensional (three-mode) matrix: sampling sites *x* variables *x* time, but examples in which two separate modes of the data array are formed by the time dimension are also widely known (e.g. months, years) [20-22]. Tucker3 model involves calculating weight matrices corresponding to each of the three modes and is one of the most basic-multi way models used in environmetrics. The model is defined by the decomposition of a three-way matrix X into a three-way *core matrix* Z and three two-way loading matrices A, B, C (one for each mode):

$$x_{ijk} \approx \sum_{l=1}^{L} \sum_{m=1}^{M} \sum_{n=1}^{N} a_{il} b_{jm} c_{kn} z_{lmn} + e_{ijk} \tag{18}$$

where e_{ijk} represents the residual error term (graphical example of decomposition realized by Tucker3 model is presented in figure 1).

Tucker3 algorithm delivers a set off possible solutions which mean large number combination of possible models with different complexities. To select a model with an optimal complexity, the variance of each combination of model complexities starting from the model with the lowest number of factor in each mode, to model with the highest complexity should be evaluated. Most often, the optimal model is the one with possibly the smallest number of factors in each of the modes but explaining a large part of data variance. In practice, a trade-off between both requirements is needed. At the same time, a set of possible Tucker3 models should be validated, e.g. using cross-validation procedure. The cross-validation is performed in such a way that part of the data are set to missing, the models are fitted to the remaining data, and the residuals between fitted and true left out elements are calculated.

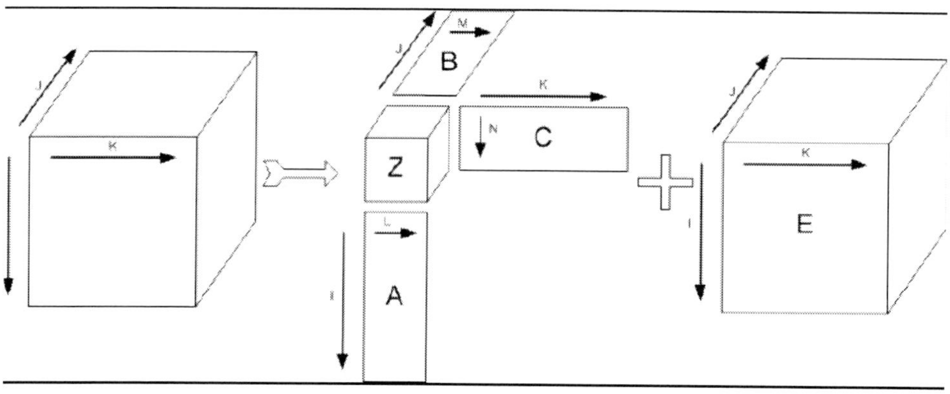

Figure 1. Tucker3 decomposition model.

On the contrary PARAFAC delivers one solution with the same number of components in every dimension. Hence the model can be defined by following equation:

$$x_{ijk} \approx \sum_{g=1}^{G} a_{ig} b_{jg} c_{kg} + e_{ijk} \tag{19}$$

where g presents the number of components.

Self-Organizing Maps

The Self-organizing map (SOM) algorithm has been proposed by Kohonen in 1980 [23]. It is a neural-network based model which shares, with the conventional ordination methods, the basic idea of displaying a high-dimensional signal manifold onto a much lower dimensional network in an orderly fashion (usually 2D space). The SOM is a competitive learning algorithm based on unsupervised learning process. This advantage causes no researcher intervention is required during the learning process and that little needs to be known about the characteristics of the input data. In the SOM algorithm, the topological relations and the number of neurons (nodes) organized on a regular low-dimensional grid are fixed from the very beginning. The number of neurons may vary from a few dozen up to several thousand. Because for SOM algorithm there are no precise rules for the choice of the various, primary defined parameters, the most common shape of the Kohonen map is a rectangular grid with the number of hexagonal nodes (n) determined using following equation:

$$n = 5 \cdot \sqrt{number\ \ of\ \ cases} \tag{20}$$

Basically, the two largest eigenvalues of the training data are calculated and the ratio between side lengths of the map grid is set to the ratio between the two maximum eigenvalues. The actual side lengths are then set so that their product is close to the determined number of map units as stated above. Hexagonal lattice is preferred because it does not favor horizontal or vertical direction. Each neutron i is represented by a d-dimensional weight vector (called also prototype vector or codebook vector) $m=[m_i,\ldots,m_d]$, where d is equal to the dimension of the input vectors. The neurons are connected to adjacent neurons by a neighborhood relation, which dictates the topology, or structure of the Kohonen map and, thus, similar objects should be mapped close together on the grid. After primary initialization, the weight vectors are characterized by random values and then SOM is trained iteratively with one of two possible algorithms: sequential or batch. A sequential training algorithm (STA) constructs the nodes in a SOM in order to represent the whole data set and their weights are optimized at each iteration step. In each step, one sample vector x from the input data set is chosen randomly and the distances between it and all the weight vectors of the SOM are calculated using some distance measure, e.g. Euclidean distance, squared Euclidean distance, Mahalanobis distance etc., thus, the optimal topology is expected. Node c (Eq. 21), whose weight vector (m) is closest to the input vector x is called the best matching unit (BMU):

$$\|x - m_c\| = \min_i \{\|x - m_i\|\} \tag{21}$$

where $\|\ \|$ is a distance measure (typically Euclidean distance). If missing data appears, they are handled by simply excluding them from the distance calculation (e.g. it is assumed that their contribution to the distance $\|x - m_i\|$ is zero). Because, the same missing value is ignored in each distance calculation (over which the minimum is taken), this is a valid solution. After finding the BMU, the weight vectors are updated in agreement with presented below update rule (Eq. 22), so that the BMU is moved closer to the input vector. The topological neighbors of the BMU are moved closer too because of their mutual connection.

$$m_i(t+1) = m_i(t) + \alpha(t)h_{ci}(t)[x(t) - m_i(t)] \tag{22}$$

where:

t – time,
m(t) – weight vector indicating the output unit's location in the space at time t,
α(t) – learning rate at time t,

h_{ci} – neighborhood non-increasing function centered in the winner unit c at the time t,
x(t) – input vector randomly drawn from the input data set at time t.

The sequential training algorithm is usually performed in two phases. In the first phase, relatively large initial learning rate $\alpha(t=0)$ and neighborhood radius σ_0 are used. In the second phase both learning rate and neighborhood radius become smaller.

The second possible training algorithm is called batch training algorithm (BTA), because instead of using a single data vector at a time, the whole data set is presented to the map before any adjustments are made. In each training step, the data set is partitioned to the Voronoi regions of the map weight vectors (e.g. each data vector belongs to the data set of the map unit to which it is closest). After that, the new weight vectors are calculated as shown in Eq. 23:

$$m_i(t+1) = \frac{\sum_{j=1}^{n} h_{ic}(t)x_j}{\sum_{j=1}^{n} h_{ic}(t)} \tag{23}$$

where $c = \arg\min_k \{\|x_j m_k\|\}$ is the index of the BMU of data sample x_j. The new weight vector is a weighted average of the data samples, where the weight of each data sample is the neighborhood function value $h_{ic}(t)$ at its BMU c. Analogous to sequential training algorithm missing values are ignored in calculating the weighted average. The quality of mapping can be quantitatively measured with the quantization error (QE) and the topographic error (TE). After competitive learning process, SOM algorithm enables graphical presentation of results in the form of set of maps (features' planes) and accomplishment of clustering tasks as well. Features' planes can be considered as a sliced version of the SOM map and provide a powerful tool to analyze the community structure. When a plenty of features is considered it is difficult to compare all maps for all features and thus becomes necessary to find similarity between them, and simultaneously, in the cases' space and classify them into clusters. Input features' planes (e.g. variables) could be visualized on a summary SOM map (called also as unified distance matrix or U-matrix) to show the contribution of each feature in the self-organization of the map. U-matrix visualizes distances between neighboring map units, and helps to identify cluster structure of the map: high values of the U-matrix indicates a cluster border, uniform areas of low values indicate clusters themselves, while each feature's plane shows the values of one feature in each map unit. In other words U-matrix expresses semi-quantitative information about the distribution of a complete set of features for a complete set of the cases while separate feature's plane visualizes distribution of a given feature for a complete set of the cases. Because of this, U-matrix joined with features' planes can be effectively applied for assessment of inter-features and inter-cases relations. According to clustering task, one of the most commonly applied algorithm is non-hierarchical K-means clustering algorithm. In this case, different values of k (predefined number of clusters) are tried and the sum of squares for each run is calculated. Finally, the best classification with the lowest Davies–Bouldin index should be chosen (it is a function of the ratio of the sum of within-cluster scatter and between-cluster separation).

Main advantages of SOM algorithm application are:

the projection of variables similarity in the form of features' planes delivers semi-quantitative information about the distribution of a given feature in the space of the cases;

SOM visualization enables presentation both similarity between positive as well as negative correlated features;

SOM visualization and SOM-supported classification is able to indicate "outliers" e.g. those features or cases which do not belong to a well-organized, homogeneous populations;

SOM is noise tolerant (this property is highly desirable when site-measured data are used).

Case Studies

The application of the multivariate statistical methods described above as one of the most important tool in assessment of lake water quality can be illustrated by specific case studies. Thus, the classification, data projection, modeling and interpretation of lake water monitoring data sets becomes understandable and turns to be a pattern to follow in lake pollution research.

Lake Pollution Control in Northern Greece

The monitoring of surface waters is an important source of information for assessment of overall pollution of the examined systems. Moreover, it is a valuable tool for better understanding of volcanic, geochemical, atmospheric and climatic processes and changes in these reservoirs [24-33]. The monitoring studies of the lakes are usually dealing with the presence of phosphorous and nitrogen, heavy metals, organic micropollutants (PAHs, PCBs, etc.) as well as the evaluation of their eutrophication status. Based on the available monitoring data, modeling studies try to predict the water quality and the available water budget aiming to the effective water management. This is not always possible since in many cases there are only few available data collected segmental and occasionally. Due to the large intra-annual variability in water chemistry, especially in low depth lakes, frequent and long term monitoring studies of the water chemistry could provide a representative and reliable estimation of the lakes. In this way the statistical treatment of the data becomes possible and valuable. Multivariate statistical modeling is a very favorable approach for classification and interpretation of the available results. These statistical approaches have been used in deep and high altitude lakes as the citations are related mostly to classical univariate statistics [34-43].

In the case study an environmetric assessment of water quality of five small to moderate sized lakes located in Northern Greece was attempted. Several multivariate statistical approaches were applied to the available datasets in order to:

detect latent factors (natural and anthropogenic) responsible for the data set structure;

identify the contribution of each factor to the total chemical concentration of each analyzed species;

classify sampling sites or sampling variables according to their similarity measure.

An extended survey aiming at the establishment of national databases concerning the surface water quality (rivers, streams, lakes) has been conducted in the area of Macedonia, Northern Greece for the period 2004-2005. Monthly sampling at selected sites and determination of various chemical and physicochemical parameters has been carried out. The speciation of nitrogen and phosphorus in rivers and steams, the distribution of heavy metals in these systems as well as the multivariate statistical interpretation of the monitoring data are presented in previous papers [44-46].

Five lakes (Volvi, Koronia, Doirani, Mikri Prespa and Megali Prespa) located in the area of Macedonia, Northern Greece, were examined in this research.

The lakes Volvi and Koronia are located in Northern Greece at approximately 11.5 km from Thessaloniki. Doirani is traversed by the border between Greece and the Former Yugoslav Republic of Macedonia (FYROM), while Mikri Prespa and Megali Prespa are two lakes in the mountainous drain basin located in Northwestern Greece at the intersection of the frontiers of Greece, Albania and FYROM (table 1).

Table 1. Main features (morphometric and general) of the lakes

Feature	Unit	Volvi	Koronia	Doirani	Mikri Prespa	Megali Prespa
Latitude (N)		40° 40'	40° 40'	40° 13'	40° 46'	40° 50'
Longitude (E)		23° 30'	23° 09'	22° 44'	21° 06'	21° 00'
Lake area	km^2	69	42	28	47.4	253.6
Catchment area	km^2	1247	350	420	260	372
Altitude	m	37	75	142	853	853
Depth (mean, max)	m	13.5 (24)	2.8 (4)	5.0 (10.4)	6.7 (7.8)	20 (55)

The major area of lakes Volvi and Koronia is protected by the Ramsar Convention as a site of international importance for the value of the wetland habitat, ideal for a variety of flora and fauna species [47]. The Greek part of the Prespa region including the lakes Mikri Prespa and a part of Megali Prespa was declared as national park because it is one of the last refuges in Europe for more than 260 different bird species, rare and/or endangered. The area is also protected by the Ramsar Convention [47].

Monthly samplings were carried out in order to monitor changes in the hydrochemical cycle and possible seasonal variations. Samples from lakes Doirani, Volvi and Koronia were gathered for 24 months whereas samples from the lakes Mikri Prespa and Megali Prespa were taken only for seven months (June - December 2005). The sampling sites were located away from quiescent sites and point sources pollution. Two sampling sites have been chosen in the lakes Volvi and Koronia in the eastern and the western part of the corresponding lake. Only one sampling site was used for the other transboundary lakes. The sampling was performed just below the water surface. The standard sampling procedure for surface waters as described in [48] was followed.

Temperature, dissolved oxygen (DO), pH and conductivity (COND.) were measured in the field. The water samples were kept cool after sampling and transported to the laboratory in a small refrigerator (4°C) within a day. The samples were filtered through a 0.45 μm membrane filter for total suspended solids (TSS) determination. Further, nitrates(III), nitrates(V), ammonia, phosphate ions, color and anionic surfactants (MBAS) were determined in filtrates, while chemical oxygen demand (COD), biochemical oxygen demand (BOD), Kjeldahl nitrogen (N_{org}) and acid-hydrolysable phosphorus (total phosphorus – TP) were determined in filtered samples. The acid-available fraction of metals (Ag, As, B, Ba, Cd, Cr, Cu, Hg, Fe, Mn, Ni, Pb, Se and Zn) was also determined. Standard analytical procedures were employed for each component [48]. The data quality was checked by careful standardization, procedure blank measurements as well as spiked and duplicate sample determination.

The environmetric approaches for the assessment of water quality are frequently employed [49-51]. In this study cluster analysis (CA), principal components analysis (PCA) and source apportioning by multiple regression analysis on principal components (PCR) were employed to the data set. Missing data were completed by mean values of the neighbor data.

Table 2. Factor loadings

	PC1	PC2	PC3	PC4	PC5	PC6
pH	0.289	0.104	-0.048	0.010	0.729	0.043
DO	0.431	0.036	-0.101	-0.075	0.695	0.078
COND.	0.806	0.336	0.367	0.023	0.145	0.147
TSS	0.732	0.438	0.265	0.094	-0.072	-0.073
COLOR	0.731	0.015	-0.173	-0.006	-0.230	0.248
BOD	0.887	0.026	-0.091	0.086	0.036	-0.057
COD	0.816	0.108	0.322	0.076	0.184	0.255
NO_2^-	0.103	0.069	0.382	0.033	0.007	0.001
NO_3^-	0.125	0.243	0.763	-0.008	0.115	-0.040
NH_4^+	0.290	0.765	0.205	0.013	0.112	-0.033
N_{org}	0.884	0.107	0.144	0.029	-0.015	0.221
PO_4^{3-}	0.236	0.123	0.495	0.048	0.053	-0.371
TP	0.892	0.007	0.182	0.092	0.028	0.121
MBAS	0.008	0.187	0.763	-0.016	-0.200	0.011
PHENOLS	-0.229	0.227	0.406	-0.192	-0.342	0.306
CN^-	0.185	-0.113	0.361	0.069	0.082	0.740
Fe	-0.074	0.762	0.479	0.128	0.139	-0.050
Mn	0.075	0.767	0.462	-0.004	-0.150	-0.022
Cu	0.085	0.798	-0.026	0.085	-0.059	-0.038
Zn	-0.201	0.057	0.081	0.613	-0.362	0.038
B	0.163	0.545	0.668	0.066	0.361	-0.054
Cd	0.262	0.013	-0.233	-0.026	0.012	0.721
Cr	-0.184	0.517	0.173	0.379	0.498	-0.009
Pb	-0.082	0.010	0.041	0.875	0.085	0.016
Ba	0.429	0.649	0.187	-0.040	-0.162	0.070
Ni	0.171	0.078	0.039	0.862	-0.044	0.013
As	0.183	0.275	-0.089	0.439	-0.421	-0.081
Explained variance [%]	29.18	17.45	13.92	8.81	7.08	5.74

Note: Marked loadings are statistically significant.

The PCA of the whole dataset (119 samples *x* 27 parameters) has indicated that up to 6 latent factors (principal components) are responsible for the general data structure. They explain over 80% of the total variance and their number was validated by the empirical test of the scree plot where only eigenvalues higher than 1 are considered as significant. In table 2 the factor loadings calculated by the Varimax approach of PCA are presented. Only statistically significant loadings (Varimax criterion) are presented.

The first latent factor accounted for 29.2% of the total variance was highly correlated with total suspended solids (TSS), conductivity, BOD, COD, organic nitrogen (N_{org}) and total phosphorus (TP). It could be considered as representing the organic dynamics of the lakes (due to natural or anthropogenic sources).

The second latent factor is associated mainly to ammonium, iron, manganese, copper, chromium and barium and explains about 17.4% of the total variance. It could be named "*sediment*" factor and is probably strongly influenced by processes of sedimentation in the lakes.

The third factor explaining 13.9% of the total variance is strongly associated to nitrates(III), nitrates(V), anion surfactants (MBAS), phenols, phosphates and boron so it could be conditionally named "*domestic wastes*" due to the role of the mentioned species in the waste waters.

The fourth factor contributing 8.8% to the total variance is mainly associated to lead, arsenic, nickel and zinc, which are products of industrial wastes ("*industrial wastes*" factor). The fifth latent factor could be referred as "*oxidation*" factor responsible to the oxidation processes in the lakes (7.1% explanation of the total variance and association to pH and dissolved oxygen). Finally, a sixth latent factor could be identified (association to cadmium and cyanide, 5.7% explanation of the total variance) as "*toxic*".

It should be kept in mind that this PCA model reflects the whole data set and represents summarized source identification for the whole lake system. It is of substantial interest to obtain similar source identification for each of the lakes included in the study. However, it seems more statistically sound to apply cluster analysis instead of PCA for the separate lake due to the limited number of samples as compared to the parameters determined.

In the next stages of the environmetric data analysis CA was applied. Several different tasks have been followed. First, it was interesting to classify all samples with respect to the objects, e.g. to the sampling periods. In figure 2a hierarchical dendrogram of the Ward's clustering algorithm is given.

The objects (totally 119, including the five lakes and the separation of samples in Koronia and Volvi finto "Eastern" and "Western" part of the lakes) are generally clustered into four big clusters classifying the lakes according to their altitude and/or sampling period. The first one contains dominantly winter and spring samples from Koronia lake (both parts); the second one - mainly Koronia samples for summer and autumn, the third cluster includes samples from Eastern and Western Volvi and the fourth - from Mikri Prespa, Megali Prespa and Doirani (high altitude lakes).

Further, clustering of the samples from the separate lakes was performed aiming identification of similarity between sampling season, on one hand, or between chemical and physicochemical parameters, on the other.

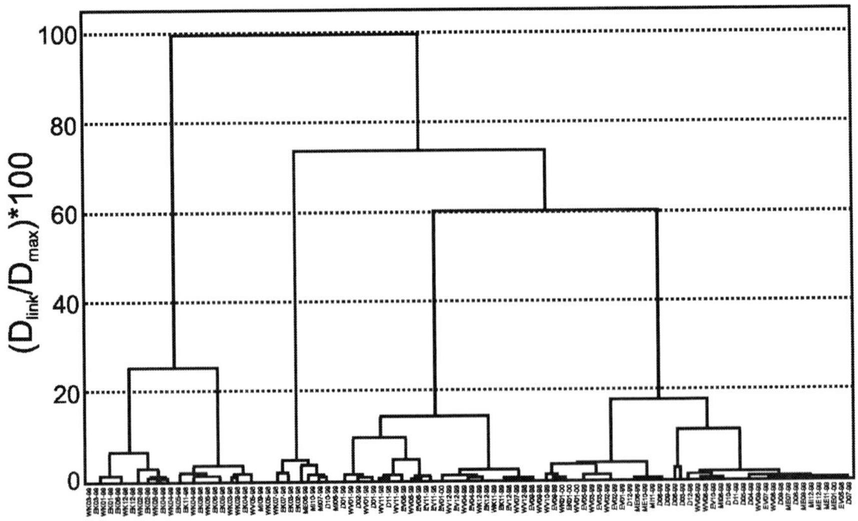

Figure 2. Hierarchical dendrogram of all sampling locations.

Koronia lake: samples are clustered into two big clusters, separating samples from 2004 from those of 2005. Lower concentrations of the analyzed species were found during 2005 than in 2004, probably due to hydrological differences. The clustering of the chemical variables confirms the relation between some of the species which was offered by the PCA of the whole set. In figure 3a hierarchical dendrogram (Ward's method, squared Euclidean distance as similarity measure) of the clustered variables is presented. Three significant clusters are detected. The first one united the latent factors "*waste inlets*" and "*sediment*" from the PCA model; the second cluster includes the "*natural*" and "*toxic*" latent factors and the third group links "*industrial wastes*" and "*oxidation*" factors. Therefore, even in case of this single lake the structural units influencing the organization of the data are the same as for the whole lake system. Probably, the environmental conditions for the whole region are quite similar and stable.

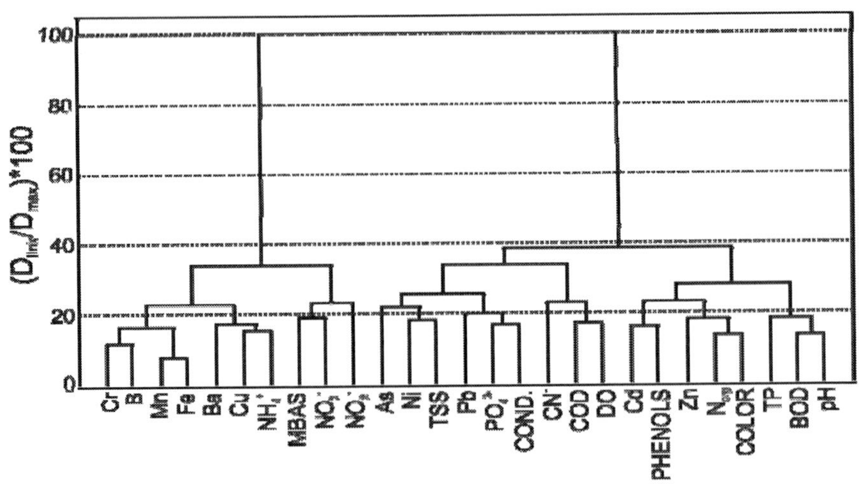

Figure 3. Hierarchical dendrogram for clustering of variables (Lake Koronia).

Volvi lake: Temporal classification is also observed in lake Volvi (two big clusters of 2004 and 2005, the same reason for separation). The clustering of the chemical variables differs from the scheme for the whole dataset. Two big clusters are divided into two subclusters each (figure 4) and the identification offered (with respect to the initial pattern for the whole region) is:

C1 (Mn, Ba, P, Zn, NH_4^+, MBAS, NO_2^-) – "*domestic wastes*" factor,
C2 (PO_4^{3-}, COD, TP, BOD, N_{org}, COLOR) – "*natural*" factor,
C3 (Ni, Cr, B, Pb, Fe, NO_3^-) – "*industrial wastes*" factor,
C4 (CN, TSS, Cd, Cu, DO, As, COND., pH) – "*acidic*" factor.

Due to the lower altitude of the Volvi lake the pollution effects are obviously stronger. This leads to a "dispersion" of the natural factors or those related to sedimentation processes into other factors where pollution events dominate. This is probably the reason leading to separation of Volvi from Koronia samples in figure 2.

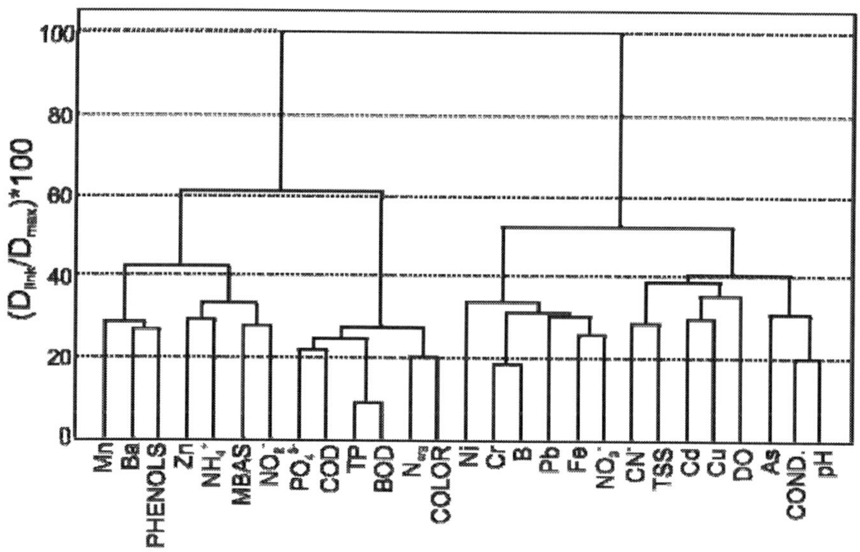

Figure 4. Hierarchical dendrogram for clustering of variables (Lake Volvi).

Doirani lake: by clustering of the sampling day, again, a temporal separation is observed as in the previous two cases (cluster consisting of 1998 samples and cluster consisting of 2005 samples). The chemical parameter clustering offers four clusters:

C1 (Ni, Fe, CN^-, NO_3^-, NO_2^-, Cr, B, DO) – "*sedimentation*" factor,
C2 (Pb, Zn, TP, As, PO_4^{3-}, Cu, MBAS, COLOR) – "*agricultural wastes*" factor,
C3 (NH_4^+, Ba, Mn, TSS, PHENOLS, N_{org}, COND.) – "*domestic wastes*" factor,
C4 (Cd, COD, BOD, pH) – "*biological*" factor.

The small area of Doirani sampled (only Greek territory) does not give probably representative results for the factors affecting the data structure. The clusters in this case differ from the general pattern and from the patterns of the two other lakes. Due to the small area checked the factors reveal a "pollution" pattern where the chemical and physicochemical variables are mixed. The names are arbitrary and only partially represent the effects of the species.

Megali and Mikri Prespa lakes: the two lakes are considered together since the monitoring of the lakes was limited to several samples taken in 2005. In the hierarchical dendrogram of the sampling days the samples are separated with respect to the season of sampling (winter 2005 and summer 2005) and not with respect to geographical or limnological reasons.

The clustering of the chemical concentrations and physicochemical parameters offers four significant clusters (figure 5):

C1 (Cr, Ni, Pb, NO_2^-, BOD, Cd, Mn, Fe, As, B, COND.) – "*sedimentation*" factor;
C2 (Zn, Cu, MBAS, TP, PO_4^{3-}) – "*domestic wastes*" factor
C3 (COLOR, N_{org}, NO_3^-, COD) – "*biological*" factors
C4 (CN^-, Ba, PHENOLS, TSS, NH_4^+, DO, pH) – "*oxidation*" factor.

The special location of the lakes is probably the reason for another pattern of structural factors, which include more specific factors.

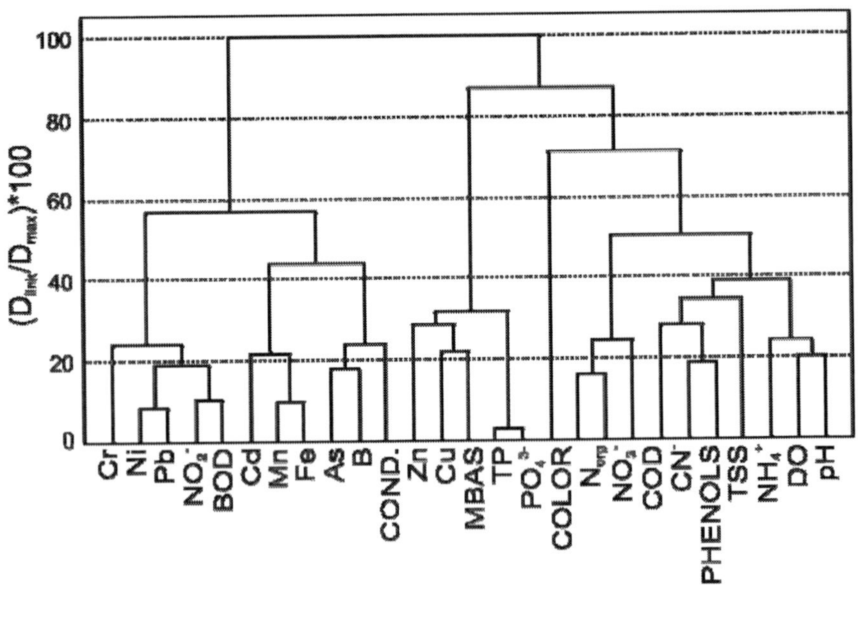

Figure 5. Hierarchical dendrogram for clustering of variables (Lake Megali Prespa and Lake Mikri Prespa).

Table 3. Element source contribution in % (whole dataset)

	Intercept	Natural factor	Sediment factor	Waste inlet	Industrial inlet	Oxidation	Toxic	R^2
DO	-	21.20 ± 8.4	-	-	-	78.8 ± 23.4	-	0.75
TSS	-	55.2 ± 16.4	25.1 ± 8.3	19.8 ± 4.4	-	-	-	0.66
COLOR	22.1 ± 7.4	52.4 ± 8.8	-	-	-	-	25.5 ± 5.4	0.77
BOD	12.2 ± 4.1	87.8 ± 23.3	-	-	-	-	-	0.82
COD	-	69.6±13.4	5.1±2.4	7.2±3.1	-	5.1±3.4	14.1±5.4	0.66
NO_2^-	49.6 ± 23.4	-	-	50.4 ± 18.7	-	-	-	0.49
NO_3^-	20.2±6.9	8.4±2.2	16.1±4.2	55.3±11.4	-	-	-	0.51
NH_4^+	12.6 ± 3.6	13.3 ± 4.7	59.4 ± 19.1	14.7 ± 3.4	-	-	-	0.66
N_{org}	-	72.0 ± 11.2	5.8 ± 2.1	6.1 ± 2.2	-	-	16.1 ± 4.3	0.72
PO_4^{3-}	18.9±8.4	19.7±7.9	19.1±10.5	43.4±9.9	-	-	-	0.86
TP	9.4 ± 4.3	90.6 ± 25.4	-	-	-	-	-	0.50
MBAS	-	-	-	100 ± 32.7	-	-	-	0.67
PHENOLS	13.5 ± 4.1	-	25.5 ± 8.6	41.6 ± 14. 1	-	-	19.9 ± 9.2	0.82
CN^-	-	5.6 ± 2.1	-	29.7 ± 10.8	-	-	64.7 ± 16.3	0.68
Fe	-	-	61.7 ± 20.7	22.2 ± 9.1	8.1 ± 4.2	7.9 ± 2.8	-	0.82
Mn	8.7 ± 4.2	-	70.3 ± 18.6	20.9± 10.8	-	-	-	\0.79

Table 3. (Continued)

	Intercept	Natural factor	Sediment factor	Waste inlet	Industrial inlet	Oxidation	Toxic	R^2
Cu	-	-	100.0 ± 14.1	-	-	-	-	0.68
Zn	15.5±18.1	-	-	-	84.5±16.3	-	-	0.71
B	-	10.0±6.1	29.9±12.9	40.4±11.6	-	20.7±9.7	-	0.77
Cd	20.3 ± 14.3	7.9 ± 4.4	-	-	-	-	72.7 ± 14.8	0.39
Cr	-	-	30.3±9.2	12.5±9.7	25.4±12.6	28.8±11.1	-	0.49
Pb	11.1 ± 8.6	-	-	-	89.9 ±25.1	-	-	0.52
Ba	9.1 ± 4.1	31.6± 14.3	52.2 ± 13.9	7.1 ± 3.3	-	-	-	0.61
Ni	27.7 ± 13.3	-	-	-	72.3 ± 15.5	-	-	0.82
As	27.7±9.9	12.4±9.7	22.7±10.1	-	38.2±8.1	-	-	0.71

Note to table 3: Intercept means the intercept of the regression and presents the portion of the unexplained concentration; R2 is the multiple correlation coefficient giving the square of the correlation coefficient between calculated (from the model) concentration and the measured one; usually it represents the percentage of the explained variance from the model and is measure for its adequateness.

In the final stage of the environmetric study one effort was undertaken to apportion each chemical and physicochemical parameter to the identified factors (considered as element sources) according to the patterns obtained for the total dataset. The apportioning was performed according to Thurston and Spengler algorithm and the results are presented in table 3. It may be seen to what extend each latent factor determines the total mass (concentration) of the parameters in consideration.

In order to check the reliability of the cluster analysis results PCA of the data from each lake with reduced number of chemical variables was performed. COLOR, dissolved oxygen, BOD and MBAS were eliminated as variables since the change of their values throughout the monitoring was quite insignificant as compared with that of the other parameters. In table 4 the summarized results of the PCA for each lake are presented.

Table 4. Principal components identification for each lake

Site	PC1	PC2	PC3	PC4	Explained total variance
Koro nia	Cr, B, Mn, Fe, Ba, Cu	NH_4^+, NO_3^-, NO_2^-	As, Ni, TSS, Pb, COND., CN, COD	Cd, TP, pH, PHENOLS, Zn, Nk, PO_4^{3-}	88.6%
	sedimentation	agricultural runoff	industrial wastes	acidic	-
Volv i	Mn, Ba, Zn, NO_3^-, NO_2^-, NH_4^+, N_{org}	PO_4^{3-}, TP	Ni, Cr, B, Pb, Fe, PHENOLS, COD	CN, TSS, Cd, Cu, pH, As, COND.	78.6%
	domestic wastes	natural	industrial wastes	acidic	-
Doir ani	Ni, Fe, CN, B, NO_3^-, NO_2^-, Cr	Pb, Zn, TP, PO_4^{3-}, Cu, As	NH_4^+, Ba, Mn, TSS, N_{org}, PHENOLS, COND.	Cd, COD, pH	83.1%
	sedimentation	agricultural runoff	domestic wastes	biological	-
Presp a	Cr, Ni, Pb, NO_2^-, NO_3^-, Cd, Mn, Fe, As, B, COND.	Zn, Cu, PO_4^{3-}, TP	N_{org}, COD, NH_4^+	CN, Ba, PHENOLS, TSS, pH	84.4%
	sedimentation	domestic wastes	Biological	acidic	-

It is readily seen that the results from PCA confirm those obtained by cluster analysis. For each lake the identified factors have a different content and significance, which is an indication for the different environmental conditions. In the table only the parameters with statistically significant factor loadings are given. In order to obtain better quantitative information about the role of each identified source apportioning models are offered for each lake (tables 5 – 8).

Table 5. Element source contribution for Koronia lake (in %)

	Intercept	Sediments	Agriculture runoff	Industrial wastes	Acidic	R^2
Mn	18.2±6.6	48.8±16.1	-	33.3±8.6	-	0.81
Ba	20.5±11.2	60.2±26.6	-	19.3±6.7	-	0.77
PHENOLS	30.8±14.2	-	5.4±1.2	-	64.8±23.3	0.65
Zn	23.5±19.2	10.5±8.2	8.8±4.2	9.5±6.2	57.7±15.1	0.76
NH_4^+	19.5±10.2	-	80.5±9.4	-	-	0.84
NO_2^-	31. 1±16.2	-	69.5±21.2	-	-	0.64
NO_3^-	26.8±10.9	-	62.8±14.3	8.4±4.2	-	0.71
PO_4^{3-}	28.5±13.7	-	31.0±12.9	10.5±6.1	-	0.75
TP	20.2±11.6	-	79.8±21.5	-	-	0.68
COD	19.5 ± 9.4	-	-	80.5±31.2	-	0.59
N_{org}	-				100.5±9.9	0.61
Ni	20.8±17.2	11.5±8.7	-	48.0±14.1	9.7±4.6	0.77
Cr	8.2±6.6	53.8±16.1	-	38.3±8.6	-	0.86
B	38.2±6.3	61.8±16.1	-	-	-	0.68
Pb	28.5±16.6	-	-	71.5±19.6	-	0.84
Fe	11.2±6.8	75.5±26.6	-	13.3±8.6	-	0.87
CN	-	-	-	100.0±9.9	-	0.63
TSS	-	43.3±9.3		56.7±18.1	-	0.72
Cd	48.3±12.7	-	9.3±7.6	-	42.4±9.1	0.66
Cu	9.2±4.2	75.5±26.6	-	15.3±8.6	-	0.77
As	-	-	-	100.0±9.9	-	0.61
H	-	-	-	-	100.0±9.9	0.75

Table 6. Element source contribution for Volvi lake (in %)

	Intercept	Domestic wastes	Natural	Industrial wastes	Acidic	R^2
Mn	34.2±11.4	36.2±9.2	11.1±6.7	18.5±11.4	-	0.81
Ba	41.7110.3	44.2±13.4	14.1±16.6	-	-	0.68
PHENOLS	-	-	-	100.0±9.4	-	0.59
Zn	31.2±19.9	38.6±11.7	-	30.2±18.1	-	0.55
NH_4^+	17.8±10.1	67.1±14.6	-	-	15.1±11.4	0.73
NO_2^-	24.3±12.3	64.6 ± 18.4	-	-	11.1±6.9	0.66
NO_3^-	27.6±13.5	66.2 ± 23.3	-	-	6.2±11.4	0.71
PO_4^{3-}	-	-	94.2±31.4	-	5.8±3.9	0.82
TP	-	-	100.0±8.9	-	-	0.59
COD	7.0±5.1	18.3±8.1	-	74.7±19.6	-	0.74
N_{org}	38.8±17.7	61.2 ± 21.9	-	-	-	0.79
Ni	15.8±12.1	-	-	84.2±26.1	-	0.64
Cr	20.2±11.7	-	-	79.8±19.4	-	0.64
B	22.9±13.5	-	22.4±9.6	54.5±17.9	-	0.59
Pb	31.3±14.1	-	-	68.7±26.2	-	0.67
Fe	34.8 ±20.3	-	12.5±8.5	63.7±17.5	-	0.76
CN	24.5±17.4	-	-	12.2±9.6	73.3±23.4	0.82
TSS	33.4±13.8	9.3±7.6	16.2±10.3	-	41.1±16.2	0.58
Cd	21.2±15.6	-	37.2±11.4	-	42.6±17.3	0.64
Cu	26.5±18.4	-	28.7±14.1	-	54.5±13.7	0.77
As	31.8±21.5	-	-	-	69.2±31.6	0.52
H	-	-	-	-	100.0±21.9	0.65

Table 7. Element source contribution for Doirani lake (in %)

	Intercept	Sediment	Agriculture runoff	Domestic wastes	Biological	R^2
Mn	32.7±11.4	29.2±16.4	-	38.1±12.6	-	0.72
Ba	22.8±13.9	-	-	77.2±19.3	-	0.57
PHENOLS	35.5±17.1	-	-	64.5±24.4	-	0.63
Zn	27.3±19.5	-	72.7±24.7	-	-	0.59
NH_4^+	13.3±6.4	-	27.1±10.5	59.6±18.7	-	0.68
NO_2^-	32.6±14.2	64.2±21.1	-	-	4.2±3.1	0.64
NO_3^-	26.1±12.4	44.7±18.2	29.2±13.4	-	-	0.68
PO_4^{3-}	28.6±14.9	-	71.4±26.6	-	-	0.73
TP	32.8±18.7	-	67.2±19.9	-	-	0.68
COD	41.6±22.8	-	-	-	58.4±26.4	0.72
N_{org}	36.7±19.2	-	-	63.3±31.1	-	0.77
Ni	18.2±8.3	81.8±32.7	-	-	-	0.81
Cr	15.3±10.1	84.7±27.8	-	-	-	0.84
B	31.4±13.9	68.6±24.4	-	-	-	0.59
Pb	43.5±16.5	-	56.5±26.8	-	-	0.57
Fe	8.3±4.9	91.7±31.5	-	-	-	0.64
CN	24.1±12.6	55.6±25.3	12.2±9.7	-	8.1±3.3	0.71
TSS	21.1±13.2	27.5±11.4	17.2±10.4	34.4±12.6	-	0.69
Cd	43.4±21.7	-	-	-	52.6±27.1	0.58
Cu	38.7±17.4	-	61.3±30.6	-	-	0.77
As	26.6±14.1	-	73.4±24.9	-	-	0.68
H	17.8±10.8	-	-	-	82.2±30.7	0.73

Table 8. Element source contribution for Prespa lake (in %)

	Intercept	Sediment	Agriculture runoff	Domestic wastes	Biological	R^2
Mn	18.8±11.4	81.2±28.7	-	-	-	0.81
Ba	26.1±14.1	21.2±10.2	-	-	52.7±34.6	0.72
PHENOLS	35.9±21.7	-	-	-	64.1±23.3	0.64
Zn	38.2±19.6	19.2±9.4	42.6±11.4	-	-	0.62
NH_4^+	24.0±11.1	-	14.4±9.7	61.6±27.4	-	0.58
NO_2^-	26.4±10.5	73.6±31.1	-	-	-	0.61
NO_3^-	42.6±22.2	51.1±20.2	-	16.3±8.8	-	0.57
PO_4^{3-}	25.8±17.3	74.2±31.5	-	-	-	0.73
TP	18.3±8.6	81.7±34.4	-	-	-	0.77
COD	12.2±6.4	-	-	78.8±13.9	9.0±4.7	0.75
N_{org}	26.7±13.3	-	-	73.3±19.1	-	0.69
Ni	30.9±11.6	69.1±27.2	-	-	-	0.74
Cr	24.4±13.1	75.6±29.9	-	-	-	0.77
B	13.5±9.5	78.3±17.4	-	8.2±6.3	-	0.66
Pb	19.3±10.2	80.8±20.0	-	-	-	0.69
Fe	12.8±7.6	87.2±19.6	-	-	-	0.84
CN	-	-	-	-	100.0±11.4	0.75
TSS	12.8±13.2	18.5±9.3	-	-	68.7±17.9	0.68
Cd	35.3±17.6	64.7±26.8	-	-	-	0.71
Cu	20.5±12.1	-	61.1±23.9	18.4±11.4	-	0.74
As	28.1±14.7	71.9±18.1	-	-	-	0.65
H	29.7±14.1	-	-	-	70.3±26.6	0.64

The apportioning models shown give enough quantitative information about the source contribution for each lake. This is a higher level of ecological information as compared with that from the whole data set when all lakes are considered as a system. The multivariate statistical data treatment of the lake water samples collected in Northern Greece made it possible to detect an adequate number of latent factors responsible for the data structure of the whole data set and the datasets for each lake. Different patterns of the data structure were found reflecting the specific lake location but, in principle, the latent factors are related to natural, sedimentation, anthropogenic and oxidation processes. The apportioning models offered give a satisfactory idea about the contribution of each factor to the formation of the total mass (concentration) of the species. Thus, a mode of assessment of the each lake water quality is established.

High Mountain Lakes Monitoring in Southern Bulgaria

The anthropogenic impact on the different natural water sources including lake ecosystems proves to be of significant importance for the ecological state of the water bodies. Even high mountain lakes are seriously influenced by anthropogenic factors and that is why the problem with lake water pollution control is involved in all scientific and social agenda on environmental protection policy.

The lake water monitoring is regularly performed if the water body serves as a drinking water source. The specific interest to high mountain lake is due to the possibility to observe and control the changes in such water quality parameters as color, acidity, turbidity, water hardness, growth of algae, ecotoxicity, etc., which could be a reason of the combined effect of various abiotic, biotic and anthropogenic factors leading finally to irreversible changes of the state of the total lake ecosystem. In principle, these lakes are located far from industrial emitters, agricultural activities or highways and it means that they do not feel the direct pollution load from urban anthropogenic sources. However, the atmospheric transfer of gaseous pollutants or air-borne particles influences the lake water quality at high altitudes. Due to this reason as well as to the relatively low water volume and low salt concentration, these lakes are extremely sensitive to atmospheric pollution. Their specific location turns them to a suitable indicator for global pollution and for the role of the transboundary transport of gases and aerosols.

The data was collected during expeditions in 2001 and 2002 for big number of lakes located in Rila mountain and Pirin mountain, which are the highest mountains in Bulgaria. The sampling sites and their heights are indicated in table 9 for both Rila (code R) and Pirin (code P) lakes. The sampling period was between May and October. The sampling itself was performed on the lake surface approximately 2 m from the costal line. The water samples (about 100 mL) were placed in polyethylene flasks. The chemical analysis was carried out within 4 days after sampling at Faculty of Chemistry, University of Sofia. Altogether eleven chemical parameters (major cations and anions like sodium, potassium, calcium, magnesium, chloride, sulfate, nitrate, hydrogen carbonate) were analysed by electrothermal atomic absorption spectrometry and ion chromatography as well as pH (potentiometrically), conductivity (conductometrically), water temperature, and dissolved matter (by summing up of chemical concentrations). Due to the rapid changes of some parameters like pH, temperature, conductivity their determination was done directly at the sampling site by the use of portable instruments. The number of lakes involved in this study was over forty. It is worth to mention that water samples were taken not only from the lakes but also from rivers and springs in the vicinity of the lakes in order to obtain a more realistic estimation of the water quality and of the various natural and anthropogenic impacts.

Table 9. Short description of the lakes in Rila and Pirin subject to assessment

Code	Sampling site	Height a.s.l. [m]
R1	Lake Granchar	2185
R1e	Drinking water from Granchar shelter	
R1f	Spring near to R1g	
R1g	Lake Higher Granchar	2224
R2	Lake Gorno Ribno	2227
R3	Lake Smradlivo	2294
R4, R5	Lake Oreshki	2275-2280
R6	Lake Dolno Ribno	2200
R7	Rila River	
R8	Lake Dolno Popovokapsko	2335
R9	Lake Gorno Popovokapsko	2352
R10	Drinking water from Strashno lake shelter	
R11	Lake Strashno	2408
R12	Lake Mineralno	2395
R13	Lake Gorno Prekorechko	2338
R15	Lake Goliamo Elensko	2472
R16	Lake Malko Elensko	2462
R17	Lake Salzata (Seven Rila Lakes group)	2535
R18	Lake Okoto (Seven Rila Lakes group)	2440
R19	Lake Babreka (Seven Rila Lakes group)	2282
R20, R21	Lake Bliznatsite(Seven Rila Lakes group)	2243
R22	Lake Trilistnika (Seven Rila Lakes group)	2216
R23	Danovisti spring	2190
R24	Lake Ribno (Seven Rila Lakes group)	2184
R25a	Lake Marichino	2410
R25b	Lake Dolno Marichino	2368
R26	Lake Ledeno	2709
R27	Lake Mussalensko	2577
R28	Lake Sedmo Mussalensko	2389
R29	Lake Gorno Chanakgiolsko	2238
R35	Drinking water from Zavrachitsa shelter	
R36	Pot water system of Zavrachitsa shelter	
R37	Tap water system near R36	
R39	Drinking water from Mussala shelter	
P1	Lake Suhodolsko	2311
P3	Lake Dolno Todorino	2510
P4	Lake Gorno Todorino	2536
P5	Lake Dalgo Banrerishko	2310
P6	Lake Djabeshko Banderishko	2322
P7	Lake Banderishko	2312
P8	Lake Muratovo	2230
P9	Lake Spanopolsko	2302
P10	Lake Dolno Georgiisko	2304

Table 9. (Continued)

Code	Sampling site	Height a.s.l. [m]
P11	LakeGorno Georgiisko	2392
P12	Sinanitsa River	
P13	Lake Sinanitsa	2181
P14	Sarchaliiska River	
Code	Sampling site	Height a.s.l. [m]
P15	Lake Chairsko	2355
P16	Lake Prevalsko	2305
P17	Lake Prevalsko	2312
P18	Lake Tevno Belmetsko	2512
P19	Lake Samnodivsko 2	2375
P20	Snow from Samodivski circus	
P21	Lake Samnodivsko 3	2372
P22	Lake Popovo	2234
P23	Lake Kremensko	2356
P24	Lake Gorno Kremensko	2352
P25	Lake Dolno Kremensko	2304
P26, P27	Lake Bezbog	2239
P28	Lake Popovo 6	2185
P29	Lake Popovo 3	2208
P30	Lake Valiavishko	2419
P31	Lake Goliamo Valiavishko	2280
P32	Lake Dolno Valiavishko	2254
P33	Spring Moiseeva cheshma	2370
P34	Lake Gorno Vasilashko	2154
P35	Lake Ribno Vasilashko	2162
P36	Lake Dolno Vasilashko	2325
P37	Lake Gorno Tipitsko	2445
P38	Lake Bashliisko	2450
P39	Lake Bashliisko 6	2430
P40	Lake Bashliisko 3	2461
P41	Lake Bashliisko 4	2313
P42	Lake Begovishko	2392

The data sets from the lakes from the two mountains were classified, modeled and interpreted by the use of cluster analysis, principal components analysis and self-organizing maps of Kohonen. The goal of the environmetric interpretation was to identify groups of similarity between the lakes, to find relationship between the chemical parameters for the lake water quality, to detect hidden factors responsible for the data structure as well as to reveal discriminating chemical parameters, which determine the separation of the lakes in different groups of similarity (or dissimilarity).

Rila Lakes Interpretation

In table 10 the basic statistics for all samples originating from Rila mountain lakes are presented. The data distribution deviates from the normal distribution for each of the variables but this is a typical case for environmental data sets. Therefore, the input data require standardization in order to apply environmetric data analysis. Altogether 106 objects (samples) from Rila lakes are considered.

Table 10. Basic statistics calculated for Rila lakes

Feature	N	Mean	Min	Max	S.D.	Skewness factor
pH	106	4.83	0.20	10.60	2.09	-0.1
DM	106	8.37	0.75	23.58	4.85	1.1
Conductivity	106	16.58	4.00	45.80	9.06	0.9
Ca^{2+}	106	17.90	5.00	60.00	9.74	1.7
Mg^{2+}	106	6.67	2.50	18.10	4.10	1.3
Na^{+}	106	26.17	11.32	91.20	11.72	2.4
K^{+}	106	3.91	0.01	41.20	4.59	5.4
HCO_3^-	106	65.84	1.28	282.80	58.26	1.8
SO_4^{2-}	106	45.02	16.92	112.83	20.19	1.5
Cl^-	106	9.09	0.28	60.80	9.27	2.2
NO_3^-	106	5.98	0.01	34.80	7.38	1.9

Note: DM – dissolved matter.

The goal of the case study is to detect similarities between the sampling locations (in different sampling periods) and relationship between the chemical parameters, which determine the lake water quality.

In figures 6 and 7 the hierarchical dendrograms for the Rila lakes objects (sampling locations) and variables (chemical parameters) are presented.

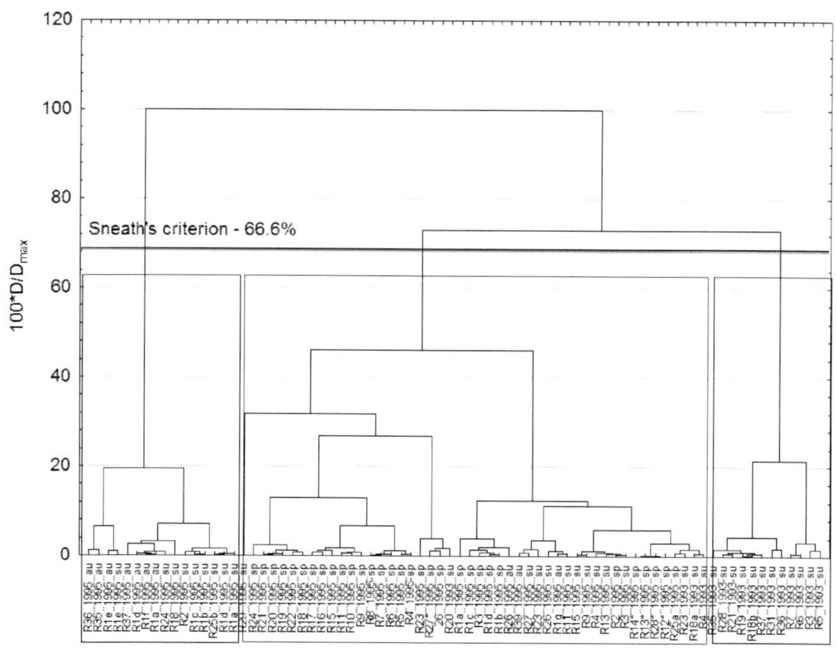

Figure 6. Hierarchical dendrogram for sampling locations, Rila lakes.

In both figures hierarchical agglomerative clustering (Ward's method of linkage, squared Euclidean distance as similarity measure, z-transformed raw data) was used. The Sneath's criterion was applied to determine the cluster significance.

Considering the sampling locations, it can be concluded that they form a quite homogeneous pattern. It means that the water quality for almost all sampling sites (lakes) is constant for the period of monitoring. There is no substantial division between the clusters for Sneath's criterion level of 1/3 D_{max} (33.3 %). Only for the less strict level of separation (66.66 %) three clusters are formed. The interpretation shows that the specific (different from others) are several samples origination from drinking water sources in the vicinity of Lake Granchar and the Zavrachitsa shelter (R29, R35, R36, R1e). It might be assumed that these are regions with slightly increased anthropogenic activity (due to tourist activity and sheep stables).

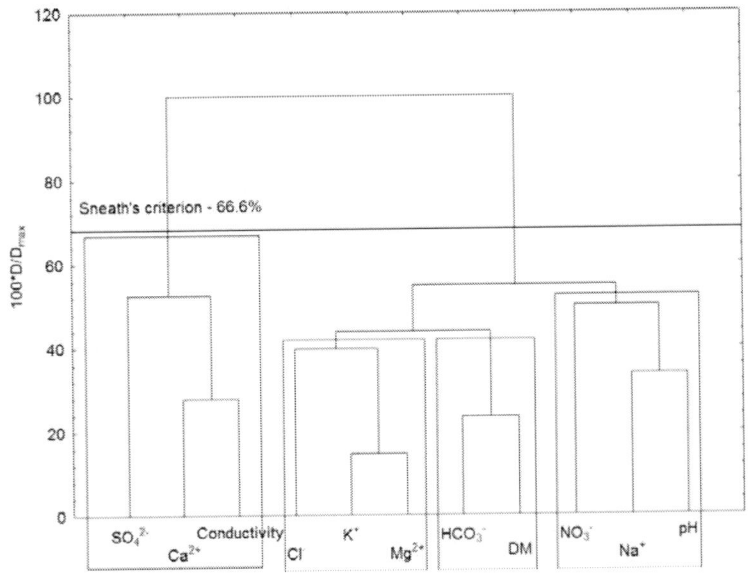

Figure 7. Hierarchical dendrogram for variables, Rila lakes.

In case of variables' similarity assessment four clusters are formed (figure 7), which indicate the influence of water hardness (Ca^{2+}, SO_4^{2-}, conductivity), water acidity (NO_3^-, pH, Na^+), salt content (K^+, Mg^{2+}, Cl^-) and bed rock (HCO_3^-, dissolved matter) impacts.

This initial classification is followed by PCA. In the next two figures (figures 8 and 9) the factor scores and factor loadings plots with highest level of explained total variance (PC1 vs. PC2) are presented.

From the score plot (figure 8) it is readily seen that the lake water homogeneity is proved by the formation of a cloud of sampling locations, which do not form separate groups of similarity. Only several outliers are found which correspond to those indicated by cluster analysis (dominantly Lake Granchar and Zavrachitsa shelter).

From figure 9 is seen that the acidic impact is mixed with the bed rock influence. The water hardness is relatively well defined. Even if the factor loading table is taken into account (table 11) one could conclude that three latent factors are responsible for the data structure and explain over 70 % of the total variance. These factors could be conditionally named: PC1 – "acidic" factor; PC2 – "*water hardness*" factor and PC3 – "*salt*" factor. They are dominantly natural factors and it proves the assumption that Rila lakes are sources of very pure water which keeps its quality for long periods.

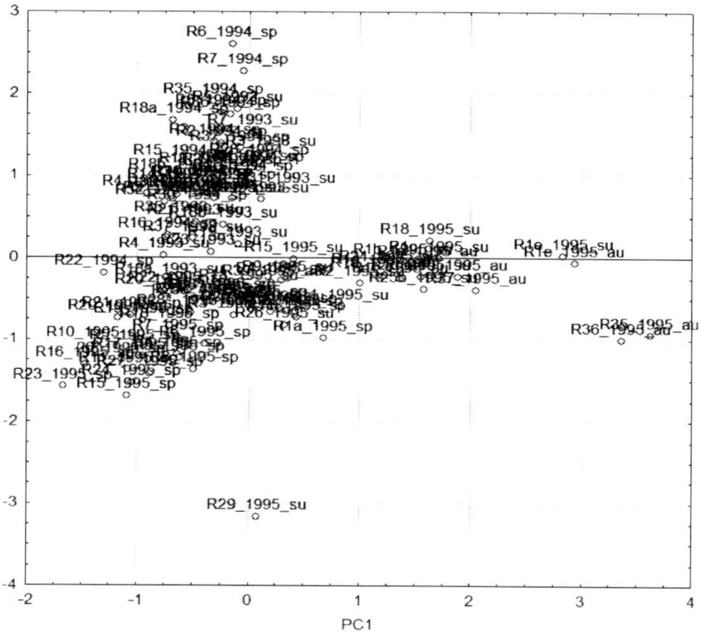

Figure 8. Factor scores plot (PC1 vs. PC 2) for Rila lakes.

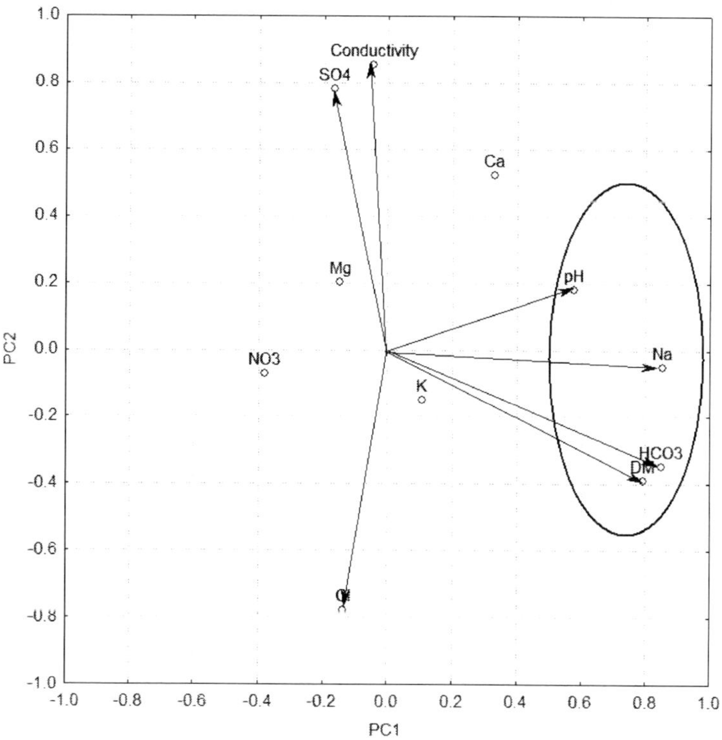

Figure 9. Factor loadings plot (PC1 vs. PC 2) for Rila lakes.

Table 11. Factor loadings (normalized Varimax rotation)

Feature	PC1	PC2	PC3
pH	0.58	0.18	-0.12
DM	0.79	-0.39	0.28
Conductivity	-0.05	0.85	0.40
Ca^{2+}	0.33	0.52	0.44
Mg^{2+}	-0.15	0.20	0.82
Na^{+}	0.85	-0.05	-0.02
K^{+}	0.11	-0.15	0.81
HCO_3^-	0.85	-0.35	0.19
SO_4^{2-}	-0.17	0.78	0.10
Cl^-	-0.14	-0.78	0.41
NO_3^-	-0.38	-0.07	0.01
% explained variance	25.2	24.6	20.8

Note: Marked loadings are statistically significant.

Since the application of SOM classification offers advantages over cluster analysis and PCA and helps in specific data interpretation, next step in the environmetric analysis was classification by SOM.

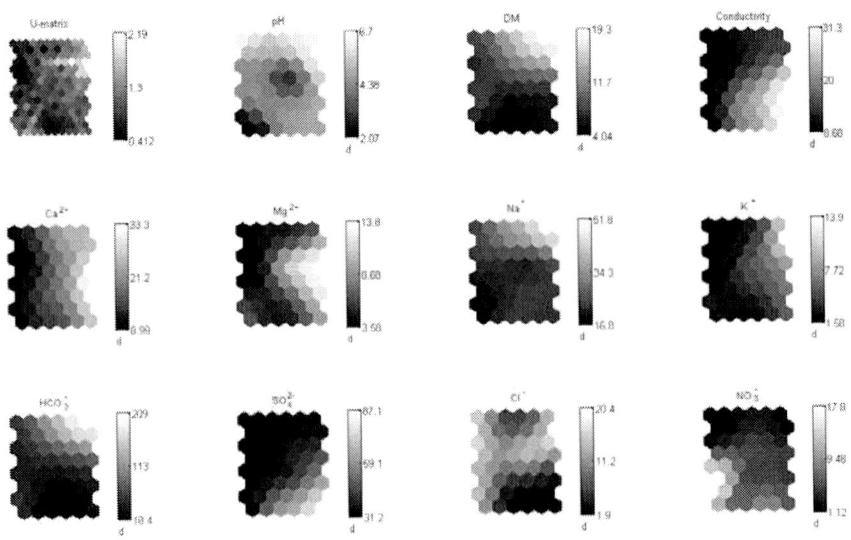

Figure 10. SOM classification for each variable at all sampling locations.

In figure 10 one can see the variables maps. It is not easy to find patterns of similarity for the different parameters of water quality (considering all sampling sites). A slight mode of similarity could be detected between Ca^{2+} and $Mg^{2+}$and between dissolved matter, Na^{+} and HCO_3^-. This is another indirect indication for the chemical homogeneity of the lake water body. Nevertheless, the next steps in SOM data interpretation give more specific information that the initial maps of the separate variables.

In figure 11 the specific grouping of all chemical parameters is presented. The relationships are given in more details and one could identify similarity pattern between SO_4^{2-} and conductivity, between Ca^{2+} and Mg^{2+}, between HCO_3^-, dissolved matter and Na^{+}. pH and NO_3^- are obviously negatively correlated and K^{+} and Cl^- are outliers. This classification does not contradict in general the outputs of CA and PCA. However, in case of SOM classification the interpretation is much more convincing as visual image.

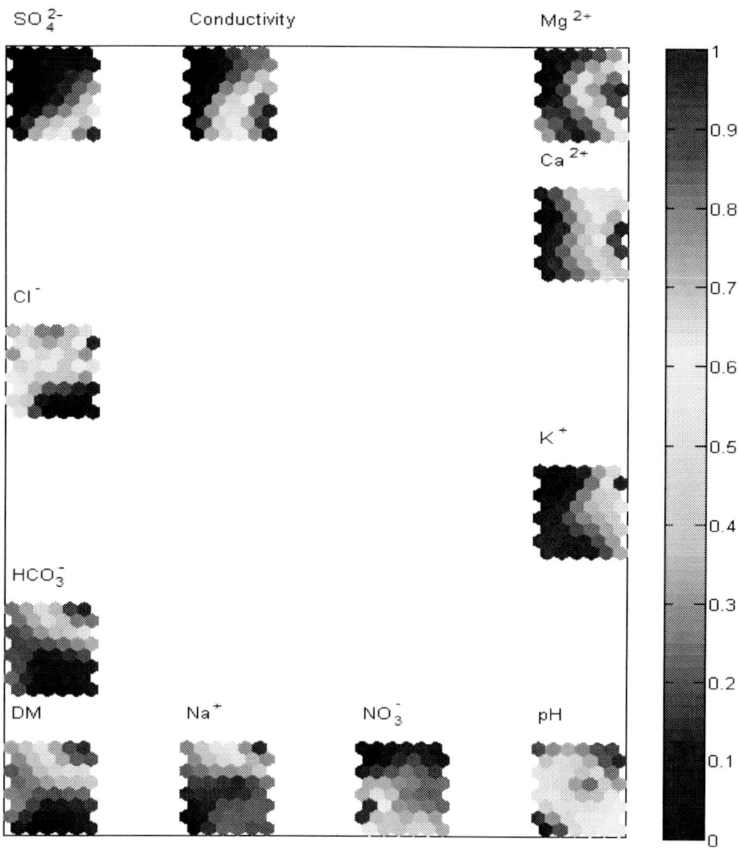

Figure 11. SOM of grouping of variables (Rila lakes).

The grouping of the sampling sites (figure 12) shows a distinct separation into six groups (clusters) being the optimal number of clusters as indicated by the minimal value of Davis – Bouldin index for this number of clusters (on the left side of the figure).

The hit diagram (right side of the figure) determines the number of samples in each node of the map. Thus, the location of each sampling site (geographical position and period of sampling) is readily determined with respect to each of the identified groups.

Cluster I is quite diffuse in its content and comprises various sampling sites from different geographical locations of Rila mountain. It means that the geographical position or sampling season is not a discriminant parameter for this group of samples. From figure 13, where the chemical parameters for each cluster are presented as average values, could be concluded that these are samples with highest pH values and lowest concentrations of NO_3^-, i.e. lake waters with very *low acidity*. All other chemical variables show their lowest level and, therefore, it could be assumed that these are *"background'* lakes with no anthropogenic impact. Typical representatives of the group are some drinking water sources in the vicinity of Lake Granchar, of Lake Gorno Ribno, of Lake Smradlivo, of Lake Marichino etc.

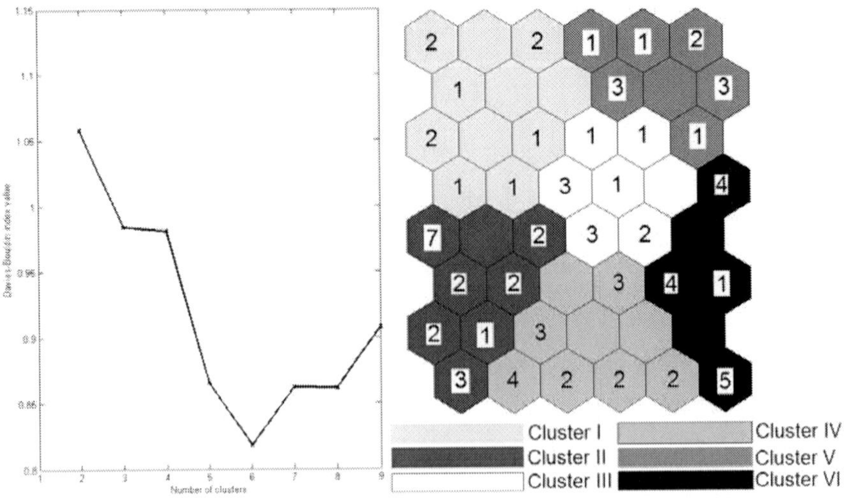

Figure 12. Groups of clusters by SOM (Rila lakes).

Cluster II includes Rila lakes located at *highest possible elevation* like Lake Moussalensko (2577 m a.s.l.), Lake Malko Elensko (2462 m a.s.l.), Lake Strashnoto (2408 m a.s.l.) etc. Except for this geographical discriminating factor, this group of lakes is characterized with very low chemical concentrations and forms another group of *background* lake systems where the anthropogenic pollution has no effect.

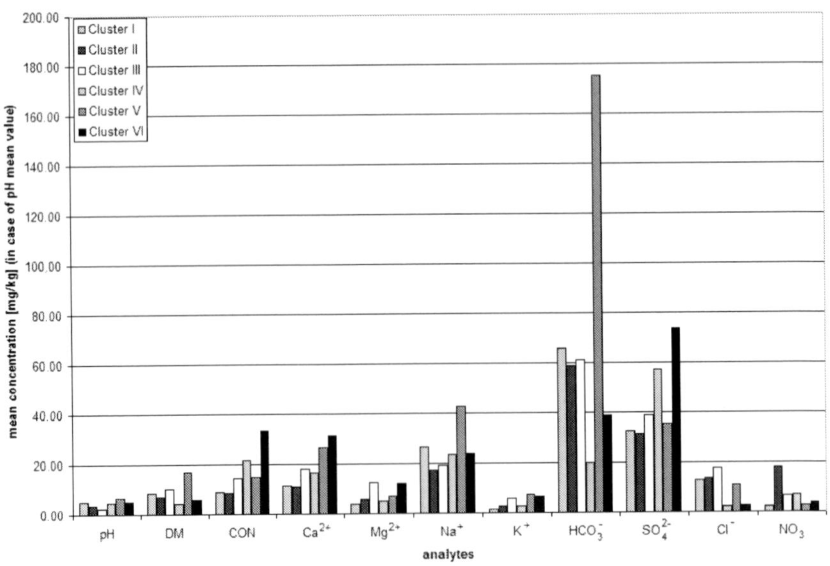

Figure 13. Discrimination diagram for all 6 SOM clusters (Rila lakes).

Cluster III comprises the entire lake group called *"Seven Rila lakes"* and therefore it represents a specific geographical pattern (all of the lakes are located in a typical circus region although at different heights). They have very similar geochemical and lithoral composition, similar hydrochemical and hydrodynamical properties, the seasonal changes are also very similar. The group is additionally characterized by its lowest pH value, highest concentrations of chloride, magnesium and potassium (specific lithoral composition).

Cluster IV is of mixed origin (Lake Sedmo Moussalensko almost 2400 m a.s.l. but also drinking water sources located at lower heights). The lake water in this cluster is characterized by very low content of dissolved matter, hydrogen carbonate and chloride. The group is a typical representative of the *water sources for human consumption* from the region of Rila mountain.

Cluster V could be conditionally attributed to lakes with *anthropogenic impact.* The average concentrations of HCO_3^-, Na^+, dissolved matter, pH are the highest ones as compared to the other groups of lakes. Lake Granchar, Lake Dolno Marichino, Rila River are typical representatives.

Cluster VI involves the group of the *Ribni lakes*. They show highest average concentrations of sulfate, calcium, magnesium and highest conductivity. The lake group is specifically located in a circus with marble and dolomite lithoral zone and contains naturally higher salt levels.

We are convinced that if the classification procedure is performed by the use of several environmetric approaches the information gained is better and allows a more reliable interpretation of the monitoring results.

Table 12. Basic statistics for all Pirin lakes

Feature	N	Mean	Min	Max	S.D.	Skewness factor
pH	74	6.7	0.50	17.0	3.0	0.8
DM	74	10.0	0.61	50.9	7.9	2.4
Conductivity	74	18.9	3.00	108.3	19.1	2.8
Ca^{2+}	74	28.8	6.00	234.0	44.3	3.7
Mg^{2+}	74	8.1	2.60	45.6	8.3	3.5
Na^+	74	23.2	5.60	65.9	11.4	1.2
K^+	74	2.2	0.00	16.0	2.2	3.7
HCO_3^-	74	76.0	0.80	623.6	95.0	3.0
SO_4^{2-}	74	23.7	1.90	104.1	17.6	1.3
Cl^-	74	7.1	0.30	39.7	7.4	1.8
NO_3^-	74	1.8	0.01	16.1	3.5	2.4

Note: DM – dissolved matter.

Pirin Lakes Interpretation

The initial data set for the samples from Pirin lakes includes 74 objects (sampling locations and sampling periods). Again, as in the case with lakes from Rila mountain no normal distribution of data (for each of the variables) is found, which means that the input data were subject of standardization before applying environmetrics.

The first step in the classification of the monitoring data was clustering of the sampling sites (Ward's method of linkage, squared Euclidean distance as similarity measure, z-transformation of data and check of the cluster significance by the Sneath's criterion). The hierarchical dendrogram is shown in figure 14.

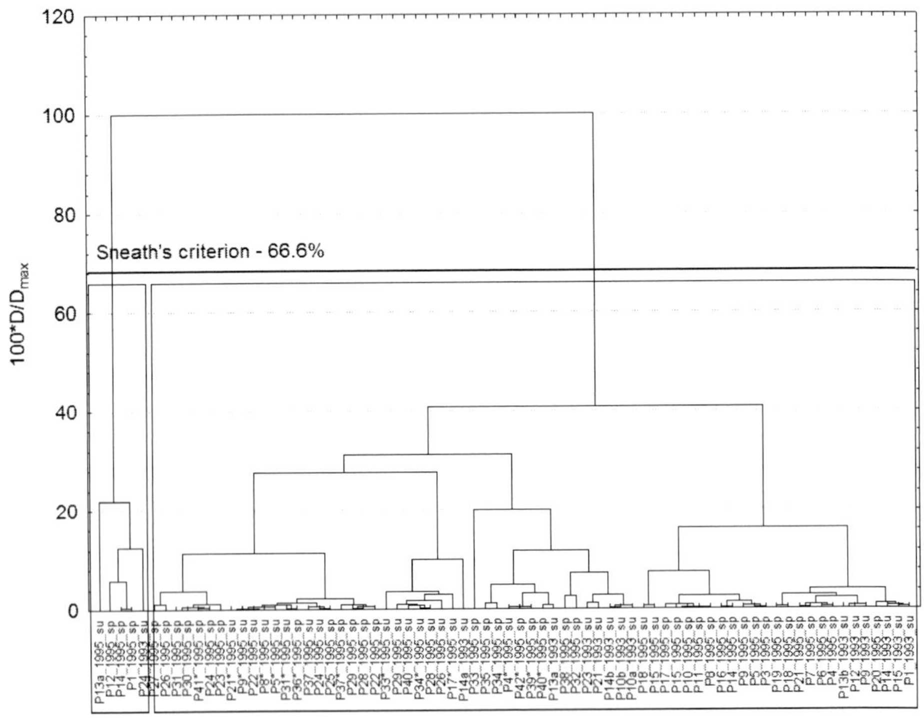

Figure 14. Hierarchical dendrogram for sampling locations, Pirin lakes.

It is readily seen that a good homogeneity of the water quality for all lakes is found. There is a tiny group of outliers (for the less strict Sneath's criterion of $2/3D_{max}$) mostly from the region of Lake Sinanitsa and Sinanitsa River characterized by quite specific lithoral composition.

In figure 15 the hierarchical dendrogram for linkage of water quality parameters is presented. Almost four clusters are formed (for the more strict Sneath's criterion of $1/3D_{max}$) as follows: (nitrates which can be recognized as connected with chlorides and sulfates), (potassium, sodium), (calcium, magnesium, conductivity), (hydrogen carbonate, dissolved matter) and pH as a outlier. It could be concluded from the cluster analysis results that factors like water hardness, water acidity, salt content and turbidity are responsible for the relative separation of the sampling events. However, this information is not convincing enough to make final decisions on the lake water quality. It is worth to mention that, in principle, these conclusions resemble those found for the Rila lakes.

In order to gain additional information from the data set PCA was performed on the normalized input data. Firstly, the factor scores plot will be considered (figure 16).

The homogeneity of the data (very close levels of most of the chemical parameters characterizing the lake water quality) is indicated again by the big cloud of similar objects. The outliers marked are from the specific dolomite circus where the lakes Sinanitsa, Suhodolsko and Gorno Kremensko are located. Again, the separation is due to a geographical (natural) rather than anthropogenic impact.

The biplot (PC1 vs. PC2) for the factor loadings (normalized data, Varimax rotation mode) indicates the grouping of the chemical variables with respect to the identified latent factors. It is seen that high correlation along PC1 axis is found for chloride, sulfate and hydrogen carbonate and along PC2 axis – for calcium, magnesium, conductivity and dissolved matter. These results slightly contradict those found by the classification with cluster analysis and that is why a more careful inspection of the PCA results is needed.

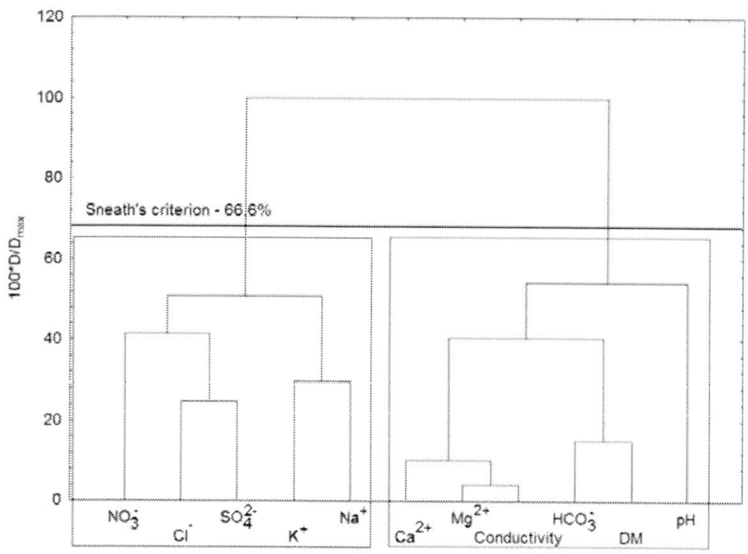

Figure 15. Hierarchical dendrogram for variables, Pirin lakes (less strict Sneath's criterion of $2/3D_{max}$ is presented).

In table 13 the factor loadings for the identified four latent factors are presented. These four factors explain over 75% of the total variance of the system. The first one could be conditionally named *"water hardness"* factor since it indicated high correlation between calcium, magnesium, conductivity and dissolved matter. The second latent factor is related to hydrogen carbonate, sulfate and chloride and, logically, its conditional name might be *"anthropogenic impact"* since it indicates possible pollution by atmospheric transfer of secondary emissions. The third factor should be attributed to processes of *lake water acidification* by the negative correlation between pH and nitrate.

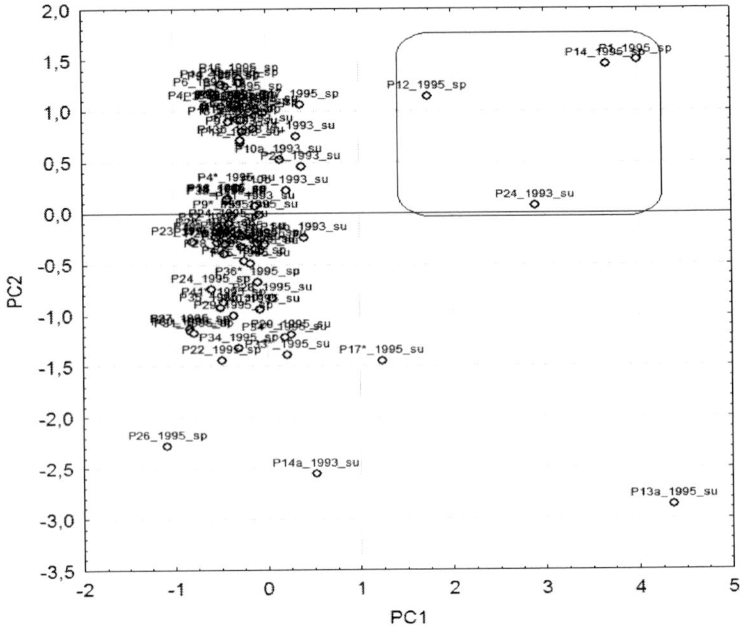

Figure 16. Factor scores plot (PC1 vs. PC2) for Pirin lakes.

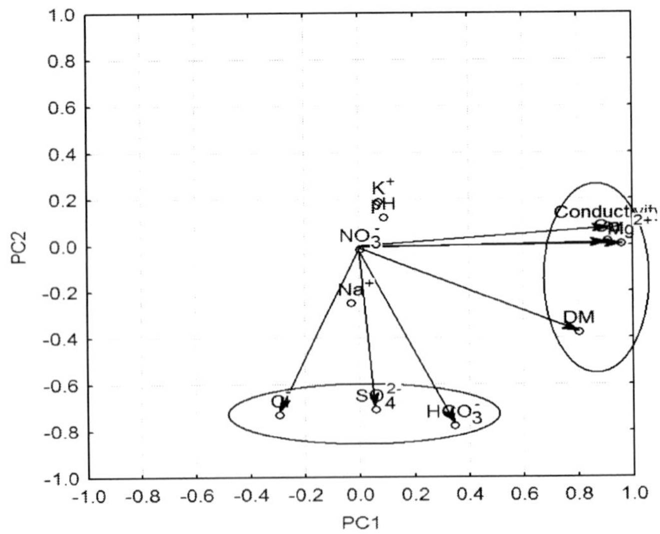

Figure 17. Factor loadings plot (PC1 vs. PC 2) forPirin lakes.

The fourth and the last latent factor is characterized by high (statistically significant loadings) for sodium and potassium and could be conditionally named *"salt"* factor.

This interpretation of the Pirin lakes data set reveals some specific hidden information in a more reliable way and helps for better understanding of the lake water quality assessment.

Table 13. Factor loadings (normalized Varimax rotation)

Feature	PC1	PC2	PC3	PC4
pH	0.093	0.119	0.737	-0.160
DM	0.805	-0.378	0.292	0.079
Conductivity	0.915	0.078	-0.161	0.117
Ca^{2+}	0.909	0.017	0.122	-0.039
Mg^{2+}	0.961	0.003	-0.002	-0.037
Na^{+}	-0.030	-0.252	0.139	0.820
K^{+}	0.072	0.184	-0.212	0.756
HCO_3^-	0.348	-0.778	0.100	0.050
SO_4^{2-}	0.059	-0.707	-0.313	0.385
Cl^-	-0.295	-0.730	-0.115	-0.199
NO_3^-	0.004	-0.018	-0.818	-0.076
Var. expl. %	31.5	17.3	13.9	13.5

Note: Marked loadings are statistically significant.

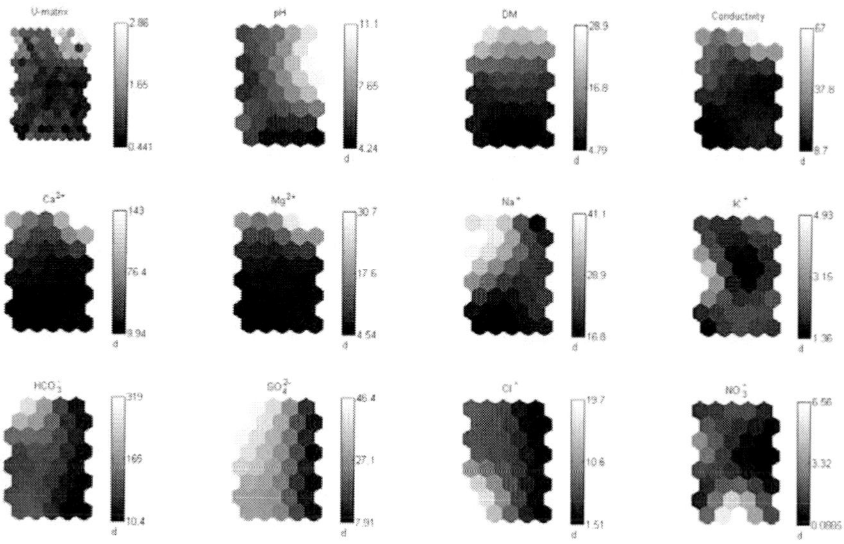

Figure 18. SOM classification for each variable at all sampling locations.

The separation of several outliers by CA and PCA and the identification of the latent factors responsible for the data structure could be improved and completed by data classification using SOM environmetrics. In figure 18 the self-organizing maps for all variables and all sampling events are shown. Similarities detected by CA and PCA are also obvious on some of the maps – e.g. conductivity, calcium, magnesium and dissolved matter maps reveal one and the same pattern – the highest concentration levels are grouped in the upper part of the maps.

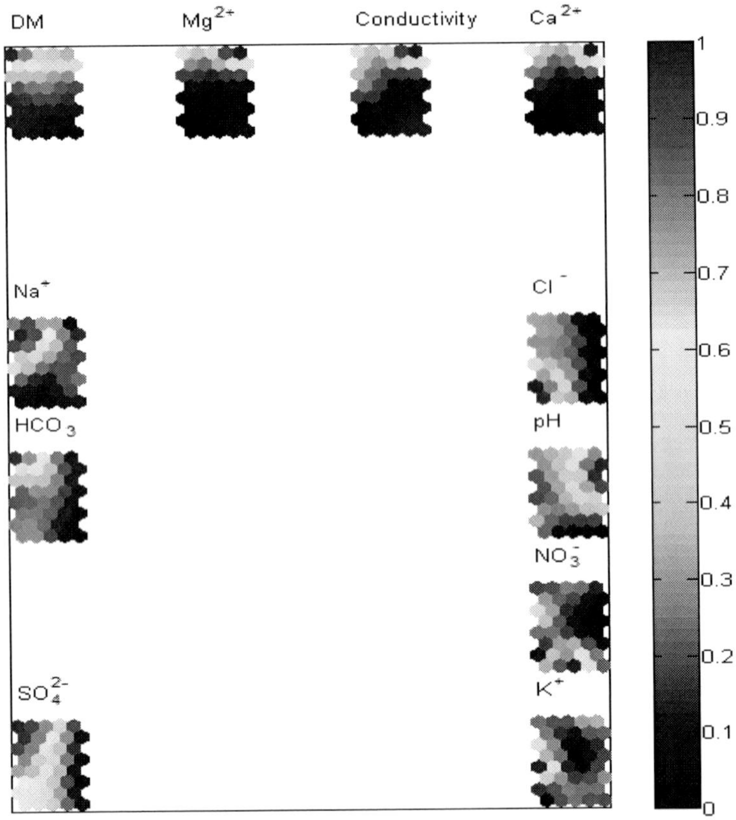

Figure 19. SOM of grouping of variables (Pirin lakes).

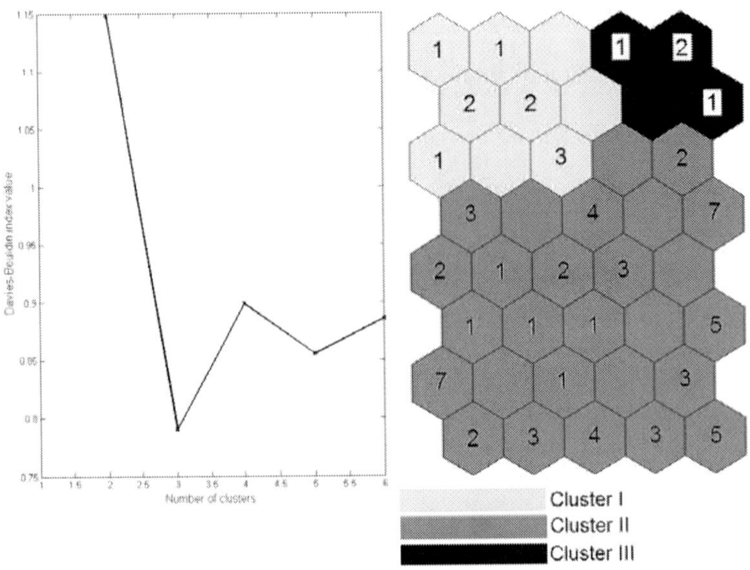

Figure 20. Groups of clusters by SOM (Pirin lakes).

In figure 19 the SOM grouping of the variables is presented. The grouping is slightly different as compared to the PCA results or CA classification. Very stable (the same for all classification approaches) is the group consisting of calcium, magnesium, dissolved matter, and conductivity. This is an important indication for the role of this relationship for the lake water quality assessment of Pirin region. In figure 19 one could detect similarity between hydrogen carbonate and sodium (not with sulfate and chloride), which is rather an indication for lithoral impact than acidification one. Thus, the different types of classification of the monitoring data indicate possibly different pathways in correlation between the water quality parameters and, parallel to it, the stable for all classification patterns.

Very important step is to divide the data set into significant clusters with their spatial vicinity (possible only by SOM classification) and, further, to detect the discriminate parameters for each one of the clusters formed. In figure 20 the optimal number of clusters and the respective hit diagram are shown.

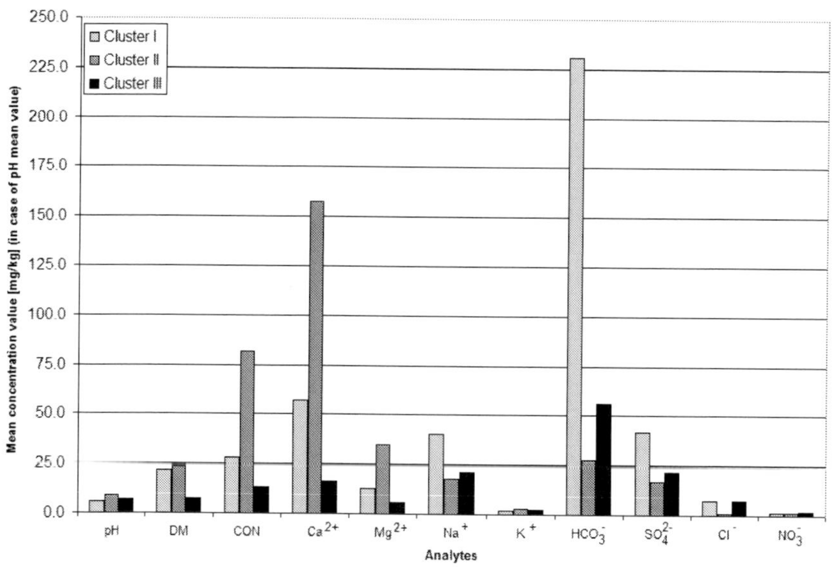

Figure 21. Discrimination diagram for all 6 SOM clusters (Pirin lakes).

For all monitoring data from Pirin lakes three clusters are formed. Their content can be easily determined along with the discriminating tracers for each one of them as presented in figure 21.

Cluster I includes dominantly lakes from the *southern part* of Pirin mountain (the group of Vasilashki lakes, which is located at relatively lower levels a.s.l. as compared to the other lakes). All objects are samples from the summer season (lowest water level) and are characterized by highest concentrations of hydrogen carbonate, sulfate, sodium and chloride. This is an indication for high salt impact but it should be attributed to natural and not to anthropogenic reasons.

Cluster II comprises a small group of lakes from the Lake Sinanitsa and Sinanitsa River vicinity. They are the outliers discussed by the interpretation of results from CA and PCA of the monitoring data. The discriminating variables in this situation are the parameters of the water hardness – calcium, magnesium, conductivity and dissolved matter. Obviously, the specific local crustal composition is the natural reason for finding similarity within this group of Pirin lakes.

Cluster III is mixed and very large. This is an indication for the homogeneity of the water quality stated before. The group of more than 60 cases is characterized by lowest levels of the concentrations of all water quality parameters.

In conclusion, it can be stated that the quality of the lake water in the high-mountain lakes in southern Bulgaria is very high and is mostly subject to natural impacts. The different environmetric data interpretations applied prove convincingly this statement.

Environmetric Assessment of Lake Water for Human Consumption

The quality of water for human consumption has always been and still is one of the most serious challenges for environmentalist. Recently, strict water quality control has been performed not only by the traditional chemical and physicochemical parameter analysis and respective comparison of the monitoring results with critical or allowable values but also by applying multivariate statistical data treatment methods such as cluster and principal components analysis, time-series analysis etc. [52–58]. This allows drawing new information from the data sets such as patterns of similarity between sampling locations, sources of pollution in the environment, seasonal behavior of chemicals, time trends etc. It is the aim of the present case study to use environmetric methods to interpret a data set of water sources for human consumption in order to assess the water quality. The efficiency of the exploratory data analysis will be shown.

Three main sources deliver potable water to the municipality of Athens, Greece: lake Iliki and lake Marathonas (close to urban environment) and lake Mornos (rural location). Water purification and control is performed by the EYDAP Company in the purification centers of Archanaes, Galatsi and Kiourka. It is of substantial interest to know how far the water from the natural sources differs from the commercial water for human consumption and if there are patterns of similarity between the natural sources and the water from the purification stations through out the year. The data set under consideration involves analytical measurements of 16 variables (turbidity (tur), pH, free acidity (fa), total hardness (th), chloride (cl), sulfate (su), nitrate (na), nitrite (ni), calcium (ca), magnesium (mg), sodium (so), chemical oxygen demand (cod), dry residue (dr), conductivity (con), phosphate (pho) and silicon dioxide (sd)) in samples from 17 sampling sites from three natural water sources (near river inlets, water collection towers and lake water – altogether 14 sites) and three purification centers (another 3 sites). Sampling was carried out in 6 bimonthly periods in 2002. The chemical analysis and the data quality check were per formed strictly according to the EU standard procedures [4] using volumetric, potentiometric, atomic and molecular spectroscopic methods in the EYDAP control laboratory.

The unsupervised data analysis using two-way PCA indicates that probably 3 latent factors determine the data structure and explain nearly 80% of the total variance of the system (PC1 – 52.8%, PC2 – 19.8%, PC3 – 7.3%). In the bivariate loadings plot (figure 23) it is readily seen that PC 1 and PC 2 could be conditionally named "*salinity factor*" (high positive loadings for free acidity, total hardness, chloride, nitrate, sulfate, calcium, magnesium, sodium, dry residue, conductivity and a high negative loading for pH, i.e. parameters responsible mainly for salinity parameters of the water) and "*turbidity factor*" (high negative loadings for turbidity, nitrite, phosphate, silicon dioxide and COD). The third latent factor (high loading for nitrate) could be conditionally named "*anthropogenic*" since it is related to the agricultural activity in the region. This latent factor is not presented in figure 22.

In the score bivariate plot (figure 23) where the sampling events (site and date, objects of interest) are grouped, only the lake Mornos events form a compact group. This group is always negative along PC1. The negative loading is due to pH, so the Mornos group is special with respect to acidity and, indeed has the highest pH values of all other sources. The events from the water purification complex are intermixed with events from the lake Marathonas and are close to the lake Mornos events, which is an important indication for the high quality of the water from these sources. The events from lake Iliki form an even more diffuse group and differ from the events from the purification complex. This plot does not provide information on any seasonal influences. The application of Tucker3 model to the data set delivers additional information. As a retreatment, the data were subjected to autoscaling. In this case the unfolding of the set was performed with respect to time as a third coordinate of the set. So, to 17 objects and 16 water quality parameters, bimonthly seasonal sampling was added as a dimension.

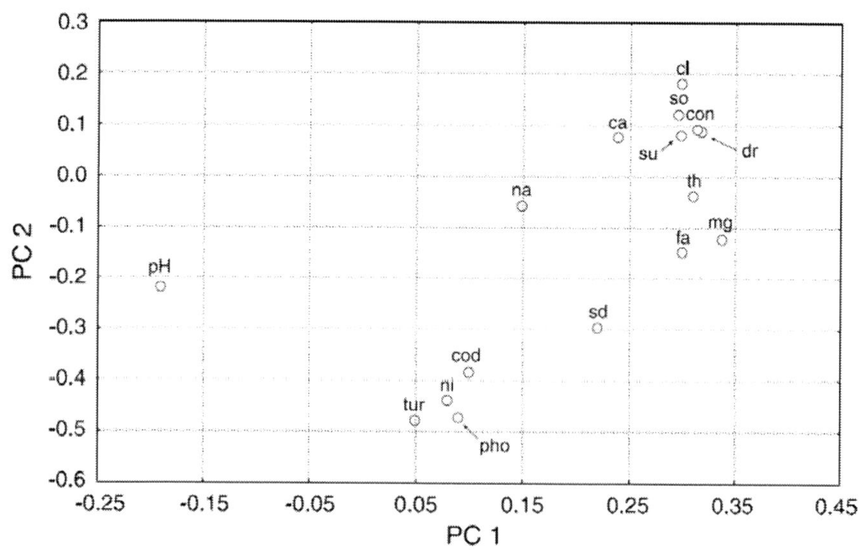

Figure 22. Factor loadings biplot (PC1 vs. PC2).

The next treatment level leads to a (2,2,2) Tucker3 model, which explains over 85% of the total variance. Again, two latent factors (score plot A1 vs. A2 as in the two-way case) might be considered responsible for the water sources' spatial separation. The first one is probably related to the absence of an urban influence on Marathonas and Mornos lakes ("*rural factor*"), which show a water quality comparable with that of the water delivered from the purification complex after purification. The second one ("*urban factor*") indicates the

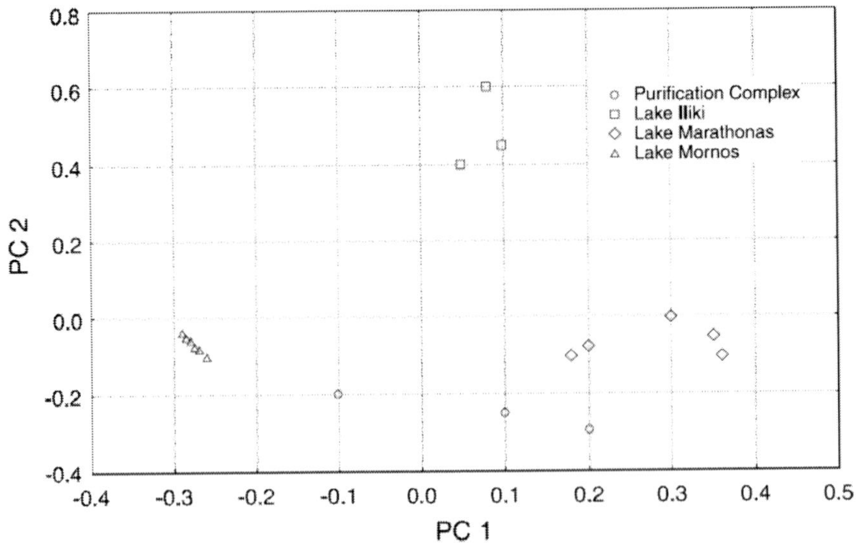

Figure 23. Factor scores biplot (PC1 vs. PC2).

possible anthropogenic influence on Iliki Lake, which is located nearer to the city. In the MPCA loadings results (B1 vs. B2, same as two-way plot) one could distinguish two hidden factors – "*salinity factor*" (high positive loadings values along B1 for free acidity, total hardness, chloride, nitrate, sulfate, calcium, magnesium, sodium, dry residue, conductivity) and "*turbidity factor*" (high negative values along B 2 for turbidity, nitrite, phosphate, silicon dioxide and COD). Nitrate loading belongs more to the first latent factor and need not be interpreted as a separate factor. It may imply that the water salinity is strongly influenced by anthropogenic sources. There are two seasonal components revealed by the MPCA (figure 24).

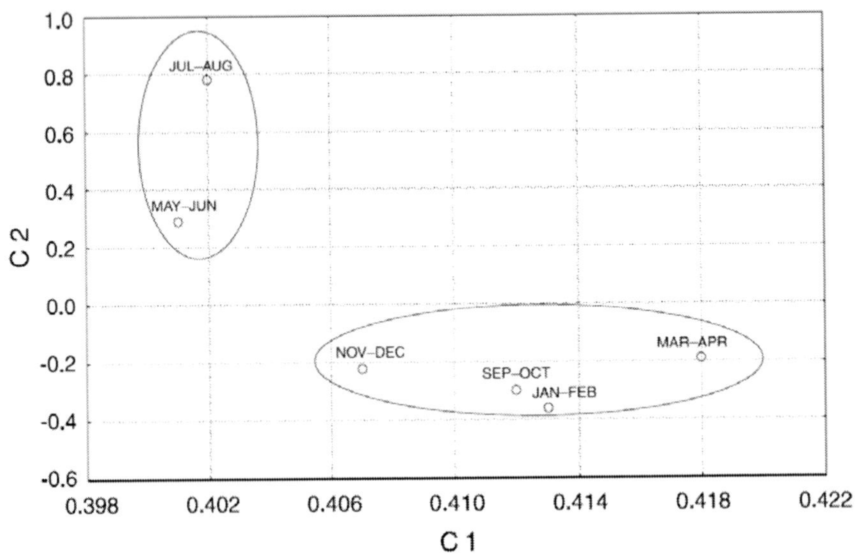

Figure 24. Bivariate MPCA scatter plot (C1 vs C2).

The first one could be conditionally named "*high water*" factor (high loadings along C1), which indicates the autumn and winter increase of the water volume of the sources (from September to February). The other seasonal pattern is represented by the conditional "*low water*" factor (May–August, high loadings along C2). Of course, this is only a hypothetical assumption since differences in temperature could also lead to changes in the water chemistry and/or biology.

The G-matrix of the Tucker 3 model (2, 2, 2) is:

$$G = \left.\begin{matrix} 27.65 & 0.13 \\ 0.16 & -6.35 \end{matrix}\right| \begin{matrix} 0.82 & -1.69 \\ 0.18 & 1.37 \end{matrix} \qquad (25)$$

From the G-matrix it was possible to determine the level of simultaneous influence of the latent factors, which determine the object, the variable and the temporal structure of the data set. It is easily seen that the interaction between A2, B2 and C1 is of great significance, i.e. the "*salinity factor*" is of significance for the "*rural*" water sources (Mornos, Marathonas, purification complex) in the "*low water*" period since the "*turbidity*" factor is of importance for the "*urban*" source (Iliki) for the same period.

The use of environmetric approach to assess the quality of several potable water sources in the region of Athens, Greece, has revealed some important hints. The superiority of three-way PCA to two-way PCA is demonstrated by the seasonal differentiation.

CONCLUSION

The recent global climate changes and increased level of pollution of many environmental compartments has led to looking for new strategies in assessment of the risk from environmental pollution. In addition, the concept of the sustainable development put on the social agenda the problem of considering the effects of many factors simultaneously on the ecological, economical, social and technological aspects of sustainability. The necessity of a new scale and metrics for risk assessment and eco-efficiency assessment is obvious because the environmental problems have to be treated in a multivariate way. The estimation of the lake water pollution in this sense is only one of the issues, which fall under the region of interest of scientists, engineers, environmentalists and politicians. Although the lakes (especially high – mountain lakes or drinking water sources) are in many cases remote places with background levels of pollutants, their quality assessment is an important part of any environmental protection strategy.

In the present chapter we have tried to prove that the lake water quality assessment possesses an extremely useful tool – the methods of chemometrics and environmetrics. If one decides to use the multivariate statistical methods for classification, projection, modeling and interpretation of lake water monitoring data in order to gain specific (often hidden, non-available from the raw data) information, then the pathway to reach it should follow several major steps:

- proving the data quality by means of metrological criteria (uncertainty, limits of detection of the monitoring methods, precision, reliability of signal, etc.),
- checking the data distribution (often non-normal) by statistical tests,

- data normalization in order to avoid problems with non – normal distribution or data dimensionality,
- classification of the monitoring data by different environmetric approaches like cluster analysis, principal components analysis, neuron net classification (self-organizing maps of Kohonen as an option of classification without training procedures),
- careful interpretation of the classification results and finding the reasons for the similarity groups,
- modeling the data in order to identify latent factors responsible for the data structure and the factor contribution to the formation of the total concentration of each of the lake water quality parameters,
- determination of seasonal patterns in lake water quality,
- comparison of the models with the real monitoring data.
- Following this mode of data interpretation we have reached to several important conclusions concerning the case studies involved:
- quantitative PCR models for five lakes located in Northern Greece are obtained and checked for adequateness; for each lake we received a reliable apportionment scheme showing the contribution of natural and anthropogenic impacts on the lake water quality; the different level of pollution for each lake was proved,
- reliable classification for high-mountain lakes in Rila and Pirin located in Southern Bulgaria is performed, which makes it possible to find patterns of similarity between the lakes in each of the mountains and to explain this similarity (or dissimilarity) by discriminating water quality parameters; it could be stated that most of the high-mountain lakes are clean with background levels of chemicals,
- the drinking water quality of the water supply source of Athens, Greece is checked as each water source is environmetrically described and modeled with respect to its water quality and seasonal changes.

REFERENCES

[1] EU Instruction 80/778. Water Analysis and Control. Brussels, Belgium, 1990.

[2] Council Directive 91/692/EEC, European Union Directive (EUD), 1991.

[3] Bulgarian Drinking Water Analysis Standard (BDWS). Ministry of Environment and Water, Sofia, Bulgaria, 1991.

[4] Greek Hygiene Office Instruction (GHOI) A5/228. Standardized methods for potable water analysis, Athens, Greece, 1986.

[5] European Commission. Directive 2000/60/EC of the European Parliament and of the Council. Community action in the field of water policy. Off.J. EC L327, 2000, 1.

[6] Simeonov, V., Wolska, L., Kuczynska, A., Gurwin, J., Tsakovski, S., Protasowicki, M. and Namiesnik, J. , Sediment-quality assessment by intelligent data analysis, *TrAC,* 2007, 26(4), 323-331.

[7] US EPA. Clarification regarding toxicity reduction and identification evaluations in the national pollution discharge elimination system program. March 2001, Washington, DC.

[8] Goodfellow G., Ausley L., Burton D., Denton D., Dorn P., Grothe, D., Heber, M., Norber-King, T. and Rogers, J.. Major ion toxicity in effluents: a review with permitting recommendations, *Environ. Toxcicol. Chem.*, 2000, 19, 175-182.

[9] Astel, A., Tsakovski, S., Barbieri, P. and Simeonov V., A comparison of SOM classification approach with cluster analysis and PCA for large environmental data sets, *Water Res.*, 41(19), 2007, 4566-4578.

[10] Simeonov, V., Wolska, L., Kuczynska, A., Gurwin, J., Tsakovski, S. and Namiesnik J., Chemometric estimation of natural water samples using toxicity tests and physicochemical parameters, *Crit. Rev. Anal. Chem.*, 2007, 37(2), 81-90.

[11] Astel, A., Tsakovski, S., Simeonov, V., Reisenhofer, E., Piselli, S. and Barbieri P., Multiway classification and modeling in surface water pollution estimation, *Anal. Bioanal. Chem.*, 2008, 390(5), 1283-1292.

[12] Simeonova, P. and Simeonov V., Chemometrics to evaluate the quality of water sources for human consumption, *Microchim. Acta*, 2006, 156 (3-4), 315-320.

[13] Astel, A., Glosinska, G., Sobczynski, T., Boszke, L., Simeonov, V. and Siepak J., Chemometrics in assessment of sustainable development rule implementation, *Centr. Europ. J. Chem.*, 2006, 4 (3), 543-564.

[14] Simeonova, P., Polluting sources apportionment for atmospheric and coastal sediments environment, *Ecol. Chem. Eng.*, 2006, 13 (10), 1021-1032.

[15] Simeonova, P., Multivariate statistical assessment of the pollution sources along the stream of Kamchia River, Bulgaria, *Ecol. Chem. Eng.*, 2007, 14 (8), 867-874.

[16] Massart, D.L. and Kaufman, L., The Interpretation of Analytical Chemical Data by the Use of Cluster Analysis. J. Wiley, NY, 1983.

[17] Vandeginste, B., Massart, D.L., Buydens, L., De Long, S., Lewi, P., Smeyers – Verbeke, J., Handbook of Chemometrics and Qualimetrics. Elsevier, Amsterdam, 1998.

[18] Einax, J.W., Zwanziger, K.H. and Geiss, S., Chemometrics in Environmental Analysis. VCH Weinheim, Germany, 1998.

[19] Thurston G. and Spengler J., A quantitative assessment of source contributions to inhalable particulate matter pollution in metropolitan Boston, *Atmos. Environ.,* 19 (1), 1985, 9-16.

[20] Tucker, L., Some Mathematical Notes on Three-Mode Factor Analysis, *Psycometrika,* 1996, 31, 279-311.

[21] Henrion, R. and Anderson, C., A new criterion for simple-structure transformations of core arrays in N-way principal components analysis, *Chemom. Intel. Lab. Syst.,* 1999, 47 (2), 189-204.

[22] Astel, A. and Małek, S., Multivariate modeling and classification of environmental n-way data from bulk precipitation quality control, *J.Chemom.*, DOI: 10.1002/cem.1156

[23] Kohonen, T., Self-organizing maps. Springer, Berlin, 2001.

[24] Aldrich, A., van der Berg, C.M.G., Thies, H. and Nickus U., The redox speciation of iron in two lakes, *Mar. Freshwat. Res.,* 2001, 52, 885-892.

[25] Vilanova, R., Fernandez, P., Martinez, C. and Grimalt, J., Polycyclic aromatic hydrocarbons in remote mountain lake waters, *Water Res.*, 35, 2001, 3916-3922.

[26] Kotti, M., Vlessidis, A., Evmiridis, N., Determination of phosphorous and nitrogen in the sediment of lake 'Pamvotis' (Greece), *Int. J. Environ. Anal. Chem.,* 2000, 78, 445-467.

[27] Fischer, E. and van der Berg, C.M.G., Determination of lead complexation in lake water by cathodic stripping voltammetry and ligand competition, *Anal. Chim. Acta,* 2001, 432, 11-20.
[28] Tokalioglu, S., Kartal, S. and Elci, L., Speciation and Determination of Heavy Metals in Lake Waters by Atomic Absorption Spectrometry after Sorption on Amberlite XAD-16 Resin, *Anal. Sci.,* 2000, 16, 1169-1178.
[29] Scott, B., Mactavish, D., Spencer, C., Strahan, W. and Muir, D., Haloacetic Acids in Canadian Lake Waters and Precipitation, *Environ. Sci. and Technol.,* 2000, 34(20), 4266-4271.
[30] Aguilera, E., Chiodini, G., Cioni, R., Guidi, M., Martini, L. and Raco, B., Water chemistry of Lake Quilotoa (Ecuador) and assessment of natural hazards, *Volcanol. Geotherm. Res.,* 2000, 97, 271-280.
[31] Tempe, R., Shevenell, L., Lecher, P., Price, J., Geochemical modeling approach to predicting arsenic concentrations in a mine pit lake, *Appl. Geochem.,* 2000, 15, 475-485.
[32] Gupta, A., Jain, R. and Gupta, K., Water quality management for the Talkatora Lake, Jaipur - a case study, *Water Sci. Technol.,* 40(2), 1999, 29-37.
[33] Klavis, M., Briede, A., Parele, E., Rodionov, V. and Klavina, I., Metal accumulation in sediments and benthic invertebrates in lakes of Latvia, *Chemosphere,* 1998, 36, 3043-3051.
[34] Stoyanova, K., Mandjukov, P., Tsakovski, S., Tcholakova, M., Karadjova, I., Gergov, G. and Simeonov, V., In: Hydrochemical and Ecological Characteristics of High-Mountain Lakes. Part I (Eds. J. Carbonnel, J. Stamenov), Bulgarian Academy of Sciences, 1995.
[35] Botova, B., Mandjukov, P., Tsakovski, S., Karadjova, I., Gergov, G. and Simeonov V., In: Hydrochemical and Ecological Characteristics of High-Mountain Lakes. Part II (Eds. J. Carbonnel, J. Stamenov), Bulgarian Academy of Sciences, 1996.
[36] Pekov, G., Mandjukov, P., Tsakovski, S. and Simeonov, V., In: Hydrochemical and Ecological Characteristics of High-Mountain Lakes. Part III (Eds. J. Carbonnel, J. Stamenov), Bulgarian Academy of Sciences, 1997.
[37] Garneli, E. and Haraldson, C., Can increased leaching of trace metals from acidified areas influence phytoplankton growth in coastal waters? *Ambio*, 1993, 22, 308-311.
[38] Marmorek, D., Bernard, D., Sutherland, G. and Wedels, C. Protocol for determining lake acidification pathways, *Water, Air and Soil Pollut.,* 1989, 44, 235-244.
[39] Rognerud, S. and Fjeld, E., Regional survey of heavy metals in lake sediments in Norway, *Ambio,* 1993, 22, 206-214.
[40] Kwiatkowski, R.E., Importance of temporal variability to the design of large lake water quality networks, *J. Great Lake Res.,* 1985, 11, 462-468.
[41] Aldenberg, T., Janse, J.H. and Kramer, P.R.G., Fitting the dynamic model PCLake to a multi-lake survey through Bayesian Statistics, *Ecol. Modell.*, 1995, 78(1-2), 83-99.
[42] Urban, R., Gorham, E., Underwood, K. and Martin, B. Geochemical processes controlling concentrations of Al, Fe and Mn in Nova Scotia lakes, *Limnol. Oceanol.,* 1990, 35, 1516-1520.
[43] Renberg, I., Korsman, T. and Anderson, N., A temporal perspective of lake acidification in Sweden, *Ambio*, 1993, 22, 264-270.

[44] Voutsa, D., Manoli, E., Samara, C., Sofoniou, M. and Stratis, J., A study of surface water quality in Macedonia, Greece: speciation of nitrogen and phosphorus, *Water, Air and Soil Pollut.,* 2001, 129, 13-23.

[45] Voutsa, D., Zachariadis, G., Kouras, A., Anthemidis, A. and Samara C., A quality study of surface waters in Macedonia, Greece. Proc. IMA'99, Halkidiki, Greece, 1999, 383-387.

[46] Simeonov, V., Stratis, J., Samara C., Zachariadis, G., Voutsa, D., Anthemidis, A., Sofoniou, M., Kouimtzis, T., Assessment of the Surface Water Quality in Northern Greece, *Water Res.,* 2003, 37, 4119-4124.

[47] Ramsar Convention Bureau. Criteria for identifying wetlands of international importance, Montrreaux, Switzerland, 1996.

[48] APHA, AWWA, WPCF. Standard methods for examination of water and wastewater, Washington, DC, 1985.

[49] Simeonov, V., Simeonova, P. and Tsitouridou, R., Chemometric Quality Assessment of Surface Waters: Two Case Studies, *Chem. Eng. Ecol.*, 2004, 11(6) , 449-469.

[50] Simeonova, P., Simeonov, D., Spasov, L., Abadzieva, N., Assessment of soil quality by the use of multivariate statistics, *Ecol. Chem. Eng.*, 2007, 14, (in press).

[51] Simeonova, P., Lovchinov, V., Dimitrov, D., Radulov, I., Quality assessment of the Yantra River Monitoring Data, *Ecol. Chem. Eng.,* 2007 14(7), 693-705.

[52] Spanos, T., Simeonov, V., Stratis, J. and Xatzixristou, X. Assessment of Water Quality for Human Consumption, *Microchim. Acta,* 2003, 141,35-40.

[53] Simeonov, V., Adam, E., Tsakovski, S., Mandjukov, P. and Stratis, J. Chemometrical Analysis and Quality Control of Pot Water and its Sources from Athens, Greece, *Tox. Envir. Chem.,* 1995, 50, 57-71.

[54] Simeonov, V., Barbieri, P., Walczak, B., Massart, D.L. and Tsakovski, S., Environmetrical Interpretation of Analytical Data of Marine Organisms from the Black Sea, *Tox. Envir. Chem.*, 2001, 79, 55-72.

[55] Simeonov, V., Sarbu, C., Massart, D.L. and Tsakovski, S., Danube River Water Data Modeling by Multivariate Data Analysis, *Mikrochim. Acta*, 2001, 137(3/4), 243-248.

[56] Simeonova, P., Simeonov, V. and Andreev, G. Water Quality Study of the Struma River Basin, Bulgaria, *Central Europ. J. Chem*. 2003, 2, 121-136.

[57] Simeonova, P., Chemometric treatment of missing elements, *Ann. Univ. Sof.Fac.Chim.,* 2007, 100, 175-186.

[58] Simeonova, P., Lovchinov, V. and Simeonov V., Data Interpretation using Multivariate Statistics for an Aerosol Sample Collection from Northern Greece. *J. Balk. Ecol.,* 2007, 10(2), 197-204.

In: Lake Pollution Research Progress
Editors: F. R. Miranda and L. M. Bernard
ISBN: 978-1-60692-106-7

Chapter 2

LAKES IN THE APULIAN KARST (SOUTHERN ITALY): GEOLOGY, KARST MORPHOLOGY, AND THEIR ROLE IN THE LOCAL HISTORY

Mario Parise*
National Research Council, IRPI, Bari, Italy

ABSTRACT

Karst environments are typically characterized by scarce presence of water at the surface: after a generally short runoff, water infiltrates underground through swallow holes and discontinuities in the rock mass to develop the subterranean complex karst systems made of variable size conduits and caves. However, locally the presence of thick residual deposits, prevailingly consisting of clays, may determine stagnancy of water at the surface, and formation of karst lakes. In Apulia, that is one of the most important regions in the Mediterranean as regards karst because of the extensive outcropping of soluble rocks, several areas show such peculiarities, that led to development of karst lakes. These landforms had a remarkable role in man's history in Apulia, since many

* m.parise@ba.irpi.cnr.it

ancient settlements were established nearby the lakes, due to availability of water. Even during more recent times, exploitation of the hydric resources contained therein resulted of extreme importance, as testified by the many dry-stone wells built within the lakes. The present article is intended to describe the geological and morphological reasons which controlled the lakes formation, the historical and social aspects related to these karst landforms, and the degradation they have been experiencing in recent years. As a consequence of the latter, at the present the lakes do not have great importance as water supplies but, nevertheless, represent habitats of great naturalistic value that are still able to support the ecological functionality and the wet environments with self-vegetation.

INTRODUCTION

Intimate connection between the surface and the subsurface is the fundamental feature of karst, that should be seen as a three-dimensional landscape where every action performed at the surface may be rapidly transferred underground, with devastating effects in terms of pollution and degradation. This derives from the peculiarity of karst, and namely its fragility, which depends upon a number of features, beginning with the main characters of hydrological transport (White, 2002): very limited water runoff at the surface, rapid infiltration of water underground through the network of fissures and conduits in the rock mass, turbulent flow, direct connection between the surface and the water table. The latter, in particular, poses high environmental problems, since a likely pollutant cannot be diluted by passing through different types of rocks in its downward movement, but rather it may reach the water table with all its potential of pollution, thus strongly impacting the water quality (Goldscheider, 2005; Van Beynen and Townsend, 2005; Gunn, 2007).

The evaluation of karst aquifers vulnerability is extremely complex, due to a number of intrinsic factors that make difficult to fully comprehend the hydrological behaviour (Bonacci, 1995; Plagnes and Bakalowicz, 2002). In the last two decades, however, a great deal of effort has been produced in Europe to develop methods to assess aquifer vulnerability in karst, and some specific methodologies have been proposed (COST Action 65, 1995; Doerfliger et al., 1999; Daly et al., 2002).

Due to the above features, karst is characterized by very limited presence of surface water, and by scarcity of runoff (Parise and Pascali, 2003). Surface water supplies in karst areas are generally more limited than in other natural settings, which results in greater human dependency on groundwater supplies than is the case in nonkarst regions with similar climates (Aley, 2000).

The few areas where there is possibility of presence of water at the surface becomes therefore of crucial importance as concerns foundation of the human settlements, and their successive developments as well. At this regard, perennial or temporary lakes in karst settings represent undoubtedly an uncommon and peculiar, if not remarkable, feature.

In karst territories, development of lakes derives essentially from three different situations:

presence of impervious residual deposits, mostly represented by clays, as filling materials of depressions. The difference in permeability between the surrounding soluble rocks, fractured and karstified, and the clays determines stagnancy of water at the surface;

colluvial deposits filling collapse sinkholes, clogging all the possible ways of underground drainage;

emergence of the water table within well-defined depressions, both of natural (sinkholes, baselevel poljes *sensu* Ford and Williams[1], 2007, etc.) and anthropogenic (quarry, open-pit mine, etc.) origin.

In addition, in karst environments several landforms exist where temporary or seasonal lakes may be formed, generally as a consequence of intense and/or prolonged rainstorms: poljes, blind valleys, endhoreic basins, etc. In some cases, it happens that man intervenes to facilitate the underground drainage, in order to gain land for agricultural practices. At this aim, swallow holes are widened, artificial channels realized, and, consequently, the lakes are canceled.

There is also another fundamental difference between non-karstic and karstic terrains: in non-karstic terrains the groundwater divides are assumed to directly underlie the surface divides, as these latter may be determined from the analysis of topographic maps. As clearly stated by Ford and Williams, *"...This approach is acceptable in karst only as an initial working hypothesis, because experience in innumerable karst catchments has shown phreatic and vadose divides to deviate significantly in plan position from overlying surface watersheds. Furthermore, phreatic divides may migrate laterally in response to changing water-flow conditions"* (Ford and Williams, 2007, page 146).

In flat karst landscape, as is the case for Apulia region of southern Italy, determining the groundwater divides is still more difficult, due to subtle topography, and uncertainties in establishing even the surface topographic divides.

Apulia is among the most important karst regions in the Mediterranean Basin: an elongated peninsula, forming the heel of the Italian boot, protruding toward the Balkans and Greece in south-eastern direction, it is constituted by carbonate rocks (figure 1) that formed the Apulia Carbonate Platform, which acted as the foreland during the phases of building-up of the Apenninic Chain in Miocene time (Doglioni et al., 1994). Its configuration, with a sub-horizontal, slightly deformed but intensely fractured, carbonate bedrock was essentially sculptured by karst processes, that started to develop after the Cretaceous age, when part of the region emerged as isolated islands from the sea (Bosellini, 2002). Later on, a wider development of karst occurred throughout the region, up to forming most of the presently

---

[1] Baselevel poljes are "... water-table dominated, occurring where dissolution has lowered the karst surface to the regional epiphreatic zone; i.e. they are windows on the water table" (Ford & Williams, 2007, page 364).

observed landforms, and the extensive network of underground caves as well (Sauro, 1991; Bruno et al., 1995; Parise, 1999).

Among the karst landforms, some areas host lakes related to the presence of impervious terrains.

These lakes had a remarkable role in man's history in Apulia, since many ancient settlements were established in their proximity, due to availability of water. Even during more recent times, exploitation of the hydric resources contained therein resulted of extreme importance, as testified by the many dry-stone wells built within the lakes. The present paper intends to delineate the main geological and morphological characters that played a role in the development of the Apulian karst lakes, by describing some peculiar situations in the different sub-regions of Apulia: Gargano, Murge and Salento (figure 1).

GARGANO

In Gargano, the promontory forming the northern sector of the region, the most interesting situation as regards karst lakes is represented by the S. Egidio polje, located between the towns of San Giovanni Rotondo and Monte Sant'Angelo. It is a wide karst valley developed along an important fault line (the Carbonara Valley transcurrent fault; Morsilli, 2008). Bounded to the north by some of the highest ridges in Gargano, including the over 1000 mt-high Mt. Calvo, the Sant'Egidio polje has been interpreted in the past by some authors as a pull apart basin, related to movement of the aforementioned fault (Guerricchio, 1986). Nevertheless, geomorphology of the area, and the related hydraulic and hydrogeological characters, cannot be fully explained without considering the karst nature of the territory, and the role played by karst processes in the development of the valley, that actually formed as a structural polje (Nicod, 1972). The main karst features of the polje, on the other hand, were already pointed out by the first scholars that analyzed the area, during the first half of the XX century (Checchia Rispoli, 1915; Baldacci, 1950).

Depending upon local topography, the "windows" may appear at the surface as a valley where a subterranean stream may be exposed and flow across the surface, or as a closed depression filled by water, that is a lake.

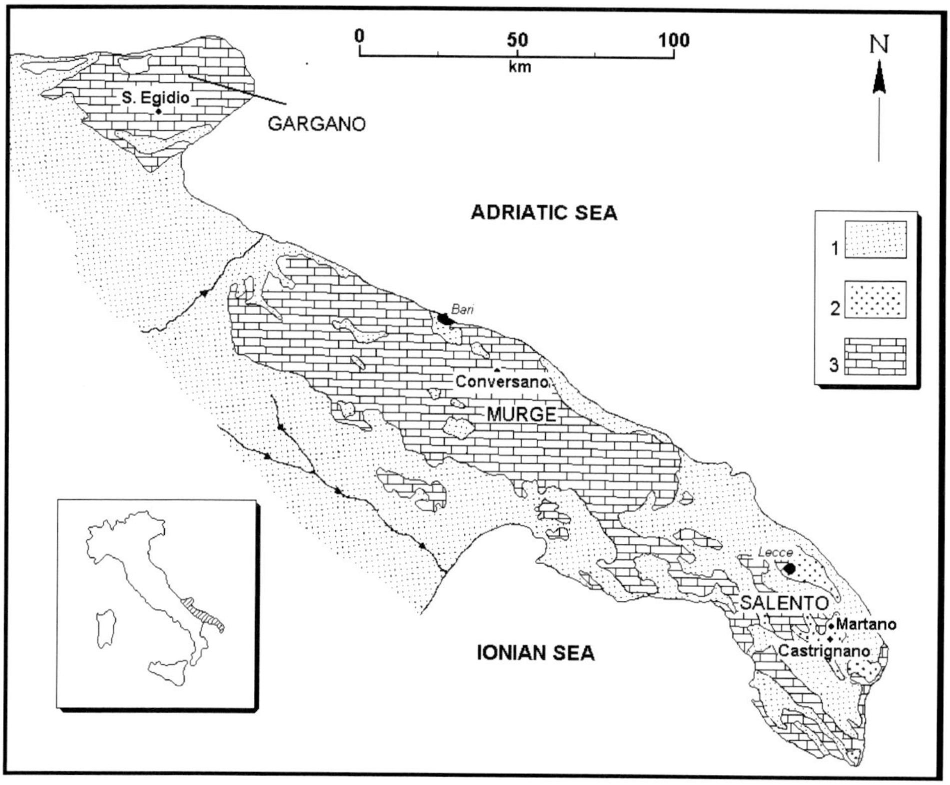

Figure 1. Geological scheme of Apulia (southern Italy), showing the localities cited in the text. Explanation: 1) recent clastic cover (Pliocene-Pleistocene); 2) bioclastic carbonate rocks (Paleogene) and calcarenites (Miocene); 3) carbonate rocks (Jurassic – Cretaceous).

The polje (figure 2) develops in SSW-NNE direction, covering some tens of hectares at elevations between 460 and 490 m a.s.l. Due to the before recalled difficulties in delineating surface watersheds in karstic territories, the limits of the hydrographic catchment are very poorly defined to the west (a morphological saddle separates it from the area where San Giovanni Rotondo is located) and to the east. The northern margin is, on the other hand, very well defined at the surface.

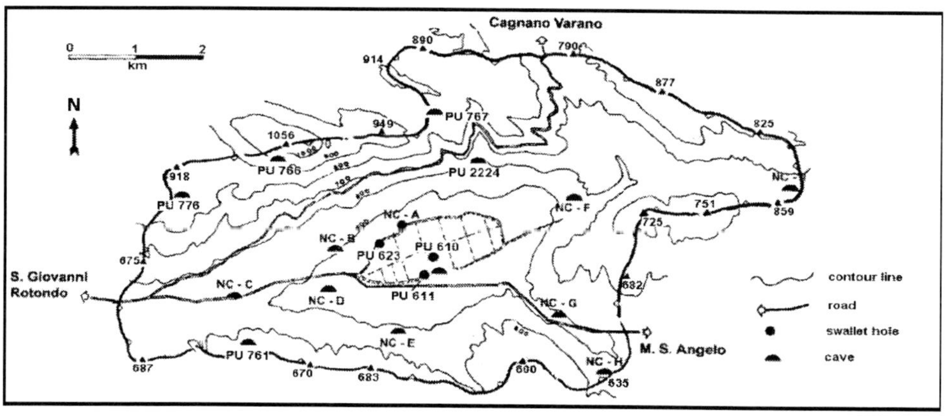

Figure 2. Sketch map of the Sant'Egidio polje (simplified after Fusilli, 2005; drawing by P. Guliani).

Several swallow holes and caves are distributed in the plain area and at its margins. Especially those located in the lowest sector of the polje played an important role in the development of the lake.

As a matter of fact, during the past centuries most of the area was occupied by a seasonal lake, one mile-long, three miles-around, and 7 palms-deep, according to the historical testimony by Manicone (1808). The water table was periodically subject to drought periods, which in some occasions were so severe to cause losses to the local economy, which was mostly based on fishing. After these events, and particularly the severe drought which occurred in 1830 (Fusilli, 2005), man, clogging the main swallow holes in the polje, was able to have again back the lake. At the turn of the XX century, a time of intense reclamation works began in many Italian regions, in the attempt of both eliminating likely sources of illness (malarial fever, especially) and of gaining new lands to agricultural practices. Such activities strongly interested the Apulia region, where several areas were object of reclamation works. Thus, the first attempts in drying up the lake at Sant'Egidio were performed through the realization of a surface channel network and the opening of the swallets. After several phases of works, the lake was actually drained.

Since 1970, a phase of degradation began for the Sant'Egidio polje. This was essentially caused by the lack of maintenance of the man-made channel network, up to that time performed by a Reclamation Society. At the same time, the liquid wastes coming from the nearby town of San Giovanni Rotondo were transferred, again by means of a surface network of open-channels, to the polje, in order to be drained at the main swallets. It is easy to understand the negative effects, in terms of pollution, that such management of the karstic territory caused in the last decades! Further, when the swallets were not able to completely drain the liquid wastes, these inundated the fields, causing pollution even at the surface.

As it can be easily appreciated from this brief description, it is clear that the environmental situation at Sant'Egidio was very bad in the time span 1970-90. Later on, after that San Giovanni Rotondo realized a depuration system, things went someways better, even though further episodes of mismanagement and degradation, mostly consisting of discharge of solid wastes into the main caves and shafts of the area, have been reported quite continuously. This, however, is a serious problem for the whole region, that is also exacerbated by lack of control from the Local Authorities (Calò and Parise, 2006). Today, many sites within the Sant'Egidio polje and at its surroundings as well, are still suffering intense degradation (Fusilli, 2005).

Conversano

The town of Conversano is located in the central part of Apulia, in the Murge sub-region, where karst affects Cretaceous limestones and dolomitic limestones, creating a flat landscape, dotted with karst landforms (figure 3) such as dolines, poljes, swallow holes (Sauro, 1991; Parise, 1999, 2006). The Cretaceous bedrock is unconformably overlain by Tertiary and Lower Pleistocene calcarenites in the proximity of the Adriatic coastline. Quaternary alluvium and eluvial deposits fill the valley floors and the topographic depressions.

Within such a context, in an overall flat environment, even slight changes in topography and little differences in elevations may become important to produce small basins or topographic depressions, produced mostly by the dissolution of soluble rocks, often working along the main discontinuity systems in the rock mass. Epikarst is limited to depth of a few meters; this means that, even though jointing in the rock mass is moderate to high, there is tapering closure of the solutionally enlarged joints rapidly with depth. In addition, these joints are also filled with red clays.

Deposition of impervious terrains within the topographic depressions, then, is the final element which concurs to the development of karst lakes, or at least to seasonal or temporary presence of water at the surface.

Infilling of the lakes consists mostly of residual deposits ranging from silty clays (*terre rosse*) to silty sands, with local intercalations of volcaniclastic materials. Interlayering of deposits with different degree of permeability, resting over an intensely fractured and karstified carbonate bedrock, makes these materials the recharge area for the underlying aquifer in the Mesozoic limestones. Poor permeability of the infilling deposits allows the rainwater to be collected and temporarily stored at the dolines and other depressions where they cover the carbonate bedrock. This determines the formation of temporary, small and shallow, karst lakes.

Ten karst lakes are distributed in the whole territory of Conversano (figures 3 and 4). The lakes may be subdivided into three groups, based upon the local morphology:

- a group having tectono-karstic origin, that includes those lakes located in valleys bounded by faults or important tectonic lineaments;
- a group of erosional-karstic origin, with the lakes located within the lines of the ancient water network;
- a group of karstic origin, that is the lakes occupying (entirely or in part) the bottom of collapse dolines.

The third group is the most numerous, counting 4 lakes (Padula, Petrullo, Chienna and Vignola), whilst the two other groups have 3 lakes each (Iavorra, Minuzzi and Agnano, as regards the first group; S, Vito, Sassano and Castiglione for the second group).

All the lakes have been historically modified by man, who tried to take advantage of the natural situation leading to presence of water at the surface or at maximum in the first meters from the ground. Until a few decades ago, the lakes were the only natural water resource easily available to the local population who, in order to use it during the dry season, built numerous bell-shaped wells in the most depressed zone of dolines, aimed at recovering the volumes of water withheld by the continental deposits (figure 5).

It was, however, since antiquity that the temporary lakes of karst origin at Conversano represented a resource of inestimable value for the local population, especially given the fact that they are located in a region where there is lack of surface water resources. In prehistoric times, the first settlements in the Conversano territory were established nearby some of the lakes: the S. Giacinto cave, near the Chienna lake, hosted Neolithic findings (Coppola, 1981), whilst remnants of a settlement dating back to several centuries before Christ have been found at the top of a ridge over the Castiglione lake (L'Abbate, 1981).

Figure 3. Geomorphological map of the Conversano area. Explanation: 1) scarp; 2) ridge; 3) old trace of river course (a), and water course in active downcutting (b); 4) valley with flat bottom; 5) doline; 6) morphological saddle. Contour interval 50 meters. Location of the ten lakes around Conversano is shown.

In the recent past, presence of the lakes was crucial even for the agricultural economy. It was then that, in order to use the water accumulated at the bottom of these depressions, the local inhabitants excavated many wells, internally coated with dry-stone walls (L'Abbate, 1992). Interesting informations about the characteristics of these works, and the technique of realization, were found in texts dating back to the second half of the XIX century by the architect Sante Simone, who in some documents describes the Conversano lakes and the hydraulic works therein realized, starting with the wells (Simone, 1880). Detailed drawings illustrate the texts, showing the inside of the wells, and their bottoms as well (figure 6).

The number of wells in each of the Conversano lakes ranges from a minimum of three at the Minuzzi lake, to the maximum number of thirty-one at the Sassano lake (table 1): the wells have a bell-shaped section, whilst their depth is variable.

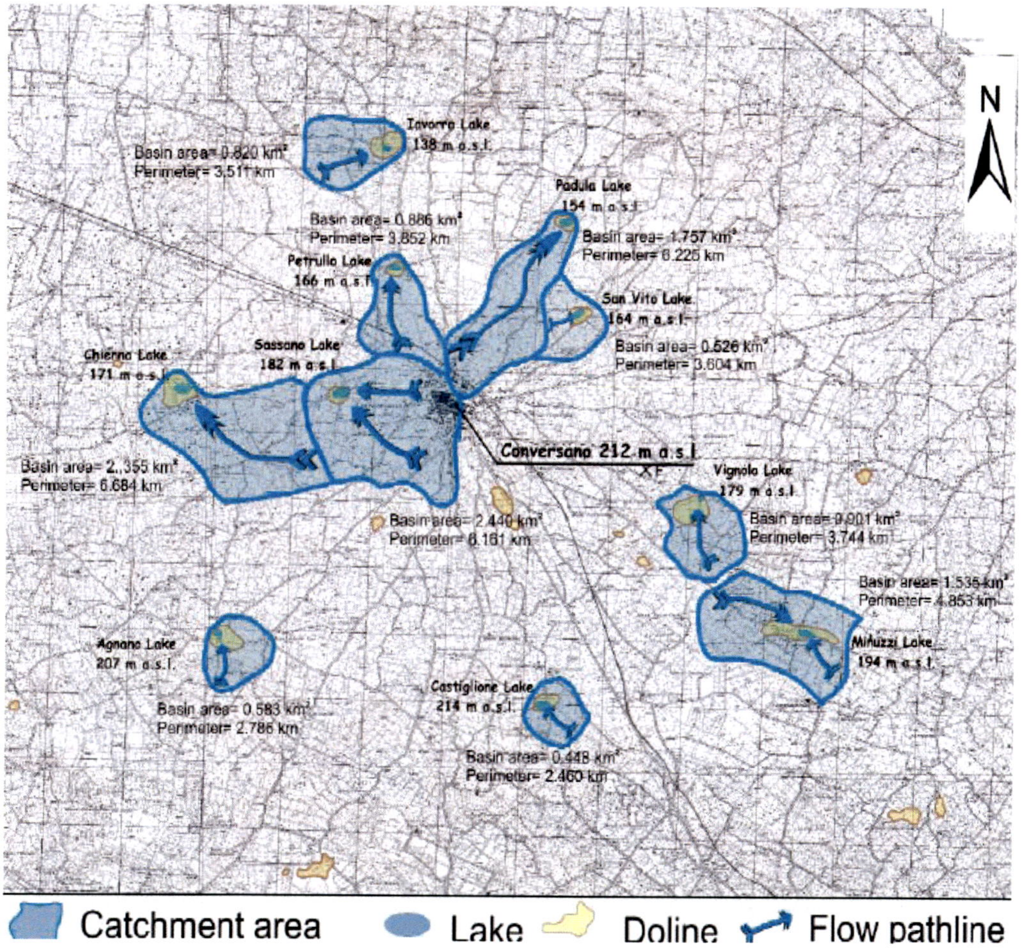

Figure 4. Catchments of the ten karst lakes in the Conversano territory (after Lopez et al., 2008).

Geomorphology of the lake catchment basins is extremely variable, ranging from more or less wide dolines to very flat valleys, discernible from the surrounding landscape with difficulty (this is the case, for example, of Iavorra lake, see also figure 7).

Notwithstanding the indubitable significance of these features for the local history, during the second half of the past century, and especially in the last decades, the lakes at Conversano had to register many episodes of degradation, starting with destruction of some wells, robbing of the coating stones (widely used for building rural houses in the area), and use of the lake area for illegal discharge of wastes. All these actions, of course, were conducive to severe pollution. During the 90's, however, something started to change, when the Province Administration of Bari funded two projects for exploitation of the lakes, and protection of the local ecosystem. These projects reached in rising again interest about the lakes, within the framework of natural, sustainable tourism exploitation, based upon respect of the wildlife and of the natural landscape.

a

b

Figure 5. Lake Sassano, at the western outskirts of Conversano. Note the difference in water levels between the wet (photo a, taken in March) and the dry (photo b, taken in September) seasons; note also the many wells in the lake. Sassano lake has the highest number of wells, counting to 31 (see also table I).

SALENTO

In Salento, the southernmost portion of Apulia, characterized by the lowest rainfall values in the region, availability of water resources at the surface was still a greater need than in the rest of Apulia (Costantini, 1988). At several sites in Salento, the presence of depressions, endhoreic basins, and valleys intervening between the slightly elevated ridges of the peninsula (the so-called *Serre Salentine*) detrmined the topographic possibilities to have stagnancy of water at the surface, or within the first meters or decimeters from the ground surface (De Giorgi, 1922).

In a way similar to that described for the Conversano lakes in the Murge, some areas became characterized by the presence of several man-made wells, built with the same technique previously described, to allow the preservation of water (figure 8), and its exploitation by man. During the wet seasons, these sites sometimes presented seasonal lakes at the surface, but for most of the year they were just the main site of collection of water resource, without the emergence of water at the surface.

a)

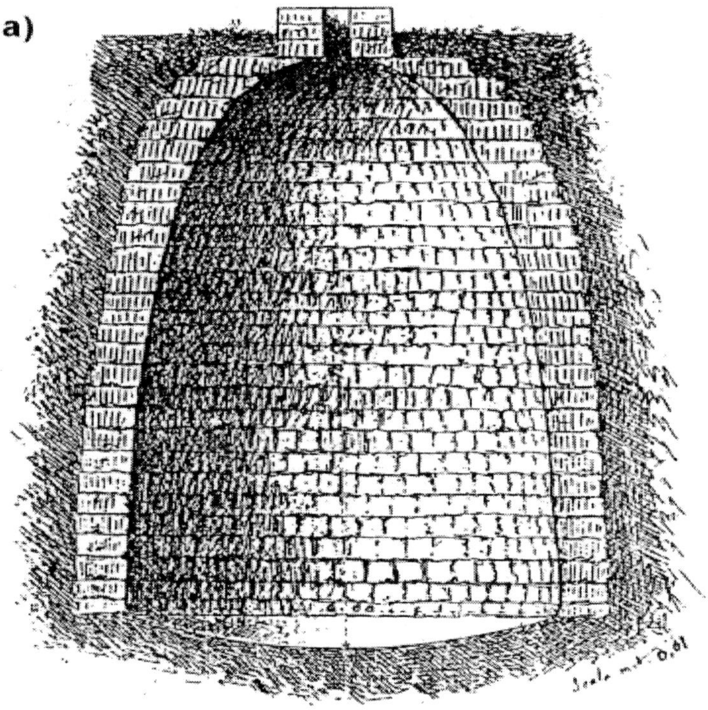

Figure 6. Continued.

Table 1. Main features of the Conversano lakes

Lake	Catchment area (km^2)	Catchment perimeter (km)	Lake area (m^2)	Average elevation (m a.s.l.)	Length main axis (m)	Direction main axis	Shape	Number of wells
Iavorra	0.820	3.511	6.025	138	110	N-S	Triangle	6
Padula	1.757	6.225	3.450	154	100	NNW-SSE	Rhomboid	4
Petrullo	0.886	3.852	4.125	166	105	E-W	Elongated	9
S. Vito	0.526	3.604	2.550	164	65	NE-SW	Circular	7
Chienna	2.355	6.684	4.000	171	115	WSW-ENE	Elongated	9
Sassano	2.440	6.161	5.250	182	130	NE-SW	Triangle	31
Vignola	0.901	3.744	1.975	179	55	NE-SW	Circular	6
Minuzzi	1.535	4.853	5.100	194	190	E-W	Triangle	4
Agnano	0.583	2.786	10.425	207	175	N-S	Rhomboid	12
Castiglione	0.448	2.460	3.175	214	65	E-W	Triangle	6

b

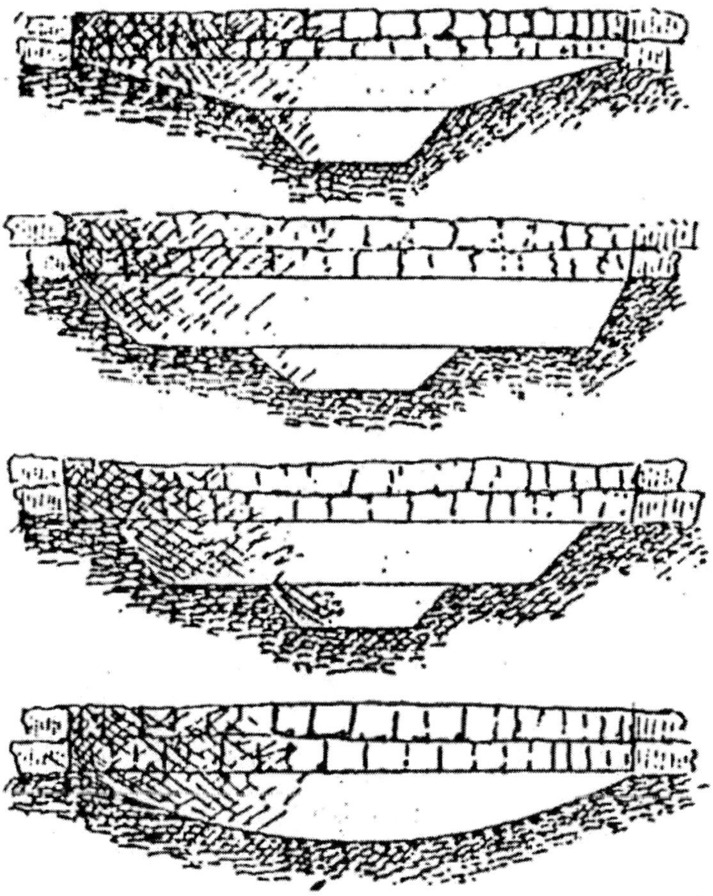

c

Figure 6. The wells at the Conversano lakes (after Simone, 1880): a) detail of a well in the Iavorra Lake; b) sections of the different types of wells; c) drawings of the bottom of the wells.

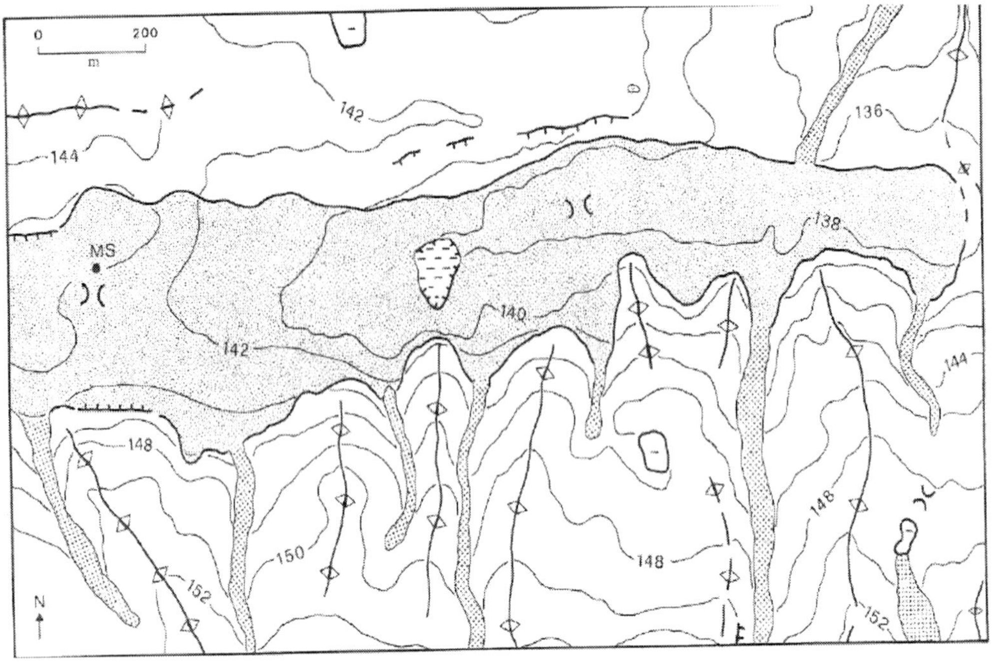

Figure 7. Geomorphological map of the Iavorra Lake. For symbols explanation, see figure 3.

Many villages in Salento still host today the remnants of ancient wells, more or less nowadays englobed within the urban areas: Martignano, Soleto, Martano, Castrignano dei Greci and Zollino (figure 9) are probably the most well-known, where tens of wells (locally called *pozzelle*) have been realized.

Figure 8. The *"pozzelle"* at Castrignano, in Salento.

The sector of the Salento peninsula where the *pozzelle* are present is the core of the so-called *Grecìa Salentina*, where a dialect strictly related to the Greek language *(griko)* is spoken still today. As shown in figure 9, originally this dialect interested wide areas of central Salento, and during the XIV-XV centuries it almost extended from the Ionian to the Adriatic coast, along a line connecting Gallipoli to Otranto. Today, it is much less limited, being restricted to an area which includes about ten villages. The origin of this ancient linguistic tradition remains uncertain, and someone reputes it an heritage of the Byzantine control. It seems, indeed, more likely that it concerns a phenomenon of progressive isolation which involved the most ancient greek-speaking population of Magna Graecia.

The persistence of the *griko* dialect is well evidenced by several names of localities in the area, and also influenced many terms describing karst features and landforms (Parise et al., 2003.)

Even in Salento, memory of the use of these man-made wells, and the historical testimony of the seasonal karst lakes are progressively being lost. The *pozzelle* are often in a very bad state, with several calcareous stones destroyed or broken, and at many sites they have completely been canceled to leave space to urban expansions of the towns. This, of course, determined a loss of memory of the long struggle of man to have available the water in semi-dry regions as this area of southern Italy.

Figure 9. Map of *Grecìa Salentina* (modified after Rohlfs, 1974). The localities with presence of *"pozzelle"* are shown in bold.

Conclusion

In territories where soluble rocks crop out, the morphology assumes peculiar aspects and the hydrogeological features are strongly affected by karst processes. Due to intrinsic fragility of karst, man can easily, and sometimes dramatically, change the natural setting, especially as regards surface and subsurface hydrology. Presence of water at the surface, up to development of lakes, is not typical of karst settings. Thus, those sites where lakes are formed in karst have a double importance, either as an uncommon situation and for the historical significance that the presence of water at the surface had for the past communities. It is not a case that most of the ancient settlements in karst territories were founded nearby the few situations where water was available at the surface or near the surface.

Even though of limited size, karst lakes can be therefore of great importance, and their study may provide interesting insights into the local history, of both the natural environment and man. At the same time, given the low relief of Apulia, and the long history of occupation of this territory by man, many ancient remnants of these landscapes have been lost, or are being lost. This is particularly true for the last decades, when an often uncontrolled urban expansions characterized wide areas of Apulia, and many traces of the historical heritage have been irrimediably destroyed. In other cases, the lakes, or the slight depressions once they occupied, and the karst caves and swallow holes in their vicinity, have become sites of frequent discharge of solid and liquid wastes, with severe consequences to the natural environment, the karst ecosystem, and the quality of groundwater. Pollution and degradation events are continuously registered in Apulia, as well as in many other italian regions. Any study regarding karst, and particularly karst groundwater, may be very useful to contribute to create an awareness of the fragility of karst, and to avoid, or at least minimize, further events of pollution in this environment of great naturalistic relevance.

References

Aley, T. (2000). Water and land-use problems in areas of conduit aquifers. In A.B. Klimchouk, D.C Ford, A.N. Palmer and W. Dreybrodt (Eds.), *Speleogenesis. Evolution of karst aquifers* (pp. 481-484). National Speleological Society, Huntsville, Alabama.

Baldacci, O. (1950) Sul carsismo di superficie del ripiano di S. Giovanni Rotondo (Promontorio Garganico). *Bollettino Società Geografica Italiana,* 6-8.

Bonacci, O. (1995) Groundwater behaviour in karst: example of the Ombla spring (Croatia). *Journal of Hydrology, 165* (1-4), 113-134.

Bosellini, A. (2002) Dinosaurs "re-write" the geodynamics of the eastern Mediterranean and the the paleogeography of the Apulia Platform. *Earth-Science Review, 59*, 211-234.

Bruno, G., Del Gaudio, V., Mascia, U., and Ruina, G. (1995) Numerical analysis of morphology in relation to coastline variation and karstic phenomena in the southeastern Murge (Apulia, Italy). *Geomorphology, 12*, 313-322.

Calò, F. and Parise, M. (2006) Evaluating the human disturbance to karst environments in southern Italy. *Acta Carsologica, 35* (2), 47-56.

Canora, F., Fidelibus, M.D., Sciortino, A., and Spilotro, G. (2008) Variation of infiltration rate through karstic surfaces due to land use changes: a case study in Murgia (SE Italy). *Engineering Geology*, *99*, 210-227.

Checchia Rispoli, G. (1915) La conca di S. Egidio sul Gargano. *Il Foglietto,* Lucera, *2*, 1-6.

Coppola, D. (1981). Le più antiche tracce di popolamento umano nel territorio: il paleolitico. In AA.VV., *Il popolamento antico nel sud-est barese* (pp. 21-35). Museo Civico di Conversano.

COST Action 65 (1995) *Karst groundwater protection.* Final Report EUR 16547, European Commission.

Costantini, A. (1988) Del modo di conservare le acque e la neve. Pozzelle e neviere nel Salento leccese. *Sallentum, 18.*

Daly, D., Dassargues, A., Drew, D., Dunne, S., Goldscheider, N., Neale, S., Popescu, I.C., and Zwahlen, F. (2002) Main concepts of the European approach for karst groundwater vulnerability assessment and mapping. *Hydrogeology Journal, 10* (2), 340-345.

De Giorgi, C. (1922) *Descrizione geologica e idrografica della Provincia di Lecce.* Tip. Ed. Salentina, Lecce, 263 pp.

Doerfliger, N., Jeannin, P.Y., and Zwahlen, F. (1999) Water vulnerability assessment in karst environments: a new method of defining protection areas using a multi-attribute approach and GIS tools (EPIK method). *Environmental Geology, 39* (2), 165-176.

Doglioni, C., Mongelli, F., and Pieri, P. (1994) The Puglia uplift (SE Italy): an anomaly in the foreland of the Apenninic subduction due to buckling of a thick continental lithosphere. *Tectonics, 13*, 1309–1321.

Ford, D., and Williams, P. (2007). *Karst hydrogeology and geomorphology.* John Wiley and Sons.

Fusilli, C. (2005) Il polje di Sant'Egidio presso San Giovanni Rotondo (FG): note storiche e descrittive. *Grotte e Dintorni, 10,* 35-40.

Goldscheider, N. (2005). Karst groundwater vulnerability mapping: application of a new method in the Swabian Alb, Germany. *Hydrogeology Journal, 13*, 555–564.

Guerricchio, A. (1986) Esempi di bacini pull apart nel Gargano (Puglia settentrionale). *Geologia Applicata e Idrogeologia, 21*, 25-36.

Gunn, J. (2007). Contributory area definition for groundwater source protection and hazard mitigation in carbonate aquifers. In M. Parise and J. Gunn (Eds.), *Natural and anthropogenic hazards in karst areas: recognition, analysis and mitigation* (pp. 97-109). Geological Society of London, spec. publ. 279.

L'Abbate, V. (1981). Il popolamento antico nell'età dei metalli. In AA.VV., *Il popolamento antico nel sud-est barese* (pp. 69-98). Museo Civico di Conversano.

L'Abbate, V. (1992) Conversano. Ricerche sui "laghi" del territorio (1877-1890). In: AA.VV., *L'Architetto Sante Simone* (pp. 305-311). Schena ed., Fasano.

Lopez, N., Spizzico, V., and Parise, M. (2008) Geomorphological, pedological, and hydrological characteristics of karst lakes at Conversano (Apulia, southern Italy) as a basis for environmental protection. *Environmental Geology.*

Manicone, M. (1808) *La fisica Appula.* Napoli, 1, 234 pp.

Morsilli, M. (2008) Il promontorio del Gargano, dal punto di vista geologico. *Proceedings Regional Meeting "Spelaion 2006"*, Borgo San Celano, 19-47.

Nicod, J. (1972) *Pays et paysages du calcaire.* Presses Universitaires de France, Paris, 244 pp.

Parise, M. (1999) Morfologia carsica epigea nel territorio di Castellana Grotte. *Itinerari Speleologici, 8*, 53-68.

Parise, M. (2006) Geomorphology of the Canale di Pirro karst polje (Apulia, southern Italy). *Zeitschrift für Geomorphologie, N.F. 147*, 143-158.

Parise, M., and Pascali, V. (2003) Surface and subsurface environmental degradation in the karst of Apulia (southern Italy). *Environmental Geology, 44*, 247-256.

Parise, M., Qiriazi, P., and Sala, S. (2004) Natural and anthropogenic hazards in karst areas of Albania. *Natural Hazards and Earth System Sciences, 4*, 569-581.

Parise, M., Qiriazi, P., and Sala, S. (2008) Evaporite karst of Albania: main features and cases of environmental degradation. *Environmental Geology, 53 (5)*, 967-974.

Parise, M., Federico, A., Delle Rose, M., and Sammarco, M. (2003) Karst terminology in Apulia (southern Italy). *Acta Carsologica, 32 (2)*, 65-82.

Plagnes, V., and Bakalowicz, M. (2002) The protection of a karst water resource from the example of the Larzac karst plateau (south of France): a matter of regulations or a matter of process knowledge? *Engineering Geology, 65* (2-3), 107-116.

Rohlfs, G. (1974) *Scavi linguistici nella Magna Grecia.* Congedo ed., Galatina, 301 pp.

Sauro, U. (1991) A polygonal karst in Alte Murge (Puglia, southern Italy). *Zeitschrift für Geomorphologie, N.F. 35* (2), 207-223.

Simone, S. (1880) *Su alcune antiche cisterne nel territorio di Conversano.* Min. Pubblica Istruzione, Direz. Gen. Antichità e Belle Arti, 1860-90, b. 7, fasc. 13, State Archive of Rome.

Van Beynen, P., and Townsend, K. (2005) A disturbance index for karst environments. *Environmental Management, 36* (1), 101–116.

White, W. B. (2002) Karst hydrology: recent developments and open questions. *Engineering Geology, 65* (2–3), 85–105.

In: Lake Pollution Research Progress
Editors: F. R. Miranda and L. M. Bernard
ISBN: 978-1-60692-106-7

Chapter 3

ECOTOXICITY AND BIOACCUMULATION OF TOXIN FROM *CYLINDROSPERMOPSIS RACIBORSKII*: TOWARDS THE DEVELOPMENT OF ENVIRONMENTAL PROTECTION GUIDELINES FOR CONTAMINATED WATER BODIES

Susan H. W. Kinnear[1], Leo J. Duivenvoorden[2] and Larelle D. Fabbro[2]

[1] Institute for Sustainable Regional Development,
CQUniversity Australia,
Bruce Highway, Rockhampton, QLD, 4702.
[2] Centre for Environmental Management,
CQ University Australia,
Bruce Highway, Rockhampton, QLD, 4702.

Keywords: aquatic ecosystem protection guidelines, blue-green algae, cyanobacteria, *Cylindrospermopsis*, cylindrospermopsin, bioaccumulation, ecological effects, ecotoxicity.

ABSTRACT

Cylindrospermopsis raciborskii is a commonly encountered cyanobacterium (blue-green algae) found in water bodies worldwide, particularly tropical and subtropical lakes and reservoirs. The changing global climate, increasing eutrophication from agricultural runoff and other factors are contributing to an increased occurrence of this alga in lakes and reservoirs worldwide, with blooms now also reported in temperate areas. Blooms pose a serious concern for water managers, since *C. raciborskii* growth is often accompanied by production of the toxin, cylindrospermopsin (CYN). This toxin clearly represents a human health concern but also targets a range of aquatic plants and animals. In recent years, research into the potential risks associated with CYN contamination of human drinking water supplies has grown considerably. However, by comparison, few studies have examined the potential for, and the influences on, ecological (environmental) effects associated with toxin-producing blooms. Consequently, whilst the World Health Organisation is working towards a provisional guideline for CYN in drinking water, the environmental management of toxic blooms remains underdeveloped: ecological risks appear to be poorly recognised, evaluated or minimised. This chapter will review the current research available on the toxicity and bioaccumulation of toxin from *C. raciborskii*, with a focus on managing the environmental toxicity of blooms. Aspects including algal growth dynamics, toxin production, modes of exposure in the aquatic environment, influences on uptake routes and potential for toxic metabolism and bioaccumulation in target organisms are discussed. Furthermore, to address the gap in environmental research concerning CYN, a summary of recent, laboratory-based ecotoxicity studies is given. Finally, a model to calculate the ecological risks associated with *Cylindrospermopsis* blooms is provided: this involves a consideration of cell concentrations, toxin concentrations and the relative proportions of cell-bound and aqueous toxin. This chapter is designed to raise awareness of the potential ecological effects associated with *C. raciborskii* in lakes and reservoirs worldwide, and to stimulate further work towards the development of ecologically relevant guidelines for toxic blooms in affected water bodies.

Abbrevations

CYNcylindrospermopsin;
CYN_{EXC} extracellular cylindrospermopsin;
CYN_{INC} intracellular cylindrospermopsin;
CYN_{TOT} total cylindrospermopsin (intracellular plus extracellular),
LPS lipopolysaccharide(s);
Q_{CYN}cylindrospermopsin cell quota,
TDI tolerable daily intake;
WSP water safety plan.

1.0. Introduction

Cyanoprokaryotes (cyanobacteria, blue-green algae) are a natural and ancient part of aquatic ecosystems. The group is represented by those non-nucleated organisms characterized by an ability to synthesize chlorophyll a [Whitton and Potts 2000] but which lack membrane-bound organelles and sexual reproduction [Komárek and Anagnostidis 1986]. The cyanoprokaryotes are a diverse and adaptive group, occupying a broad range of habitats including Antarctic ice shelves and volcanoes. However, they are most commonly of interest as a component of the phytoplankton populations of freshwater, estuarine and marine environments. Of particular importance are those species known to form blooms, especially where these are associated with the production of potent toxins.

Some cyanoprokaryote genera have superior growth abilities due to the presence of buoyancy-regulating gas vacuoles, often combined with the capability to capture a variety of light wavelengths and undertake nitrogen fixation [Bowling 1994]. Cyanoprokaryote dominance has been recorded under conditions of low light penetration, water column stability, warm water temperatures and high nutrient availability but low nitrogen to phosphorus ratios [Fabbro and Duivenvoorden 1996; Jacoby *et al.* 2000; Hesse and Kohl 2001; Saker and Griffiths 2001]. Such conditions are often characteristic of tropical and subtropical riverine impoundments and inland lakes. Here, rapid cell division often leads to cell concentrations reaching 'bloom' proportions, currently defined in Australia as > 15,000 cells mL^{-1} [Jones et al. 2002].

Sometimes, cyanoprokaryote blooms are associated with the production of poisonous compounds known as cyanotoxins. Toxin production is highly variable within and between blooms according to speciation, genetic composition, and the prevalence of conditions that allow maximal toxin production by individual cells [Carmichael 2001b].

Toxic algal blooms exert acute and chronic lethal and sublethal effects on a range of terrestrial and aquatic organisms. The reviews of Schwimmer and Schwimmer [1968], Beasley *et al.* [1989], Carmichael and Falconer [1993] and Duy *et al.* [2000] provide substantial evidence of animal deaths associated with intoxication by cyanoprokaryotes. Mortalities have been recorded from livestock, domestic fowl, pets and wildlife; extensive fish kills are also common. Several accounts have linked adverse effects in terrestrial species with intoxication by algal blooms (for example, Matsunaga *et al.* [1999]). However, the accuracy of these might be questioned, particularly prior to the 1980s when understanding of cyanotoxins was limited due to non-rigorous reporting of bloom events. Effects on humans have also been comprehensively reviewed [see Falconer 2001; Carmichael *et al.* 2001], including (most recently), the risks for public health due to consumption of contaminated seafood [Ibelings and Chorus 2007].

2.0. *Cylindrospermopsis Raciborskii* and Its Toxin

2.1. *Cylindrospermopsis raciborskii*

Cylindrospermopsis raciborskii Woloszynska Seenayya et Subba Raju 1972 is a filamentous and heterocystous cyanobacterium [Komárek and Anagnostidis 1989]. Blooms of *C. raciborskii* have been reported from rivers, impoundments, lakes, ponds and dams, and are frequently accompanied by toxin production [Branco and Senna 1994; Shaw *et al.* 1999; Bouvy *et al.* 2000; McGregor and Fabbro 2000; Fastner *et al.* 2003; Saker *et al.* 2003; Bouaïcha and Nasri 2004]. In 1979, *C. raciborskii* was implicated in the 'Palm Island mystery disease', where toxin exposure was responsible for the hospitalisation of over one hundred children suffering acute gastroenteritis [Byth 1980; Hawkins *et al.* 1985]. Australia appears to be a hotspot for *C. raciborskii*, with blooms frequently documented from rivers, impoundments, lakes, ponds and dams in Queensland [Saker *et al.* 1999a; Saker and Eaglesham 1999; Shaw *et al.* 1999; McGregor and Fabbro 2000] as well as in other states [Baker 1996]. Recently, global climate change has been examined as a trigger for the widespread distribution, frequency and duration of toxic *C. raciborskii* blooms, particularly in temperate areas [Garnett et al. 2003; Neilan *et al.* 2003; Briand *et al.* 2004; Chonudomkul *et al.* 2004; Wiedner *et al.* 2007].

2.2. Cylindrospermopsin

Cylindrospermopsin (CYN) is a tricyclic alkaloid cytotoxin produced by *C. raciborskii* [Ohtani et al. 1992] and several other blue-green species including *Umezakia natans, Aphanizomenon ovalisporum, Aph. flos-aque, Anabaena bergii* var *limnetica, Raphidiopsis curvata* and *Lyngbya wollei* [Shaw *et al.* 1999; Fergusson and Saint 2000; Schembri *et al.* 2001; Preußel *et al.* 2006; Seifert *et al.* 2007]. Structurally, this highly water-soluble molecule is a sulfated guanidinium zwitterion with relatively low molecular weight [Sivonen and Jones 1999; Shaw *et al.* 2000]. CYN inhibits both protein synthesis and glutathione synthesis in mammalian hepatocytes [Runnegar *et al.* 1994; Runnegar *et al.* 1995] and is associated with genotoxic action such as DNA fragmentation, strand breakage and mutation [Shen *et al.* 2002; Reisner *et al.* 2004; Humpage *et al.* 2005]. The nucleotide structure of CYN together with the presence of the guanidine and sulphate groups have aroused suspicion that CYN may also be a carcinogen [Ohtani *et al.* 1992; Humpage *et al.* 2000; Shen *et al.* 2002].

2.3. Ecological Effects of *C. raciborskii* and CYN

CYN is of particular environmental concern since its primary mechanism of action – protein synthesis inhibition – applies to all aquatic species. In zooplankton and phytoplankton, toxic *C. raciborskii* blooms have been implicated in local, transient and episodic changes in diversity, abundance, community composition and size classes [Bouvy *et al.* 2000; Bouvy *et al.* 2001; Fabbro *et al.* 2001; Nogueira *et al.* 2004a; Leonard and Paerl 2005; Nogueira *et al.* 2006]. Whole cell extracts or live cultures of *C. raciborskii* containing CYN have caused tissue injuries and changes to growth, behaviour, mortality and reproduction in aquatic animals and plants [White 2006; White *et al.* 2006; Kinnear *et al.* 2007; White *et al.* 2007; Kinnear *et al.* 2007 ; Kinnear *et al.* 2008]. In addition, bioaccumulation of CYN has been shown in *Melanoides tuberculata* snails [White *et al.* 2006]; *Bufo marinus* tadpoles [White *et al.* 2007]; *Anodonta* mussels [Saker *et al.* 2004]; and *Cherax* crayfish and *Melanotaenia* rainbowfish [Saker and Eaglesham 1999]. Conversely, bioconcentration of toxin has not yet been recorded from aquatic plants [White *et al.* 2005a; Kinnear *et al.* 2008].

The adverse ecological effects of *Cylindrospermopsis* and CYN exposures, combined with the increased occurrence of toxin-producing blooms, highlights an urgent need to create suitable environmental management strategies for toxic blooms. The development of ecosystem guidelines requires an understanding of the factors that influence the ecotoxicity of *C. raciborskii* blooms: these are discussed in detail below.

3.0. Factors Influencing the Potential for Ecotoxicity

There is a multiplicity of factors the influence the extent of ecological effects caused by *C. raciborskii* blooms. These include: the occurrence and cell concentrations of a bloom; if toxin is produced, and, if so, its concentration and bioavailability; which organisms are likely to be exposed; how toxin is taken up; individual susceptibilities; the potential for bioaccumulation;

whether other toxins are present and the changes in water quality that accompany a bloom (figure 1).

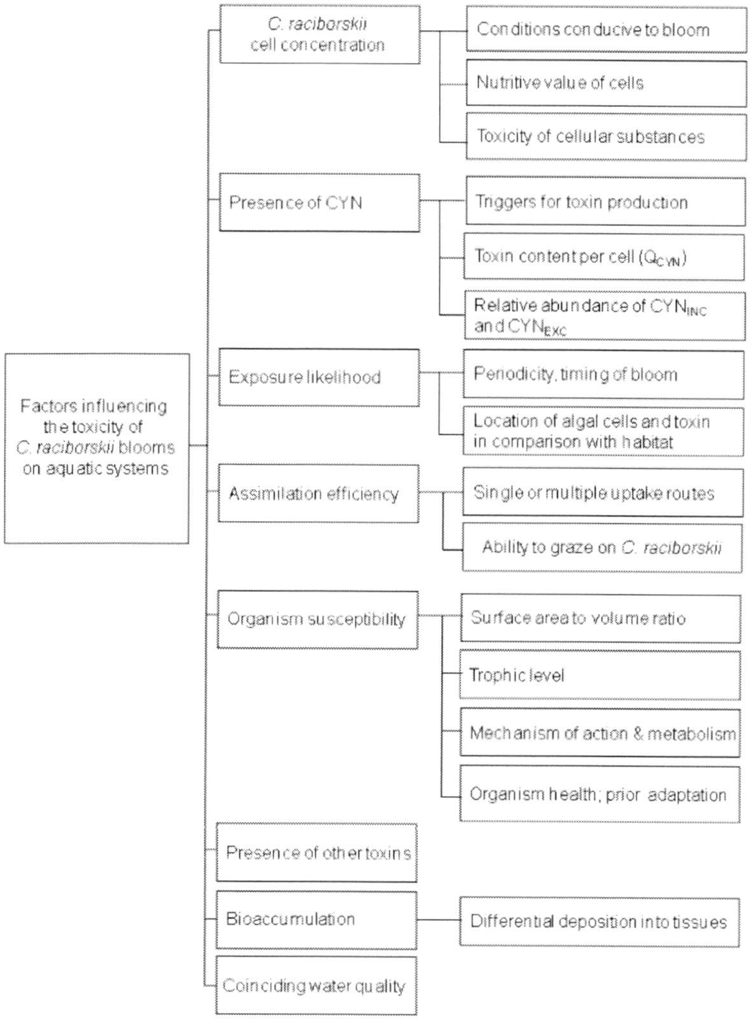

Figure 1. Summary of factors influencing the toxicity of Cylindrospermopsis raciborskii on aquatic ecosystems.

3.1. Growth of *C.raciborskii* in the Water Column

C. raciborskii occurs year-round in deep stratified lakes and rivers from tropical to subtropical climates, though it may also bloom seasonally in shallow temperate waters [Fabbro and Duivenvoorden 1996; Padisák 1997; Bormans *et al.* 2004]. The ecology of *C. raciborskii* has been extensively reviewed [Chiswell *et al.* 1997; Padisák 1997]. In brief, *C. raciborskii* prefers stable, stratified and warm (>25°C) water columns with low light availability and slight alkalinity [Padisák 1997]. Since it is heterocystous, *C. raciborskii* can fix nitrogen and therefore may dominate in waters with N-deficiency.

C. raciborskii is now considered an adaptive species capable of invading new environments [Neilan *et al.* 2003]. This has important implications in terms of the number of aquatic species likely to experience toxic *C. raciborskii* blooms in their habitat. Exposure to *C. raciborskii* cells may be associated with adverse effects in two ways: one, toxicity from substances in the cells themselves (other than CYN, which is discussed later); and two, the potential for monospecific blooms to cause changes in the diet of aquatic grazers.

Some *C. raciborskii* isolates exert toxicities that are unrelated to CYN concentrations [Hawkins *et al.* 1997; Carmichael 2001a]. These additional toxicities probably result from unidentified *C. raciborskii* cell substances [Hawkins *et al.* 1997; Falconer *et al.* 1999; Norris *et al.* 1999; Froscio *et al.* 2001; Saker *et al.* 2003; Falconer 2005]. For example, lipopolysaccharides (LPSs) are found in the cell wall of all cyanobacteria and may be highly inflammatory substances [Raziuddin *et al.* 1983]. However, the LPS content in *Cylindrospermopsis* is not known; therefore, attributing the toxicity of non-toxic producing *Cylindrospermopsis* solely to LPSs seems inappropriate. Many blue-green algae are also capable of producing other allelopathic compounds (also known as 'secondary metabolites'): the ecological role of these remain unclear and desperately understudied [Leflaive and Ten-Hage 2007]. Recently, it has been suggested that allelopathy in *C. raciborskii* is a possible explanation for its invasion into mid-latitudes [Figueredo *et al.* 2007].

C. raciborskii cells may also cause adverse impacts via the nutritive status of the food web. Like most cyanobacteria, cells of *C. raciborskii* represent poor nutritional value for grazers [Lirås *et al.* 1998; Leonard and Paerl 2005]. The cells may also be difficult to ingest, particularly coiled morphotypes or especially large filaments, though some zooplankters can graze these successfully [Fabbro *et al.* 2001]. Finally, if *C. raciborskii* dominates the water column, other highly palatable and nutritious plankton (e.g., green algae) may be absent. This combination of factors might mean that grazing animals (and, in turn, their predators) are faced with shortages in suitable food supply during especially large or prolonged bloom events. Thus, potential 'toxicities' could include poor growth rates, reduced fecundity and even starvation. Hence, *C. raciborskii* blooms may be associated with a number of ecological effects even in the absence of CYN.

3.2. Dynamics of Toxin Production by *C. raciborskii*

3.2.1.Genetics and Triggers for CYN Production

Few studies have examined the percentage of natural *C. raciborskii* blooms that produce toxin [McGregor and Fabbro 2000]. Toxin production does not appear related to the Australian geographical origin of the species [Schembri *et al.* 2001, and different strains have different toxicities – even if they are genetically similar [Saker and Neilan 2001; Bernard *et al.* 2003; Fastner *et al.* 2003; Fergusson and Saint 2003]. There is some evidence that coiled morphotypes have less CYN content than straight filaments [McGregor and Fabbro 2000].

Factors influencing the toxin content of natural and cultured *C. raciborskii* populations were recently reviewed by Griffiths and Saker [2003] and include pH, light, temperature, nutrient and mixing regimes [Branco and Senna 1994; Fabbro and Duivenvoorden 1996; Chiswell *et al.* 1997; Chiswell *et al.* 1999; Saker and Griffiths 2000; Bormans *et al.* 2004]].

Growth rates also appear coupled with toxin production [Hawkins *et al.* 2001] and the size of vegetative cells may provide an indication of likely toxin content [Hawkins *et al.* 2001; Saker and Neilan 2001].

The gene cluster responsible for CYN biosynthesis has recently been characterised [Mihali *et al.* 2008] and this will aid future research work focussing on the toxigenic strains of *C. raciborskii*.

3.2.2. Toxin Content Per Cell (QCYN)

Toxin quotas (Q_{CYN}) describe the quantity of toxin (e.g., of CYN) produced in a single cell [e.g., of *C. raciborskii*]. Different algal blooms may represent different ecotoxicity risks due to cell quota characteristics. For example, consider two blooms of equal toxin concentration, but different cell concentrations. The bloom with high cell density (but low Q_{CYN}) might represent less ecological risk to a grazer of *C. raciborskii* than would one with low cell density [but high cell quota]. This is because grazing on the latter may lead to increased toxin loading. Of course, this is only true where grazers consume equal quantities of cells in each scenario. Moreover, there is limited information regarding how grazing pressure itself may affect toxin cell quotas [Fabbro *et al.* 2001; White *et al.* 2006].

Mesocosm experiments have demonstrated that the Q_{CYN} of *C. raciborskii* is higher in tropical rather than sub-tropical environments [Garnett *et al.* 2003]. The toxin producing capability of vegetative *C. raciborskii* cells compared with specialised cells (akinetes and heterocytes) has not been examined. *Microcystis* cells have a genetically fixed toxin quota [Orr and Jones 1998], hence, since akinetes and heterocytes in *C. raciborskii* begin life vegetatively [Chiswell *et al.* 1997], their genetic capability for toxin production could equal that of vegetative cells. On the other hand, resting stages and cysts of some algae – such as *Alexandrium* – are many times more toxic than their vegetative counterparts [Landsberg 2002]. The possibility for akinetes to have high Q_{CYNs} may be important ecologically during bloom senescence, when they are produced in large numbers before drifting into the sediments [Chiswell *et al.* 1999]. This could signal potential toxicity to the benthic (possibly even hyporheic) fauna that might otherwise avoid toxin exposure.

3.2.3. Relative Abundance of Intracellular and Extracellular Toxin

If CYN is held within the algal cells where it was manufactured, it is known as the cell-bound or intracellular toxin fraction (CYN_{INC}). However, if CYN is secreted or transported out of the cell, or if the cells lyse and liberate toxin, the toxin may also be present in the dissolved form (aqueous or extracellular component, CYN_{EXC}). The significance of these fractions to bioavailability and uptake routes in aquatic organisms was described in White *et al.*, [2005b]. Briefly, toxin exposure may occur via transdermal uptake of CYN_{EXC}, oral consumption of toxin-laden cells (CYN_{INC}) and/or (accidental) drinking of suspended particles and aqueous concentrations.

The relative abundance of CYN_{EXC} and CYN_{INC} is critical in affecting toxicity as even small concentrations of CYN_{INC} can dramatically increase the severity of toxic responses in organisms like snails and tadpoles [White 2006; White *et al.* 2006; White *et al.* 2007]. The ratio of CYN_{INC}: CYN_{EXC} is highly variable during *C. raciborskii* blooms, but most CYN is often dissolved [Carson 2000], a pattern which contrasts almost all other algal toxins. CYN_{EXC} is particularly dominant during the postexponential growth phase [Saker and Griffiths 2000]. Hawkins *et al.* [2001] speculated that older *C. raciborskii* cells may be associated with CYN_{EXC} due to increased cell wall permeability. However, it is conceivable that *C. raciborskii* could also transport CYN out of the cell: this has the ecological advantage of exerting toxicity without cells being consumed. Since CYN is small and hydrophilic, passive transport across the cytoplasmic membrane could be possible. Unfortunately, fluxes in $CYN_{INC:}$ CYN_{EXC} ratios, and why and how these occur, have yet to be closely mapped.

3.3. Likelihood of Toxin Exposure

3.3.1. Periodicity and Timing of Bloom

Cylindrospermopsis is rapidly invading new habitats [Neilan *et al.* 2003; Briand *et al.* 2004]. Hence, an increasingly broad range of aquatic species from tropical, subtropical and temperate environments are likely to encounter toxic blooms. Globally, *C. raciborskii* cell concentrations appear to peak in (late) summer [McGregor and Fabbro 2000; Saker and Griffiths 2001; Briand *et al.* 2002; Fastner *et al.* 2003; Hamilton *et al.* 2005]; however, more toxin may actually be produced when cooler temperatures prevail [Saker and Griffiths 2000]. Therefore, if aquatic species reproduce in summer, embryonic or larval stages could be exposed to high cell concentrations of *C. raciborskii,* but not necessarily high cell quotas. The opposite problem has been identified for tropical *Microcystis* blooms, which are highly toxic during the summer months that coincide with critical breeding periods for many fish species [Wiegand and Pflugmacher 2001].

3.3.2. Location of Cells and Toxin in Relation to Target Species

The positioning of algal cells within an aquatic habitat may be critical in affecting toxin exposure and uptake rates. Spatial variation of *Cylindrospermopsis* may allow different habitats to experience different cell and toxin concentrations (for example, planktonic environments might feature surface scums and CYN_{INC}, compared with benthic areas harbouring senescent cells and CYN_{EXC}) [see White *et al.* 2005b]. Nonetheless, once in the extracellular form, CYN may diffuse into almost all aquatic habitats [Christoffersen 1996]. This again highlights the importance of the proportion of the CYN_{EXC} and CYN_{INC} fractions, along with total toxin concentrations. New research from Australia shows that the zonation of toxin in the water column may be linked with the form being CYN or deoxy-CYN [Everson *et al.*, in preparation]: this too would ultimately affect the nature and extent of toxin exposure and the resultant effects in aquatic organisms.

3.3.3. Persistence of Toxin

CYN is a highly stable compound with respect to heat, light and pH [Chiswell *et al.* 1999], although pure CYN breaks down rapidly in sunlight when cell pigments are present (90% complete in two to three days). Photocatalytic degradation of CYN appears to be pH dependent, being most efficient at pH 9.0 [Chiswell *et al.* 1999]. In addition, recent work has suggested that once produced, CYN may remain in the water column for extended periods (> 40 days), since microbial degradation appears negligible [Wormer *et al.* 2008]. This characteristic increases the toxicity and bioaccumulation potential of CYN, since prolonged exposure times could result from even a short-lived bloom. It is also conceivable that longevity of toxin fractions (CYN_{INC} and CYN_{EXC}) may vary throughout a bloom: Lahti *et al.*

[1997b] reported that dissolved MC was more persistent compared with MC in particulate material. Again, this may have a bearing on the ultimate toxicity of a *C. raciborskii* bloom.

3.4. Assimilation Efficiency

3.4.1. Method(S) of Uptake

The pathway(s) for, and extent of, CYN uptake is critically important in affecting end toxicities in aquatic organisms. Uptake of CYN is poorly understood. The sulfate group on the CYN molecule is not required for cell entry or toxicity [Runnegar *et al.* 2002]. Uptake via passive diffusion might be responsible, given the relatively small molecular weight of CYN [Chong *et al.* 2002], however Runnegar *et al.* [2002] considered that CYN was unlikely to permeate cell walls since it is so hydrophilic. Since CYN affects the liver, the toxin could require similar transport carriers as those for microcystin (a hepatotoxin). It seems that bile acid transporters are utilized during the initial stages of CYN uptake, but a secondary system may also be present since bile acid inhibition provides protection against toxicity for only 72h [Chong *et al.* 2002].

Regarding CYN uptake in aquatic plants, some authors have concluded that the effects of cyanotoxins would be minimal and short lived, given that plant-bioavailable toxin (the extracellular form) is of generally low concentration and is transient during a bloom [Jones *et al.* 1994; Jones and Orr 1994; Casanova *et al.* 1999]. However, CYN is usually extracellular [Saker and Griffiths 2000; Norris *et al.* 2001; Metcalf *et al.* 2002a], so plants may instead experience a high CYN uptake risk. To date, *Hydrilla verticillata* and *Spirodela olighorriza* are the only aquatic macrophytes for which CYN uptake has been examined: both plants appear to adsorb, but not accumulate, toxin [White *et al.* 2005a; Kinnear *et al.* 2007]. However, these results could have been confounded by the detection technique (high performance liquid chromatography), which cannot detect bound CYN.

Transdermal uptake, accidental drinking and grazing probably all contribute to tissue contamination in aquatic animals, but it is uptake via grazing on CYN-laden cells that has been linked with the highest concentrations of tissue toxins in gastropods and tadpoles [White *et al.* 2006; White *et al.* 2007]. It is also possible that uptake via grazing is accelerated by CYN-mediated tissue injuries, which allow greater CYN adsorption/absorption rates across the digestive epithelia [Seawright *et al.* 1999; Kinnear *et al.* 2007].

3.4.2. Ability to Graze on C. raciborskii

Grazing on toxin-laden *C. raciborskii* cells appears to increase CYN toxicity dramatically, thus, factors that affect grazing rates will also affect toxin uptake. For example, organism size, metabolism and nutritive needs all influence grazing rates. Preferential selection of cells with reduced cell quotas may also be used as a mechanism to reduce or avoid toxicity. In studies with the aquatic snail *Melanoides tuberculata*, *C. raciborskii* cell quotas almost tripled following snail grazing, which may represent a competitive response by the alga to increase toxin load and lower cell palatability [White *et al.* 2006]. The role of other food sources should not be discounted, since the availability of a non-toxic alternative might also cause reduced grazing on toxic cells [Soares *et al.* 2004].

3.5. Organism Susceptibility

3.5.1. Size and Surface Area to Volume Ratios

The size of an organism is important in determining ecological risks because of the relationship between surface area to volume ratios and transdermal toxin uptake. Both Christoffersen [1996] and Carmichael [1996] concluded that there is a relationship between organism size and algal toxin sensitivity, with smaller organisms at greater risk. However, surface area to volume ratios may be significant only in organisms that receive toxin primarily from the transdermal route. By contrast, organisms with multiple uptake routes seem like to obtain most of their toxin load from cell grazing, meaning that minor changes in transdermal uptake due to surface area are insignificant.

3.5.2. Trophic Level

Relationships between CYN sensitivity and trophic levels have not been studied. However, if biomagnification of toxin is possible, consumer organisms located high in the food web may experience very concentrated toxin doses.

3.5.3. Mechanism of Action and Metabolism

The susceptibility of aquatic plants to CYN toxicity via protein synthesis inhibition and/or the production of toxic CYN metabolites remains poorly understood. In animals, production of metabolites during CYN breakdown in the liver results in additional toxicity, since the breakdown compounds are more toxic than the parent molecule [Runnegar *et al.* 1994]. Therefore, it is conceivable that aquatic organisms with less evolved liver apparatus have some protection from CYN toxicity. For example, the hepatopancreas found in snails may provide only poor toxin metabolism compared with the liver of amphibians. In addition, cytochrome P-450 enzymes, which are responsible for the bioactivation of CYN necessary to cause toxic effects [Runnegar *et al.* 1994; Shaw *et al.* 2000], are especially concentrated in vertebrate livers. This could explain the marked differences in the sensitivities recorded in conjunction with *C. raciborskii* exposures, for example, in *Melanoides* snails and *Bufo* tadpoles [White *et al.* 2007; Kinnear *et al.* 2007].

Whether there is a trend for biologically advanced animals to experience greater toxicity from CYN due to the functioning of the liver (or other organs responsible for toxin metabolism) remains unresolved. Quite severely toxic effects of CYN exposure have been observed in simple invertebrates [Hiripi *et al.* 1998; Metcalf *et al.* 2002b]. In contrast, Saker *et al.* [2004] and Saker and Eaglesham [1999] recorded high levels of accumulated CYN from more advanced organisms [mussels, crayfish, rainbow fish] with no mortality. There is also considerable variability in the toxicity of CYN amongst mammalian subjects such as mice and cattle [Saker *et al.* 1999], and even between individuals of the same species [Seawright *et al.* 1999]. Taken together, these studies suggest that the mode and extent of CYN toxicity is complex and involves more than the detoxification processes in the liver (or equivalent organ).

3.5.4. Organism Health and Prior Adaptation

Many other factors could influence the responses of aquatic organisms to CYN exposure, including sex, life cycle stage, disease, degree of parasitism, nutritional status and history. Few studies have examined these factors with respect to CYN toxicity.

Evolutionary adaptation of aquatic organisms to CYN and *C. raciborskii* seems likely. Both Ressom *et al.* [1994] and Lirås *et al.* [1998] suggested that the shared environment between cyanobacteria and aquatic organisms is likely to have encouraged the development of coexistence strategies. There have been several studies to supporting this idea. Shaw *et al.* [2002] speculated that tolerance in mice repeatedly exposed to CYN for 90 days could result from the inhibition of enzymes required for metabolic activation of CYN, or from induction of enzymes capable of CYN degradation. Such enzymes may be either constantly expressed or switched on and off in response to toxin exposures [Beattie et al. 2003]. Recently, Gustafsson *et al.* [2005] demonstrated that increased tolerance to toxic *Microcystis* could be induced in *Daphnia magna*, and that this tolerance could be passed on to successive generations.

3.6. Production of Toxins and Substances Other than CYN

C. raciborskii is known to produce saxitoxin, although this usually does not occur in conjunction with CYN production [Pomati *et al.* 2003; Castro *et al.* 2004; Pomati *et al.* 2004a; Pomati *et al.* 2004b]. There are also a number of other CYN analogs produced by *C. raciborskii,* including deoxy-CYN, the toxicity of which remains unclear [Norris *et al.* 1999; Looper *et al.* 2005]. However, it appears that in *C. raciborskii* at least, deoxy-CYN represents only a small proportion (10%) of total toxin content. Nonetheless, thorough risk assessments reflecting the likely effects associated with *C. raciborskii* blooms will need to consider whether the strain in question is a producer of CYN, its analogs, saxitoxin or a combination of these.

Some studies suggest that synergism between CYN, other toxins and cell components would increase the overall toxicity in a bloom event [Norris *et al.* 1999]. However, the mechanism of CYN toxicity is unusual: byproducts produced during CYN metabolism are more toxic than the parent molecule [Runnegar *et al.* 1994]. This being the case, LPSs (or other allelopathic cell substances) which reduce the ability of an organism to metabolise CYN could in fact offer protection against toxicity [Metcalf *et al.* 2002b]. For example, pre-exposure to LPS from *Microcystis* cells increases the LC_{50} of cylindrospermopsin to *Artemia salina* by 8 units, indicating decreased toxicity [Lindsay *et al.* 2006].

3.7. Bioaccumulation

CYN bioaccumulation has been demonstrated in several aquatic organisms (figure 2). This is ecologically significant as increased tissue toxin loads may increase lethal and sublethal toxicities. Tissue toxins may also represent a source of toxin contamination for predators that are otherwise unexposed to CYN (and/or *C. raciborskii*). Of the animals studied for CYN bioaccumulation, gastropods and bivalves have recorded the highest tissue toxin concentrations; followed by crustacean, amphibian and fish species (figure 2). Thus, the biological complexity of each organism appears to decrease its susceptibility to bioaccumulation. This contrasts with patterns of toxicity (where more advanced organisms experience greater toxicities); however, the influences of other factors (grazing rates, toxicity, toxin metabolism) on accumulation rates have yet to be properly studied.

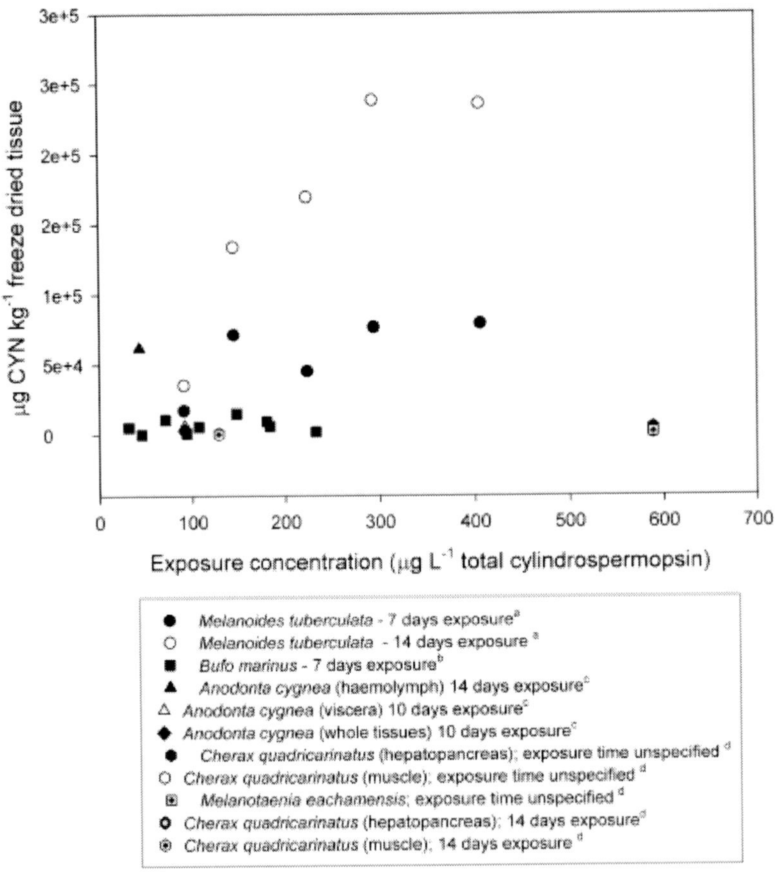

Figure 2. Cylindrospermopsin concentrations of aquatic organisms as reported in published literature. a [White *et al.* 2006]; b [White *et al.* 2007]; c [Saker *et al.* 2004]; d [Saker and Eaglesham 1999]. Values from published literature are as reported or closest approximation from figures provided. Note: Data from Nogueira *et al.* [2004a] not included since values were reported as fresh weights.

3.7.1. Differential deposition into tissues

Organotropy occurs when accumulated substances are attracted towards, or deposited into, specific tissues or organs. This is known to occur during CYN bioaccumulation: Saker and Eaglesham [1999] reported that crayfish hepatopancreas contained five times the toxin concentration of muscle tissues; and Saker *et al.* [2004] reported that 68.1% of the total CYN load in swan mussels came from the haemolymph. This may indicate that CYN has an affinity for blood or lymph. Such toxin partitioning might influence the ecological effects of CYN: for example, differential toxin deposition could allow toxin to be stored away from organs that are targets for toxicity. However, this seems unlikely, given that CYN is often recovered from the liver (or hepatopancreas), a primary site for toxicity.

3.8. Coinciding Water Quality

One of the most difficult issues to address in ecotoxicity testing is the confounding of results by deteriorated water quality. Alkaline pH values, oxygenation supersaturation and low light conditions often accompany a cyanobacterial bloom. Decomposition of blooms may also lead to deoxygenation and the production of nitrogenous compounds [Carmichael 1996; Oberemm 2001]. Thus, the probability for additive or synergistic effects must be recognised when extrapolating results from laboratory-based trials into the field.

4.0. Towards the Development of Aquatic Ecosystem Protection Guidelines

Regulatory options for human health risks associated with toxic algal blooms have been extensively discussed [Chorus and Bartram 1999; Chorus 2005; Chorus 2005b]. However, quite different approaches are needed to address ecosystem health risks. Few nations have regulatory strategies that specifically target the environmental impacts of *C. raciborskii* (or any other cyanobacteria). Some countries instead rely on basic catchment management plans to manage ecological risks, such as eutrophication action plans in the UK and South Africa [Ferguson 1997] and 'total maximum daily loads' for nutrients and chlorophyll-a in the United States [Burns 2005]. There is an urgent need for a greater recognition of the ecological effects associated with toxic algal blooms, and for the development of appropriate management strategies for the protection of aquatic ecosystem health.

4.1. Guidelines Based Solely on Toxin Concentrations

Currently, the World Health Organisation (WHO) drinking water guidelines are based on toxin concentrations. Environmental guidelines could also be based on toxin concentrations, since these are clearly important in influencing the extent of ecological effects in aquatic organisms. However, toxin concentrations alone may not give a complete picture of risk. Firstly, pollutant values used in such guidelines typically rely on huge quantities of laboratory and field-based ecotoxicity data [ANZECC 2000]. This kind of research information is not yet widely available for *Cylindrospermopsis* and CYN. Secondly, guidelines based on toxin concentrations are often very prescriptive, involving the calculation of lifelong tolerable daily intakes (TDI). In drinking water, human TDIs are calculated for different age and exposure scenarios. When additional species are considered, new TDIs are calculated for each: Duy *et al.* [2000] calculated sixteen TDIs for CYN contamination of livestock drinking water: the consequent guideline values ranged from 0.6 – 9.1 $\mu g\ L^{-1}$. It would be impossible to set individual guidelines for an entire aquatic ecosystem that contains hundreds of different species, often with hugely different risk scenarios and life spans.

A third problem is that a guideline based on the CYN concentration alone does not address the possibility of more than one CYN analog being present (for example, CYN and deoxy-CYN), or even the presence of other toxins and cell compounds such as anatoxin and LPSs.

Fourthly, if environmental standards are set using only toxin concentrations, repeated toxin testing may be required to ensure compliance. Since *Cylindrospermopsis* prefers low-light environments [Padisák 1997], there is considerable difficultly in visually detecting subsurface blooms [McGregor and Fabbro 2000]. Occasional, random sampling is unlikely to capture peaks and troughs in toxin concentrations, particularly in the case of short bloom intervals. Few nations have official methods for CYN detection: in Australia, high performance liquid chromatography-mass spectrometry is the method of choice [Nicholson and Burch 2001; NHMRC and NRMMC 2004]; the draft guidelines for New Zealand also nominate liquid chromatography-mass spectrometry [Wood and Holland 2005]. Certified quantitative analytical standards for CYN do not yet exist, which prevents interlaboratory calibrations [Chorus 2005b] although current techniques seem generally reproducible and comparable [Törökné *et al.* 2004].

Finally, the existing studies have shown that intracellular toxin concentrations are critical in influencing toxic response and bioaccumulation potential associated with exposure. Since environmental risk scenarios may vary according to the relative abundance of CYN_{INC}: CYN_{EXC}, guideline values should not rely on total toxin concentrations (as currently done for drinking water) but, ideally, on all available fractions.

4.2. Guidelines Based Solely on Cell Concentrations

Cell concentrations (or cell biomass) are a valuable way to measure exposure risks associated with a bloom because these allow an assessment of toxicities related to *C. raciborskii* cell substances other than CYN. Since cell concentrations are necessary to determine cell quotas (Q_{CYN}s), they may also help determine the likely toxin loads ingested by grazer species. In the case of *Cylindrospermopsis*, however, basing an environmental standard solely on cell counts is not appropriate. Most obviously, not all strains of *C. raciborskii* produce CYN. For those that do, the relationship between *C. raciborskii* cell concentrations and toxin concentration are not consistent [Eaglesham *et al.* 1999; Griffiths and Saker 2003]. Thus, using cell concentrations as a surrogate for toxin concentrations is not reliable.

4.3. A Possible Solution

On their own, neither *C. raciborskii* cell concentrations nor total CYN concentrations are adequate indicators of the ecological risks posed by toxic blooms. Instead, a combination of critical factors is needed to properly quantify environmental risks related to a *C. raciborskii* bloom. The management of ecological risks seems best approached by combining estimates of risk associated with *C. raciborskii* (via cell concentrations, biomass or chlorophyll measurements) and those for CYN (via toxin concentrations). A model that uses this combination is shown in table 1. In addition, the importance of CYN_{INC} is highlighted by a third factor, the percentage of CYN_{INC} in the bloom. Multiplication of the three factors A, B and C gives a final risk value that indicates the likelihood for adverse ecological effects, ranging from 'low' to 'extreme'.

Table 1. Possible method for calculating and interpreting an ecological risk value for blooms of Cylindrospermopsis raciborskii. Shaded cells show an example

C. raciborskii (cells mL^{-1})[a]	Risk factor (A)
≤ 20, 000	1
20,001 – 50, 000	2
50,001 – 100,000[b]	3
100,001 – 200, 000[b]	4
≥ 200, 001[b]	5
Toxin concentration (µg L-1) CYNTOT	Risk factor (B)
≤ 0.5	1
0.6 – 0.99	2
1.0 – 9.99	4
10 – 99.99	8
≥ 100	10
% of toxin as CYNINC	Risk factor (C)
≤ 25%	1
26 – 49 %	1.5
50 – 75%	2
≥ 76%	2.5
FINAL RISK VALUE ((A*B*C)	(18)
Final Risk value	Final Risk Rating
0.5 – 2.49	Low
2.5 – 4.99	Medium
5.0 – 19.99	High
≥ 20	Extreme

[a] or surrogate measure of cellular mass such as biovolume or chlorophyll units;
[b] add 0.5 to risk factor A if scums accompany the bloom.

The shaded cells in table 1 indicate the risk calculation for a *C. raciborskii* bloom of 60,000 cells mL^{-1} and 3 µg L^{-1} of toxin with 30% of the toxin being CYN_{INC}, which results in a high risk of ecological effects. This example is relevant to reservoirs in Queensland (Australia) in which toxic *C. raciborskii* is commonly reported [McGregor and Fabbro 2000]. Unfortunately, there are not enough data regarding the toxicity of *C. raciborskii* and CYN mixtures to aquatic organisms to enable a close consideration of risks. Thus, the risk factor values in the table are arbitrary, as are the ranges of cell and toxin concentrations that occupy each category.

Once an ecological risk rating has been calculated, appropriate management actions could be triggered based on the perceived severity of the environmental risk. This may introduce a problem regarding connotations of 'low', 'medium', 'high' and 'extreme' risk ratings. For example, a 'low' risk rating could be interpreted as 'low probability of adverse effects is occurring'; or 'that risks will occur, but only to selected species in the system'; or 'that risks will occur in many species, but on a low scale of severity' (e.g., minor changes in zooplankton diet compared with mass fish mortalities). The final interpretation (and management response) is likely to depend on precise details of particular cases, especially the numbers and types of different species present in a water body, their likely sensitivities (as discussed earlier) and past bloom history (e.g., prolonged or recurrent bloom events).

Cell biomass or equivalent chlorophyll *a* values could replace cell concentration data in the model: both these have been used in regulatory approaches in the past. However, there is some doubt regarding the accuracy of biomass measurements, particularly where sample preservatives cause inaccurate measurements [Hawkins *et al.* 2005]. In addition, the use of chlorophyll *a* units to characterise risks associated with a bloom must assume *C. raciborskii* dominance; and this may not always be the case.

This model has some limitations. The contribution of deoxy-CYN has not been addressed, since there are contrasting reports of its toxicity [Norris *et al.* 1999; Neumann *et al.* 2005]. Knowledge of CYN_{INC} concentrations is required, which requires testing beyond that of simply total toxins. This could be problematic as many parameters can be time consuming and expensive to determine [e.g., *C. raciborskii* cell density, CYN concentrations and the percentage of CYN occurring in the intracellular form]. However, this additional cost appears to be warranted given the linkages between CYN_{INC} and increased toxicities and bioaccumulation [Kinnear *et al.* 2007; White *et al.* 2007; Kinnear *et al.* 2007]. Furthermore, since blooms may experience radical changes in cell and toxin concentrations and/or fluxes between intracellular and extracellular toxin dominance, repeated testing would be required during a bloom. However, most environmental guidelines would require ongoing monitoring. Finally, any comprehensive estimation of ecological risks would require a concomitant consideration of water quality (pH, alkalinity, temperature) and the potential synergies or antagonisms between *C. raciborskii*, CYN and other toxicants present in natural water bodies (e.g., herbicides, other toxins).

4.4. Guidelines as Part of a Wider Environmental Management Approach

Environment guidelines, such as those suggested in table 1, have an inherent problem in that they are reactive by nature. Drinking water guidelines work (at least in theory) because they prevent toxin contaminated water supplies from reaching a human population. In contrast, in the natural environment, exposure occurs simultaneously with bloom development and the production of toxins: whilst managers are assigning a risk rating, ecological effects are already occurring. Consequently, well developed monitoring and early warning systems are likely to be especially useful in overcoming the 'reactiveness' of guidelines. This may include the use of rapid toxin detection methods (e.g. ELISA), the use of PCR to detect *C. raciborskii* strains that are genetically capable of toxin production [Rasmussen et al. 2005], and/or the development of bioindicators or biomarkers for CYN intoxication.

A relatively new approach to toxic algal bloom management, as introduced by Chorus [2005b], is Water Safety Plans (WSPs). Here, a management 'loop' enables users to define a problem, configure a solution and check the success of that solution Ecological risk values (as suggested in table 1) could help to identify at-risk water bodies requiring implementation of a WSP. This combined approach would encourage a greater recognition and understanding of factors contributing to toxic *Cylindrospermopsis* blooms, compared with sole use of guidelines that encourage compliance but not necessarily an understanding of hazards [Chorus 2005b]. However, the trade-off of the WSP approach remains that, in order to achieve a greater level of water [and food] security, managers and technical operators must invest a good deal of time in both conducting comprehensive risk analyses and ensuring that the resulting management procedures are adhered to [Ibelings and Chorus 2007].

5.0. Future Work

Future studies should focus on providing the information needed to progress guidelines for environmental protection of water bodies affected by toxic *C. raciborskii* blooms. In the laboratory, testing of live cultures containing environmentally relevant cell and toxin concentrations similar is useful. This approach is limiting in terms of pinpointing CYN-related (compared with *Cylindrospermopsis*-related) effects, since live cultures contain other bioactive compounds [Sivonen and Jones 1999; Saker *et al.* 2003; Nogueira *et al.* 2004a] which means toxic effects cannot be ascribed solely to CYN. However, the live-culture approach has greater environmental relevance compared with laboratory studies using purified toxin, because ecological effects occurring during natural blooms will reflect the synergisms and antagonisms of coinciding water quality, multiple toxins and toxic *C. raciborskii* cellular substances. Of course, field studies are the ultimate approach, although the logistics of monitoring and sampling can sometimes be difficult.

An entirely different question, but one worthy of study, is that of the natural role of cyanobacterial toxins in aquatic ecosystems generally. This subject has been examined by few authors [e. g. Babica *et al.* 2006], and none appear to have examined CYN specifically. In terms of survival, the benefits of toxin production by blue-greens are multiple – for example, the ability to reduce grazing pressure by eliminating zooplankton and other predators, or to reduce competition for nutrients, light and other resources by reducing other phytoplankton and macrophytes. Both of these scenarios seem likely: if so, toxin production would be advantageous (for the alga) in almost all settings – although at a trade-off from the metabolic cost of production. A greater understanding of the possible evolutionary advantages of CYN production by *C. raciborskii* – and, by extension, the triggers for toxin production and cessation – may provide insights into a better approach for the ecological management of toxic blooms.

6.0. Conclusion

A multiplicity of factors can influence the ecological effects associated with CYN-producing *C. raciborskii* blooms. A range of adverse effects have already been recorded following exposure to this alga and its toxin. Bioaccumulation of CYN in aquatic organisms has been barely studied; however, existing studies have shown that both invertebrates and vertebrates are capable of toxin accumulation. Guidelines for ecological risks could combine a consideration of cell concentrations, toxin concentrations and toxin fraction (CYN_{INC}), and may achieve greater environmental protection if integrated into a wider management approach such as a water safety plan. The model suggested in table 1 is a demonstrative only; however, it could be further developed into an aquatic ecosystem protection guideline for *C. raciborskii* blooms as new, detailed ecotoxicological data become available.

Since cyanobacteria are a natural component of aquatic ecosystems; it would be folly to introduce a management framework that attempts to prevent blooms from occurring. Unfortunately, it would seem that the existing, higher-level measures designed to curb the 'artificial' increase in blooms – those resulting from human activity, climate change, or a combination of these – fail to recognise the ecosystem effects linked with CYN toxicities in natural environments. Aquatic ecosystems are immensely complex and have unpredictable sensitivities, and their proper study and management is a challenging task. The model suggested in this chapter is the first to direct addresses environmental effects of CYN blooms and therefore represents an important step forward in achieving a more holistic management of water bodies polluted by toxic blooms.

REFERENCES

ANZECC (2000). *Australian and New Zealand Guidelines for Fresh and Marine Water Quality*. Canberra: Australian and New Zealand Conservation Council.

Babica, P. , Bláha, L. and Maršálek, B. (2006). Exploring the natural role of microcystins - a review of effects on photoautotrophic organisms. *Journal of Phycology 42(1)*: 9-20.

Beattie, K.A. , Ressler, J. , Wiegand, C. , Krause, E. , Codd, G.A. , Steinberg, C. and Pflugmacher, S. (2003). Comparative effects and metabolism of two microcystins and nodularin in the brine shrimp *Artemia salina*. *Aquatic Toxicology 62*: 219-226.

Bernard, C. , Harvey, M. , Biré, R. , Krys, S. and Fontaine, J.J. (2003). Toxicological comparison of diverse *Cylindrospermopsis raciborskii* strains: Evidence of liver damage caused by a French *C. raciborskii* strain. *Environmental Toxicology 18*: 176-186.

Bormans, M. , Ford, P.W. , Fabbro, L.D. and Hancock, G. (2004). Onset and persistence of cyanobacterial blooms in a large impounded tropical river, Australia. *Marine and Freshwater Research 55*: 1-15.

Bouaïcha, N. and Nasri, A.-B. (2004). First report of Cyanobacterium *Cylindrospermopsis raciborskii* from Algerian Freshwaters. *Environmental Toxicology 19*: 541-543.

Bouvy, M. , Falcão, D. , Marinho, M. , Pagano, M. and Moura, A. (2000). Occurrence of *Cylindrospermopsis* (Cyanobacteria) in 39 Brazilian tropical reservoirs during the 1998 drought. *Aquatic Microbial Ecology 23*: 13-27.

Bouvy, M. , Pagano, M. and Troussellier, M. (2001). Effects of a cyanobacterial bloom (*Cylindrospermopsis raciborskii*) on bacteria and zooplankton communities in Ingazeira reservoir (northeast Brazil). *Aquatic Microbial Ecology 25*: 215-227.

Bowling, L. (1994). Occurrence and Possible Causes of a Severe Cyanobacterial Bloom in Lake Cargelligo, New South Wales. In: Jones, G (editor). *'Cyanobacterial Research in Australia', Australian Journal of Marine and Freshwater Research*. pp. 737-745

Branco, C.W. and Senna, P.A. (1994). Factors influencing the development of *Cylindrospermopsis raciborskii* and *Microcystis aeruginosa* in the Paranoá Reservoir, Brasília, Brazil. *Algological Studies 75*: 85-96.

Briand, J.F. , Leboulanger, C. , Humbert, J.-F. , Bernard, C. and Dufour, P. (2004). *Cylindrospermopsis raciborskii* (Cyanobacteria) Invasion at Mid-Latitudes: Selection; Wide Physiological Tolerance, or Global Warming? *Journal of Phycology 40*: 231-238.

Briand, J.F. , Robillot, C. , Quiblier-Llobéras, C. , Humbert, J.F. , Couté, A. and Bernard, C. (2002). Environmental context of *Cylindrospermopsis raciborskii* (Cyanobacteria) blooms in a shallow pond in France. *Water Research 36*: 3183-3192.

Burns, J. (2005). United States of America: Cyanobacteria and the status of regulatory approaches. In: Chorus, I (editor). *Current approaches to cyanotoxin risk assessment,* *risk management and regulations in different countries, WaBoLu 02/05.*Umweltbundesamt, Dessau. pp.

Byth, S. (1980). Palm Island Mystery Disease. *Medical Journal of Australia 2*: 40-42.

Carmichael, W.W. (1996). Toxic *Microcystis* and the environment. In: Watanabe, M, KI Harada, WW Carmichael and H Fujiki (editor). *Toxic Microcystis.*Boca Raton: CRC Press. pp. 1-12

Casanova, M.T. , Burch, M.D. , Brock, M.A. and Bond, P.M. (1999). Does toxic *Microcystis aeruginosa* affect aquatic plant establishment? *Environmental Toxicology 14*: 97-109.

Castro, D. , Vera, D. , Lagos, N. , García, C. and Vásquez, M. (2004). The effect of temperature on growth and production of paralytic shellfish poisoning toxins by the cyanobacterium *Cylindrospermopsis raciborskii* C10. *Toxicon 44*: 483-489.

Chiswell, R.K. , Shaw, G.R. , Eaglesham, G. , Smith, M.J. , Norris, K.R. , Seawright, A.A. and Moore, M.R. (1999). Stability of cylindrospermopsin, the toxin from the cyanobacterium, *Cylindrospermopsis raciborskii*: Effect of pH, temperature and sunlight on decomposition. *Environmental Toxicology 14*: 155-161.

Chiswell, R.K. , Smith, M. , Norris, R. , Eaglesham, G. , Shaw, G. , Seawright, A.A. and Moore, M. (1997). The cyanobacterium, *Cylindrospermopsis raciborskii*, and its related toxin, cylindrospermopsin. *Australasian Journal of Ecotoxicology 3*: 7-23.

Chong, M.W.K. , Wong, B.S.F. , Lam, P.K.S. , Shaw, G.R. and Seawright, A.A. (2002). Toxicity and uptake mechanism of cylindrospermopsin and lophyrotomin in primary rat hepatocytes. *Toxicon 40*: 205-211.

Chonudomkul, D. , Yongmanitchai, W. , Theeragool, G. , Kawachi, M. , Kasai, F. , Kaya, K. and Watanabe, M.M. (2004). Morphology, genetic diversity, temperature tolerance, and toxicity of *Cylindrospermopsis raciborskii* (Nostocales, Cyanobacteria) strains from Thailand and Japan. *FEMS Microbiology Ecology 48*: 345-355.

Chorus, I., (editor). (2005). Current approaches to cyanotoxin risk assessment, risk management and regulations in different countries. WaBoLu 02/05. Umweltbundesamt, Dessau, Germany: Federal Environmental Agency.

Chorus, I. (2005b). Water Safety Plans A better regulatory approach to prevent human exposure to harmful cyanobacteria. In: Huisman, J, HCP Matthijs and PM Visser (editor). *Harmful Cyanobacteria.*Dordrecht, The Netherlands: Springer. pp. 201-226

Chorus, I. and Bartram, J., (editor). (1999). Toxic cyanobacteria in water: a guide to their public health consequences, monitoring and management. London: Published on behalf of the World Health Organisation by E and FN Spoon.

Christoffersen, K. (1996). Ecological implications of cyanobacterial toxins in aquatic food webs. *Phycologia 35(6 Supplement)*: 42-50.

Fabbro, L.D. , Baker, M. , Duivenvoorden , L.J. , Pegg, G. and Shiel, R. (2001). The Effects of the Ciliate *Paramecium* cf. *caudatum* Ehrenberg on Toxin Producing Cylindrospermopsis Isolated from the Fitzroy River, Australia. *Environmental Toxicology 16*: 489-497.

Fabbro, L.D. and Duivenvoorden, L.J. (1996). Profile of a bloom of the cyanobacterium *Cylindrospermopsis raciborskii* (Woloszynska) Seenaya and Subba Raju in the Fitzroy River in tropical central Queensland. *Marine and Freshwater Research 30*: 579-595.

Falconer, I.R. (2005). *Cyanobacterial Toxins of Drinking Water Supplies: Cylindrospermopsins and Microcystins*. Boca Raton: CRC Press.

Falconer, I.R. , Hardy, S.J. , Humpage, A.R. , Froscio, S.M. , Tozer, G.J. and Hawkins, P.R. (1999). Hepatic and renal toxicity of the blue-green alga (Cyanobacterium) *Cylindrospermopsis raciborskii* in Male Swiss Albino Mice. *Environmental Toxicology 14*: 143-150.

Fastner, J. , Heinze, R. , Humpage, A.R. , Mischke, U. , Eaglesham, G.K. and Chorus, I. (2003). Cylindrospermopsin occurrence in two German lakes and preliminary assessment of toxicity and toxin production of *Cylindrospermopsis raciborskii* (Cyanobacteria) isolates. *Toxicon 42*: 313-321.

Ferguson, A.J.D. (1997). The role of modelling in the control of toxic blue-green algae. *Hydrobiologia 349*: 1-4.

Fergusson, K.M. and Saint, C.P. (2000). Molecular phylogeny of *Anabaena circinalis* and Its identification in environmental samples by PCR. *Appl. Environ. Microbiol. 66(9)*: 4145-4148.

Fergusson, K.M. and Saint, C.P. (2003). Multiplex PCR assay for *Cylindrospermopsis raciborskii* and cylindrospermopsin-producing cyanobacteria. *Environmental Toxicology 18*: 120-125.

Figueredo, C. , Giani, A. and Bird, D.F. (2007). Does Allelopathy contribute to *Cylindrospermopsis raciborksii* (Cyanobacteria) bloom occurrence and geographic expansion? *Journal of Phycology 43(2)*: 256-265.

Froscio, S.M. , Humpage, A.R. , Burcham, P.C. and Falconer, I.R. (2001). Cell-free Protein Synthesis Inhibition Assay for the Cyanobacterial Toxin Cylindrospermopsin. *Environmental Toxicology 16*: 408-412.

Garnett, C. , Shaw, G. , Moore, D. , Florian, P. and Moore, M. (2003). Impact of climate change on toxic cyanobacterial (blue-green algal) blooms and algal toxin production in Queensland. Queensland Department of Natural Resources and Mines, the National Research Centre for Environmental Toxicology, and Environmental Health Unit, Queensland Health, Rocklea. 111.

Griffiths, D.J. and Saker, M.L. (2003). The Palm Island Mystery Disease 20 Years on: A review of research on the cyanotoxin cylindrospermopsin. *Environmental Toxicology 18(2)*: 78-93.

Gustafsson, S. , Rengefors, K. and Hansson, L.-A. (2005). Increased consumer fitness following transfer of toxin tolerance to offspring via maternal effects. *Ecology 86(10)*: 2561-2567.

Hamilton, P.B. , M., L.L. , Dean, S. and Pick, F.R. (2005). The occurrence of the cyanobaterium *Cylindrospermopsis raciborskii* in Constance Lake: an exotic cyanoprokaryote new to Canada. *Phycologia 44(1)*: 17-25.

Hawkins, P.R. , Chandrasena, N.R. , Jones, G.J. , Humpage, A.R. and Falconer, I.R. (1997). Isolation and toxicity of *Cylindrospermopsis raciborskii* from an ornamental lake. *Toxicon 35*: 341-346.

Hawkins, P.R. , Putt, E. , Falconer, I.R. and Humpage, A.R. (2001). Phenotypical variation in a toxic strain of the phytoplankter, *Cylindrospermopsis raciborskii* (Nostocales, Cyanophyceae) during batch culture. *Environmental Toxicology 16(6)*: 460-467.

Hawkins, P.R. , Runnegar, M.T.C. , Jackson, A.R.B. and Falconer, I.R. (1985). Severe hepatotoxicity caused by the tropical cyanobacterium (blue-green alga) *Cylindrospermopsis raciborskii* (Woloszynska) Seenaya and Subba Raju isolated from a domestic water supply reservoir. *Applied and Environmental Microbiology 50(5)*: 1292-1295.

Hesse, K. and Kohl, J. (2001). Effects of light and nutrient supply on growth and microcystin content of different strains of *Microcystis aeruginosa*. In: Chorus, I (editor). *Cyanotoxins Occurrence, Causes, Consequences.*Berlin: Springer-Verlag. pp.

Humpage, A.R. , Fenech, M. , Thomas, P. and Falconer, I.R. (2000). Micronucleus induction and chromosome loss in transformed human white cells indicate clastogenic and aneugenic action of the cyanobacterial toxin, cylindrospermopsin. *Mutation Research 472*: 155-161.

Humpage, A.R. , Fontaine, F. , Froscio, S.M. , Burcham, P.C. and Falconer, I.R. (2005). Cylindrospermopsin genotoxicity and cytotoxicity: role pf cytochrome P-450 and oxidative stress. *Journal of Toxicology and Environmental Health Part A 68*: 739-753.

Ibelings, B.W. and Chorus, I. (2007). Accumulation of cyanobacterial toxins in freshwater "seafood" and its consequences for public health: A review. *Environmental Pollution 150(1)*: 177-192.

Jacoby, J.M. , Collier, D.C. , Welch, E.B. , Hardy, F.J. and Crayton, M. (2000). Environmental factors associated with a toxic bloom of *Microcystis aeruginosa. Canadian Journal of Fisheries and Aquatic Sciences 57(1)*: 231-240.

Jones, G. , Baker, P.D. , Burch, M.D. and Harvey, F.L. (2002). National Protocol for the Monitoring of Cyanobacteria and their Toxins in Surface Waters. Draft V5.0 for consideration LWBC, July 2002. Agriculture and Resource Management Council of Australia and New Zealand National Algal Management, Canberra.

Jones, G.J. , Bourne, D.G. , Blakely, R.L. and Doelle, H. (1994). Degradation of the cyanobacterial hepatotoxin microcystin by aquatic bacteria. *Natural toxins 2*: 228-235.

Jones, G.J. and Orr, P.T. (1994). Release and degradation of microcystin following algicide treatment of a *Microcystis aeruginosa* bloom in a recreational lake, as determined by HPLC and protein phosphatase inhibition assay. *Water Research 28(4)*: 871-876.

Kinnear, S. , Duivcnvoordcn, L. and Fabbro, L. (2007). Growth and bioconcentration in Spirodela oligorrhiza following exposure to Cylindrospermopsis raciborskii whole cell extracts. *Australasian Journal of Ecotoxicology* 12:19-31Kinnear, S. , Fabbro, L. and Duivenvoorden, L. (2008). Variable Growth Responses of Water Thyme (Hydrilla verticillata) to Whole-Cell Extracts of Cylindrospermopsis raciborskii. *Archives of Environmental Contamination and Toxicology 54(2)*: 187-194.

Kinnear, S.H.W. , Duivenvoorden, L.J. and Fabbro, L.D. (2007). Sublethal responses in Melanoides tuberculata following exposure to Cylindrospermopsis raciborskii containing cylindrospermopsin. *Harmful Algae 6(5)*: 642-650.

Kinnear, S.H.W. , Fabbro, L.D. , Duivenvoorden, L.J. and Hibberd, E.M.A. (2007). Multiple-organ toxicity resulting from cylindrospermopsin exposure in tadpoles of the cane toad (Bufo marinus). *Environmental Toxicology 22(6)*: 550-558.

Komárek, J. and Anagnostidis, K. (1986). Modern approach to the classification system of cyanophytes. *Archiv für Hydrobiologie Supplement 73(2) Algological Studies 43*: 157-226.

Komárek, J. and Anagnostidis, K. (1989). Modern approach to the classification system of cyanophytes 4 - Nostocales. *Archiv für Hydrobiologie Supplement 82, 3 Algological Studies 56*: 247-345.

Lahti, K. , Rapala, J. , Färdig, M. , Niemelä, M. and Sivonen, K. (1997b). Persistence of cyanobacterial hepatotoxin, Microcystin-LR in particulate material and dissolved in lake water. *Water Research 31(5)*: 1005-1012.

Landsberg, J.H. (2002). The effects of harmful algal blooms on aquatic organisms. *Reviews in Fisheries Science 10(2)*: 113-390.

Leflaive, J. and Ten-Hage, L. (2007). Algal and cyanobacterial secondary metabolites in freshwaters: A comparison of allelopathic compounds and toxins. *Freshwater Biology 52(2)*: 199-214.

Leonard, J.A. and Paerl, H.W. (2005). Zooplankton community structure, micro-zooplankton grazing impact, and seston energy content in the St. Johns river system, Florida as influenced by the toxic cyanobacterium *Cylindrospermopsis raciborskii. Hydrobiologia 537*: 89-97.

Lindsay, J. , Metcalf, J.S. and Codd, G.A. (2006). Protection against the toxicity of microcystin-LR and cylindrospermopsin in *Artemia salina* and *Daphnia* spp. by pre-treatment with cyanobacterial lipopolysaccharide (LPS). *Toxicon 48*: 995-1001.

Lirås, V. , Lindberg, M. , Nystrom, P. , Annadotter, H. , Lawton, L. and Graf, B. (1998). Can ingested cyanobacteria be harmful to the signal crayfish (*Pacifastacus leniusculus*)? *Freshwater Biology 39*: 233-242.

Looper, R.E. , Runnegar, M.T.C. and Williams, R.M. (2005). Synthesis of the Putative Structure of 7-Deoxycylindrospermopsin: C7 Oxygenation Is Not Required for the Inhibition of Protein Synthesis. *Angewandt Chemie International Edition 44(25)*: 3879-3881.

McGregor, G.B. and Fabbro, L.D. (2000). Dominance of *Cylindrospermopsis raciborskii* (Nostocales, Cyanoprokaryota) in Queensland tropical and subtropical reservoirs: Implications for monitoring and management. *Lakes and Reservoirs: Research and Management 5*: 195-205.

Metcalf, J.S. , Beattie, K.A. , Saker, M.L. and Codd, G.A. (2002a). Effects of organic solvents on the high performance liquid chromatographic analysis of the cyanobacterial toxin cylindrospermopsin and its recovery from environmental eutrophic waters by solid phase extraction. *FEMS Microbiology Letters 216*: 159-164.

Metcalf, J.S. , Lindsay, J. , Beattie, K.A. , Birmingham, S. , Saker, M.L. , Törökné, A.K. and Codd, G.A. (2002b). Toxicity of cylindrospermopsin to the brine shrimp *Artemia salina*: comparisons with protein synthesis inhibitors and microcystins. *Toxicon 40*: 1115-1120.

Mihali, T.K. , Kellmann, R. , Muenchhoff, J. , Barrow, K.D. and Neilan, B.A. (2008). Characterization of the gene cluster responsible for cylindrospermopsin biosynthesis. *Applied and Environmental Microbiology 74(3)*: 716-722.

Neilan, B.A. , Saker, M.L. , Fastner, J. , Törökné, A.K. and Burns, B.P. (2003). Phylogeography of the invasive cyanobacterium *Cylindrospermopsis raciborskii. Molecular Ecology 12*: 133-140.

Nogueira, I.C.G. , Lobo-da-Cunha, A. and Vasconcelos, V.M. (2006). Effects of *Cylindrospermopsis raciborskii* and *Aphanizomenon ovalisporum* (cyanobacteria) ingestion on *Daphnia magna* midgut and associated diverticula epithelium. *Aquatic Toxicology 80*: 194-203.

Nogueira, I.C.G. , Saker, M.L. , Pflugmacher, S. , Wiegand, C. and Vasconcelos, V.M. (2004a). Toxicity of the Cyanobacterium *Cylindrospermopsis raciborskii* to *Daphnia magna. Environmental Toxicology 19*: 453-459.

Norris, R.L. , Eaglesham, G. , Pierens, G. , Shaw, G. , Smith, M.J. , Chiswell, R.K. , Seawright, A.A. and Moore, M.R. (1999). Deoxycylindrospermopsin, an analog of cylindrospermopsin from *Cylindrospermopsis raciborskii. Environmental Toxicology 14*: 163-165.

Norris, R.L. , Eaglesham, G.K. , Shaw, G.R. , Senogles, P. , Chiswell, R.K. , Smith, M.J. , Davis, J.A. , Seawright, A.A. and Moore, M.R. (2001). Extraction and Purification of the Zwitterions Cylindrospermopsin and Deoxycylindrospermopsin from *Cylindrospermopsis raciborskii. Environmental Toxicology 16*: 394-396.

Oberemm, A. (2001). Effects of cyanotoxins on early life stages of fish and amphibians. In: Chorus, I (editor). *Cyanotoxins: Occurrence, Causes, Consequences.*Berlin: Springer-Verlag. pp. 240-248

Ohtani, I. , Moore, R.E. and Runnegar, M.T.C. (1992). Cylindrospermopsin: a potent hepatotoxin from the blue-green alga *Cylindrospermopsis raciborskii. Journal of the American Chemical Society 114*: 7941-7942.

Orr, P. and Jones, G. (1998). Relationship between microcystin production and cell division rates in nitrogen-limited *Microcystis aeruginosa* cultures. *Limnology and Oceanography 43(7)*: 1604-1614.

Padisák, J. (1997). *Cylindrospermopsis raciborskii* (Woloszynska) Seenayya et Subba Raju, an expanding highly adaptive cyanobacterium: worldwide distribution and review of its ecology. *Archiv für Hydrobiologie. Supplement 107 (Monographic Studies) 4*: 563-593.

Pomati, F. , Moffitt, M.C. , Cavaliere, R. and Neilan, B.A. (2004a). Evidence for differences in the metabolism of saxitoxin and C1+2 toxins in the freshwater cyanobacterium *Cylindrospermopsis raciborskii* T3. *Biochimica et Biophysica Acta 1674*: 60-67.

Pomati, F. , Neilan, B.A. , Suzuki, T. , Manarolla, G. and Rossetti, C. (2003). Enhancement of intracellular saxitoxin accumulation by lidocaine hydrochloride in the cyanobacterium *Cylindrospermopsis raciborskii* T3 (Nostocales). *Journal of Phycology 39*: 535-542.

Pomati, F. , Rossetti, C. , Manarolla, G. , Burne, B.P. and Neilan, B.A. (2004b). Interactions between intracellular Na^+ levels and saxitoxin production in *Cylindrospermopsis raciborskii* T3. *Microbiology 150*: 455-461.

Preußel, K. , Stüken, A. , Wiedner, C. , Chorus, I. and Fastner, J. (2006). First report on cylindrospermopsin producing *Aphanizomenon flos-aquae* (Cyanobacteria) isolated from two German lakes. *Toxicon 47*: 156-162.

Rasmussen, J.P., Campbell, R., Monis, P.T. and Saint, C.P. (2005). The genetic determinants of cylindrospermopsin production and their detection using real-time PCR. 5th Workshop of the Australian Research Network on Algal Toxins (Oral session abstracts), 9 - 11 July, 2005, Moreton Bay Research Station, Queensland. Australia.

Raziuddin, S., Siegelmann, H.W. and Tornabene, T.G. (1983). Lipopolysaccharides of the cyanobacterium *Microcystis aeruginosa. European Journal of Biochemistry 137*: 333-336.

Reisner, M., Carmeli, S., Werman, M. and Sukenik, A. (2004). The Cyanobacterial Toxin Cylindrospermopsin Inhibits Pyrimidine Nucleotide Synthesis and Alters Cholesterol Distribution in Mice. *Toxicological Sciences 82(8)*: 620-627.

Ressom, R., San Soong, F., Fitzgerald, J., Turczynowicz, L., El Saadi, O., Roder, D., Maynard, T. and Falconer, I. (1994). *Health Effects of Toxic Cyanobacteria (Blue Green Algae).* Canberra: National Health and Medical Resources Council.

Runnegar, M.T., Kong, S., Zhong, Y. and Lu, S.C. (1995). Inhibition of reduced glutathione synthesis by cyanobacterial alkaloid cylindrospermopsin in cultured rat hepatocytes. *Biochemical Pharmacology 49(2)*: 219-225.

Runnegar, M.T., Kong, S.M., Zhong, Y.Z., Ge, J.L. and Lu, S.C. (1994). The role of glutathione in the toxicity of a novel cyanobacterial alkaloid cylindrospermopsin in cultured rat hepatocytes. *Biochemical and Biophysical Research Communications 201(1)*: 235-241.

Saker, M.L. and Eaglesham, G.K. (1999). The accumulation of cylindrospermopsin from the cyanobacterium *Cylindrospermopsis raciborskii* in tissues of the redclaw crayfish *Cherax quadricarinatus*. *Toxicon 37*: 1065-1077.

Saker, M.L. and Griffiths, D.J. (2000). The effect of temperature on growth and cylindrospermopsin content of seven isolates of the cyanobacterium *Cylindrospermopsis raciborskii* (Woloszynska) Seenaya and Subba Raju from water bodies in northern Australia. *Phycologia 39(4)*: 349-354.

Saker, M.L. and Griffiths, D.J. (2001). Occurrence of blooms of the cyanobacterium Cylindrospermopsis raciborskii (Woloszynska) Seenayya and Subba Raju in a north Queensland domestic water supply. *Marine and Freshwater Research 52*: 907-915.

Saker, M.L., Metcalf, J.S., Codd, G.A. and Vasconcelos, V.M. (2004). Accumulation and depuration of the cyanobacterial toxin cylindrospermopsin in the freshwater mussel *Anodonta cygnea*. *Toxicon 43*: 185-194.

Saker, M.L. and Neilan, B.A. (2001). Varied diazotrophies, morphologies and toxicities of genetically similar isolates of *Cylindrospermopsis raciborskii* (Nostocales, Cyanophyceae) from Northern Australia. *Applied and Environmental Microbiology 67(4)*: 1839-1845.

Saker, M.L., Nogueira, I.C.G., Vasconcelos, V.M., Neilan, B.A., Eaglesham, G.K. and Pereira, P. (2003). First report and toxicological assessment of the cyanobacterium *Cylindrospermopsis raciborskii* from Portuguese freshwaters. *Ecotoxicology and Environmental Safety 55*: 243-250.

Schembri, M.A., Neilan, B.A. and Saint, C.P. (2001). Identification of genes implicated in toxin production in the cyanobacterium *Cylindrospermopsis raciborskii*. *Environmental Toxicology 16*: 413-421.

Seawright, A.A., Nolan, C.C., Shaw, G.R., Chiswell, R.K., Norris, R.L., Moore, M.R. and Smith, M.J. (1999). The oral toxicity for mice of the tropical cyanobacterium *Cylindrospermopsis raciborskii*. *Environmental Toxicology 14*: 135-142.

Seifert, M., McGregor, G., Eaglesham, G., Wickramasinghe, W. and Shaw, G. (2007). First evidence for the production of cylindrospermopsin and deoxy-cylindrospermopsin by the freshwater benthic cyanobacterium, Lyngbya wollei (Farlow ex Gomont) Speziale and Dyck. *Harmful Algae 6(1)*: 73-80.

Shaw, G., McKenzie, R.A., Wickramasinghe, W.A., Seawright, A.A., K., E.G. and Fabbro, L.D. (2002). Comparative toxicity of the cyanobacterial toxin, Cylindrospermopsin, between mice and cattle: human implications. 4th Workshop of the Australian Research Network for Algal Toxins (ARNAT), Australian Institute of Marine Science, Townsville.

Shaw, G. , Seawright, A.A. , Moore, M.R. and Lam, P.K.S. (2000). Cylindrospermopsin, A Cyanobacterial Alkaloid: Evaluation of Its Toxicologic Activity. *Therapeutic Drug Monitoring 22*: 89-92.

Shaw, G., Sufenik, A., Livne, A., Chiswell, R.K., Smith, M.J., Seawright, A.A., Norris, K.R., Eaglesham, G. and Moore, M.R. (1999). Blooms of the cylindrospermopsin containing cyanobacterium, *Aphanizomenon ovalisporum* (Forti) in newly constructed lakes, Queensland, Australia. *Environmental Toxicology 14*: 167-177.

Shen, X., Lam, P.K.S., Shaw, G.R. and Wickramasinghe, W. (2002). Genotoxicity investigation of a cyanobacterial toxin, cylindrospermopsin. *Toxicon 40*: 1499-1501.

Sivonen, K. and Jones, G. (1999). Cyanobacterial toxins. In: Chorus, I and J Bartram (editor). *Toxic cyanobacteria in water.* London: E and FN Spoon. pp.

Soares, R.M., Magalhães, V.F. and Azevedo, S.M.F.O. (2004). Accumulation and depuration of microcystins (cyanobacteria hepatotoxins) in *Tilapia rendalli* (Cichlidae) under laboratory conditions. *Aquatic Toxicology 70*: 1-10.

White, S.H. (2006). Cylindrospermopsin in whole cell extracts and live cultures of *Cylindrospermopsis raciborskii*: ecotoxicity, bioaccumulation and management. School of Biological and Environmental Sciences, Faculty of Sciences, Engineering and Health. Rockhampton, Central Queensland University: 371.

White, S.H., Duivenvoorden, L.J. and Fabbro, L.D. (2005a). Absence of Free-Cylindrospermopsin Bioconcentration in Water Thyme (*Hydrilla verticillata*). *Bulletin of Environmental Contamination and Toxicology 75(3)*: 574-583.

White, S.H., Duivenvoorden, L.J. and Fabbro, L.D. (2005b). A decision-making framework for ecological impacts associated with the accumulation of cyanotoxins (cylindrospermopsin and microcystin). *Lakes and Reservoirs: Research and Management 10*: 25-37.

White, S.H., Duivenvoorden, L.J., Fabbro, L.D. and Eaglesham, G.K. (2006). Influence of intracellular toxin concentration on cylindrospermopsin bioaccumulation in a freshwater gastropod (*Melanoides tuberculata*). *Toxicon 47(5)*: 497-509.

White, S.H., Duivenvoorden, L.J., Fabbro, L.D. and Eaglesham, G.K. (2007). Mortality and toxin bioaccumulation in Bufo marinus following exposure to Cylindrospermopsis raciborskii cell extracts and live cultures. *Environmental Pollution 147(1)*: 158-167.

Whitton, B.A. and Potts, M., (editor). (2000). The Ecology of Cyanobacteria Their Diversity in Time and Space. Dordrecht: Kluwer Academic Publishers.

Wiedner, C., Rücker, J., Brüggemann, R. and B, N. (2007). Climate change affects timing and size of populations of an invasive cyanobacterium in temperate regions. *Oecologia 152(3)*: 473-484.

Wiegand, C. and Pflugmacher, S. (2001). Uptake of Microcystin-LR in Aquatic Organisms. In: Chorus, I (editor). *Cyanotoxins Occurrence, Causes, Consequences.* Berlin: Springer-Verlag. pp. 249 -252

Wormer, L., Cires, S., Carrasco, D. and Quesada, A. (2008). Cylindrospermopsin is not degraded by co-occurring natural bacterial communities during a 40-day study. *Harmful Algae 7(2)*: 206-213.

In: Lake Pollution Research Progress
Editors: F. R. Miranda and L. M. Bernard
ISBN: 978-1-60692-106-7

Chapter 4

FOCUS ON UNDERSTANDING THE RELATION BETWEEN LAKES AND POLLUTION – MODEL-BASED APPROACH AND CASE STUDY OF SUBARCTIC LAKE

Ryunosuke Kikuchi*[*1] and Tamara T. Gorbacheva[2]
[1] ESAC - Polytechnic Institute of Coimbra, Coimbra, Portugal
[2] INEP - Kola Science Center; Russian Academy of Sciences, Apatity, Russia

ABSTRACT

Fresh surface water accounts for just 1/10,000 of the total water available on the planet, and lakes contain almost all of the fresh surface water. Once contaminated, it is difficult to restore fresh-water quality: lakes therefore require special protection from contamination. This chapter aims to (i) understand the basic relation between lakes and pollution and (ii) answer a fundamental question: "How can we manage lakes effectively and properly?"

Models allow administrators to predict changes within the natural lake system resulting from management actions; however, extensive data input and a large number of

---

* Corresponding author: Tel.: +351 239 802287; fax: +351 239 802979. *E-mail address*: kikuchi@mail.esac.pt

parameters are required to simulate complex lake ecosystems. Heat loss from Arctic/subsractic lakes tends to be rapid in late summer and results in complete mixing of the water. Based on such subarctic shallow lakes, a case study (68°02'N and 33°11E) was carried out to investigate water quality, sediment profile and bulk deposition. Although there are various sources of heavy metals - weathering, industrial processes and so on, the presented study shows that there are prime (cautious) factors responsible for changing the lake/sediment quality. This finding suggests that the combination of some key factors with a simple model may be practical in daily lake management.

INTRODUCTION

Fresh surface water accounts for just 1/10,000 of the total water available on the planet [Baugartner, 1996], but this amount seems large when the volume is expressed as ~125,000 km^3. On a global scale, the amount of fresh surface water is almost constant year to year, being replenished by water precipitation previously evaporated from the ocean (~350,000 km^3) and land areas (~70,000 km^3). Most precipitation falls back into the ocean, and only ~110,000 km^3 falls on the land; that is, fresh water is a scarce resource. Once contaminated, it is difficult and expensive to restore fresh-water quality [Baugartner, 1996]. Thus, the study of fresh-water pollution has focused primarily on streams and lakes.

Lakes contain almost all of the fresh surface water on the planet. The water in rivers and streams makes up less than 1% of the volume in lakes [Baugartner, 1996]. This fact indicates that lakes require special protection from contamination. Furthermore, it takes many years to replenish lakes owing to the relatively small amount of precipitation that falls on lakes and the small amount of stream water that runs directly into lakes. On average, lake replenishment takes 100 years, whereas the replacement time for water in streams and rivers is about ten days [Baugartner, 1996]. If contaminants are disturbed throughout an average lake, the incoming water cannot restore the lake to its initial quality for a long time.

Heavy metals or trace metals are defined as metals with a density >5 g cm^{-3}. These metals form oxides and sulfides which are very hard to dissolve, and they tend to be bound in stable complexes with organic and inorganic particles. Heavy metals are generally found in small amounts in sediments (<0.1%). The great interest in heavy metals derives from the fact that these elements are supplied to aquatic systems in great excess by man. Furthermore, some heavy metals are hazardous to the aquatic ecosystem. The sources of heavy metals can be divided into five categories [Forstner and Wittman, 1979]: geologic weathering, industrial processes of ores and metals, the use of metals and metal components, leaching of metals from garbage and solid waste dumps, and animal and human excretions.

The pathways of pollutants to lake ecosystems are mainly divided into two groups [Hakanson and Jansson, 2002]: (i) water pathways – surface runoff, groundwater and waste outlets, and (ii) atmospheric pathways – wet deposition (rain and snow), dry deposition (falling particles) and gas absorption [U.S. EPA, 2007]. In addition, pollutants may also come out of the lakes through volatilization of gases from the water; together, gas absorption and volatilization are called gas exchange [U.S. EPA, 2007].

The purpose of lake control is to supervise the status of the received input substances and see whether no inappropriate contamination takes place. The ultimate aim should be to diagnose possible negative ecological effects, but this is a difficult task: on what species? on what organ? sediments or biota themselves and relative to other contaminants? and so on.

These types of questions are generally impossible to answer in a strictly scientific way, since present knowledge of pathways and effects of toxic substances in lake ecosystems is quite limited [Hakanson and Jansson, 2002]. It is obvious that basic and applied research and better methods for lake pollution control are needed. The purpose of this chapter is to basically understand the relationship between input substances and lake quality. The qualitative and quantitative relationship is discussed from two different viewpoints – the modeling approach and a field survey performed in a subarctic shallow lake.

Basic Information

Multidisciplinary knowledge may be required to understand the relation between lake quality and pollution – limnology, ecology, sedimentology, mathematics, atmospheric chemistry, meteorology, pollution science, geoscience, hydrology/hydraulics etc. The overall aim of this chapter is to present a comprehensive outline on the relation of lakes with pollution; therefore, basic information about key parameters is briefly reviewed first, followed by a description of the main discussion (modeling and field survey).

Sedimentation

The classic theory (see [Simons, 1980]) underlines the notion that particles heavier than water fall at a constant velocity in calm waters, where the force of downward motion is equal to the drag force resisting motion. The setting force for a sphere is given by;

$$Fs = (1/6) \cdot (\pi \cdot d^3) \cdot (p_s - p) \cdot g$$

where Fs = the force causing motion, d = particle diameter, p_s = particle density, p = water density and g = acceleration due to gravity.

There is a difference between laminar flow and turbulent flow [Smith, 1995]: if Reynolds Number (Re) is less than 0.5, the flow around the falling particle is laminar; and if Re is greater than 0.5, the flow is turbulent. The known Stokes' law for spherical particles is derived by putting Fs equal to the total drag force (= mass × acceleration) when the drag coefficient is equal to 24/Re, so the setting velocity is given by;

$$v = (p_s - p) \cdot g \cdot d^2 / (18 \cdot \mu)$$

where v = setting velocity and μ = viscosity coefficient. This equation describes the setting velocity up to Reynolds Number (Re) of 0.5. At a large Re value, the form resistance and inertia become important, and the drag coefficient decreases. For large spherical particles, the settling velocity is approximately proportional to the square root of the particle diameter.

Factors Influencing Sedimentation in Lakes

Some examples are given to show the linkage between lakes and sedimentation. On the lake surface, scum rows (ripples) sometimes appear to be pointing in the same direction as the wind and parallel to one another; these scum rows are caused by a rather complex water movement known as "Langmuir Circulation" [Thorpe, 2002]. Figure 1 shows a blend of wind speed and wave movement.

As the wind blows across a lake, a unit of water is moved from point A to point B. As this unit of water leaves point A, more water rushes up from beneath to occupy the space left behind. This net movement of water creates an upwelling. At point B, there is now more water than before. A downwelling occurs as the excess water pushes downward. As this continues to happen, spiraling cells are established in the water. The arrows indicate the direction of the water movement within each cell. Wherever the cells touch the surface, scum resting on top of the surface tension is pushed from the upwelling point to the downwelling point.

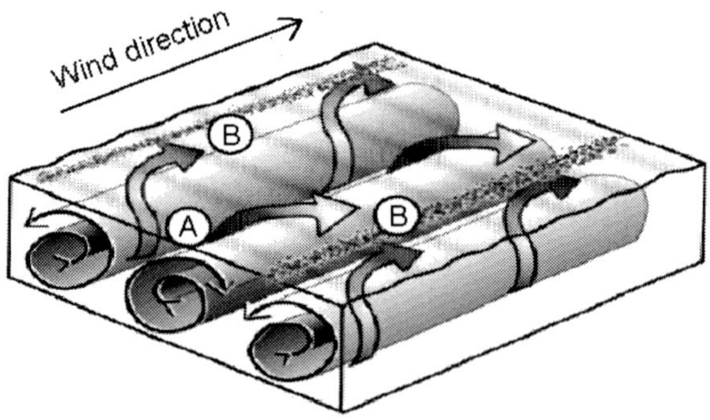

Figure 1. Schematic illustration of Langmuir circulation on the lake surface (redrawn from [Thrope, 2002]). Point A = upwelling and point B = downwelling.

Wind-induced waves in lakes significantly affect the sediment, even at great water depth [Hakanson and Jansson, 2002]; the orbital motion (U_m) of the water becomes oscillating with a horizontal displacement (l_n) near the lake bottom, the maximum horizontal velocity is associated with this motion (U_m), and the empirical relationship between the wave motion and sediment grain size (< 0.5 mm) moved by the passive wave is given as follows;

$$p \cdot U_m^2 \cdot (p_s - p) \cdot g \cdot d = C \cdot (l_n/d)^{1/2}$$

where C = empirical constant (≈0.13 in 1.0 g cm^{-3} water density [Sternberg and Larsen 1975]).

The following factors influence the dominant dynamics in the lake bottom: (i) an energy factor related to the wind/wave and the wave position (i.e. critical depth); (ii) a lake form factor related to the hypsographic curve (convex or concave); (iii) a slop factor related to the fact that fine deposits seldom stay permanently on slopes of about 5% inclination. Figure 2 illustrates the above-mentioned factors.

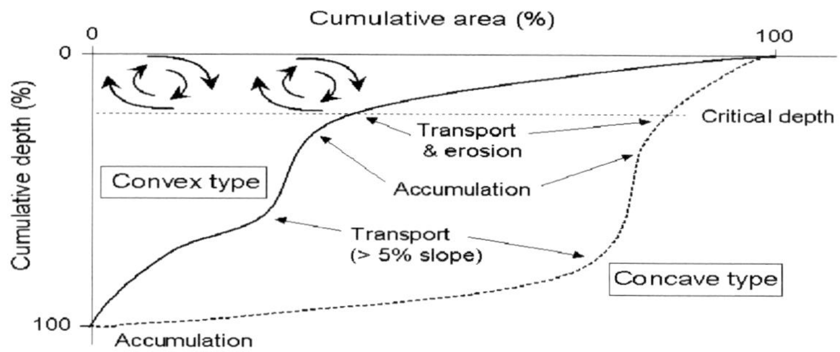

Figure 2. Conceptual illustration for factors influencing bottom dynamics with lake hypsographic curve as parameter (convex type and concave type) (based on [Hakanson, 1982]).

Grasshopper Effect

The atmospheric pathway is important for some toxic substances since they can remain in the environment long after they have been used or produced, and these substances make their way into the environment through a cycle of long-range air transport and deposition known as the "grasshopper effect" [Environment Canada, 1998; U.S. EPA, 2007]: persistent and volatile pollutants (e.g. certain pesticides, industrial chemicals and heavy metals) evaporate out of the soil where they are used, and travel in the atmosphere, condensing out again when the temperature drops. The process repeated in "hops" can carry them thousands of kilometers in a matter of days [Environment Canada, 1998].

Cold climate puts it at the receiving end of this process, with measurable concentrations of DDT, toxaphene, chlordane, PCBs and mercury found in both the Great Lakes (Canada) and the Arctic [Environment Canada, 1998]. Ingested by fish and other species, the chemicals travel up the food chain, accumulating in the fatty tissue of predatory animals, including people. Some of these pollutants are linked with developmental and reproductive impacts on wildlife, and may have similar effects on humans. The consequences may be serious for native people in the North, because traditional food sources are being contaminated [Environment Canada, 1998].

Arctic/Subarctic Lakes

Arctic/subarctic lakes are typically prevalent on low-lying landscapes, such as coastal and interior plains. There are many kettle (produced by the melting of buried glacial ice), moraine, and ice-scour lakes on the undulating terrain of postglacial arctic landscapes (e.g., the Canadian Shield, Fennoscandia, and the Kola Peninsula) [Korhola and Weckström, 2005]. Thermokarst lakes are also quite common in the Arctic (e.g., along the Alaskan coast and in Siberia), developing in depressions formed by thawing permafrost. Small ponds also dominate portions of the arctic/subarctic landscape (e.g., the low-lying terrain of Fennoscandia); typically less than 2 m deep, these freeze solid over the winter.

Local catchments are typically the primary source of water for arctic lakes [Woo and Xia, 1995]. Spring runoff originates from snow accumulation on lake ice, hillslope runoff, and lateral overflow from wetlands and streams [Woo, Heron and Steer, 1981; Marsh and Hey,

1989]. Outlets of small lakes may be snow-dammed, and eventually release rapid and large flows downstream [Woo, 1980].

The hydro-ecology of many small arctic/subarctic lakes is intimately linked with climatic conditions. The timing and speed of lake-ice melt depend on the rate of temperature increase in late spring and early summer, wind, and inflow of basin meltwater and terrestrial heat exchanges.

In northern Fennoscandia, for example, lakes >10 m deep are usually stratified during the summer and have well-developed thermoclines [Korhola et al., 2002]. In contrast, many high-arctic lakes mix vertically, thereby reducing thermal stratification [Welch, Legault and Bergmann, 1987]; similarly, small shallow lakes do not stratify because they warm quickly and are highly wind-mixed. Heat loss from arctic lakes tends to be rapid in late summer and early autumn and often results in complete mixing.

Atmospheric Deposition of Particles

Atmospheric deposition refers to the removal of pollutants from the air to soil, water and other surfaces. Deposition to water bodies can occur directly to the surface or indirectly, when material deposited to the land surface enters a water body through runoff [Blumberg et al., 2000].

Wet deposition refers to the incorporation of both gases and particles into all types of precipitation: rain, fog and snow. Pollutants may be removed from air by wet deposition through three main mechanisms [Blumberg et al., 2000]: (i) small particles can serve as cloud condensation nuclei and become entrapped in raindrops; (ii) particles can be incorporated into falling raindrops (i.e. particle scavenging); and (iii) gaseous pollutants can be dissolved into cloud droplets and falling rain or snow.

Dry particle deposition is broadly defined as the transport of particles and the contaminants associated with them onto surfaces. In general, the amount of contaminants deposited depends on concentrations in the air mass. The relationship is complex, however, depending on such physical factors as wind speed, the area of the receiving surface and whether that surface is water or land, and the properties of the contaminant, such as reactivity and the size of the particle with which it is associated [Blumberg et al., 2000]. For large particles the deposition velocities approach their gravitational settling velocities. For particles less than about 0.1 μm in size, the deposition velocity increases with decreasing particle diameter because of Brownian diffusion [Cohen, 1998]. Thus, particles in this size range will reside in the atmosphere for long time periods and can be transported over long distances. The flux of particle-bound pollutants from the atmosphere can be represented by the following equation [Cohen, 1998];

$$N_a = V_a \cdot C_a$$

where N_a is the flux of the particle pollutant, C_a is the mass of the pollutant in the particle phase per unit volume of air, and V_a is the overall deposition velocity.

Background Levels

It is important to establish a natural reference (i.e. background level) in order to qualify the contamination degree and compare the specific metals which have different concentrations in lake sediments. Figure 3 illustrates the schematic distribution of metal content in a sediment core.

According to the proposed definition [Lee, 1970; Hakanson and Jansson, 2002], the natural background level is the lowest value obtained in a sediment core. The natural background distribution depends on compaction, diffusion and bioturbation (i.e. layer disturbance by biological activity) in the sediments. Different metals have differential vertical mobility; e.g, lead is regarded as rather immobile whereas iron and manganese are known to be mobile and highly dependent on chemical character in the sediments.

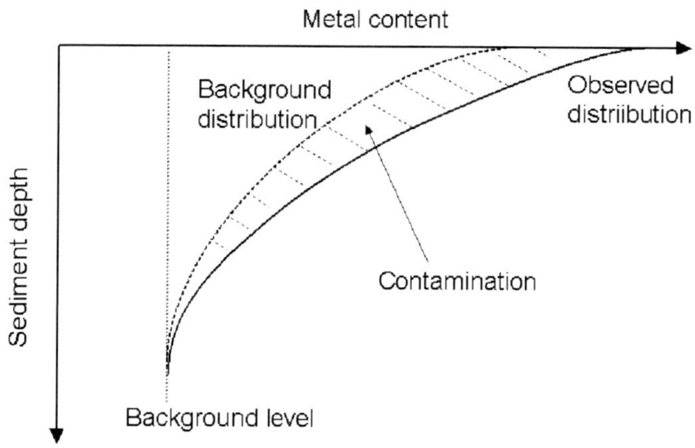

Figure 3. Schematic drawing of background level, background distribution and observed distribution of metal in sediment core (redrawn from [Hakanson and Jansson, 2002]).

Data Analysis Techniques

Sets of real field data contain not only information useful for quality assessment but also confusing noise [Praus, 2005]. Mostly, measured variables are not normally distributed, often co-linear or autocorrelated, containing outliers, erroneous or nonsense values. In order to reveal mutual dependence or logical structures of data, there are several procedures generally called data mining techniques. Some of them are based on the reduction of data dimensionality, such as principal component analysis, factor analysis, independent component analysis, independent factor analysis, generative topographic mapping and so on [Praus, 2005]. Two typical methods for data analysis are briefly presented here.

The contamination factor can be expressed for any given site or sediment core and as a mean for basins or lakes, and this factor expresses the relation between the pollutant concentration and the natural background concentration. The contamination factor accounts for the contamination of a single element [Hakanson and Jansson, 2002], and it is given by;

$$C_f = C_{0\text{-}x} / K$$

where C_f = the contamination factor of a given substance, $C_{0\text{-}x}$ = the substance concentration determined near the sediment surface (0 to x cm depth) - the top layer would be preferable (i.e. 0-1 cm) in most cases, and K (or B) = lake background level (or B = average geochemical background [Turekian and Wedepohl, 1961]). As shown in the above formula, the C_f value enables direct comparisons to be made between the measured contaminants and background levels. The following terminology can be used to describe the C_f values [Hakanson and Jansson, 2002]: low contamination = $C_f<1$; moderate contamination = $1\leq C_f<3$; considerable contamination = $3\leq C_f<6$; and extreme contamination = $6\leq C_f$.

The basic goal in Principal component analysis (PCA) is to reduce the data dimension. PCA is mathematically defined as an orthogonal linear transformation that transforms the data to a new coordinate system such that the greatest variance by any projection of the data comes to lie on the 1st coordinate (called the 1st principal component), the 2nd greatest variance on the second coordinate, and so on. PCA is theoretically the optimum transformation for a given data item in least square terms [Jolliffe, 2002]. The direction of the 1st principal component w_1 is defined by;

$$\mathrm{w}_1 = \underset{|\mathrm{w}|=1}{\arg\max} E\{(\mathrm{w}^T \cdot x)^2\}$$

where w_1 is of the same dimension m as the random data vector x; w^T is the matrix of basic vectors; the function of arg max stands for the argument of the maximum, and E is the operator of expectation value. The k-th principal component is determined as the principal component of the residual;

$$\mathrm{w}_k = \underset{|\mathrm{w}|=1}{\arg\max} E\{[\mathrm{w}^T - \sum_{i=1}^{k=1} \mathrm{w}_i \mathrm{w}_i^T\, x)]^2\}$$

The principal components are given by;

$$\mathrm{Si} = \mathrm{W}_i^{\mathrm{T}} \cdot x$$

The matrix value (W_i^T) is accomplished using the covariance matrix $E\{x \cdot x^T\} = C$ where x^T is a data matrix, and the W_i values are singular eigenvectors of C that correspond to the n largest eigenvectors of C [Jolliffe, 2002]. The results of a PCA are usually discussed in terms of component loadings and scores [Dunteman, 1989]: the component loadings are the correlation coefficient between the variables and factors, and; the principal component score is calculated for the standardized variable, i.e. the original datum minus the mean of the variable then divided by its standard deviation. [Krzanowski, 1988]. Multivariate statistical methods such as PCA have been employed in the following fields: quality assessment of surface water [Vega et al., 1998; Wunderlin et al., 2001], groundwater [Reghunath et al., 2002], and environmental research [Ceballos et al., 1998; Lambrakis et al., 2004].

POTENTIALLY TOXIC ELEMENTS

The term "heavy metals" is applied to a group of trace metals with a density greater than 6 gm cm^{-3}. This group was previously called toxic metals; however, other trace elements are also toxic to living organisms when present in excess. Such toxic trace elements are now collectively given the holistic term 'potentially toxic elements (PTEs)' [Alloway, 1995].

The natural background value for heavy metals in sediments is highly variable, depending on the bedrock structure (e.g. clay and shale have high content of metal); however, the increased supply of heavy metals to lakes and lake sediments is often connected with human activities [Hakanson and Jansson, 2002]. The content of heavy metals in lake sediment offers a key to the pollution history of the lake, and the horizontal distribution of metals in lake sediment can be used to assess the transportation and sedimentation rate of metals. Table 1 summarizes the classification of naturally-occurring metals according to their toxicity in the hydrologic environment. This classification may be used as criteria for comparison of metal-containing sediments.

Table 1. Classification of naturally-occurring metals in the hydrologic environment

Extremely low (or null) toxicity	Low toxicity	Moderate to high toxicity
Al, Bi, Ca, Cs, Fe, Li, K, Mn, Mg, Mo, Na, Sr and Rb	Au, Ba, Ce, Dy, Er, Eu, Gd, Ge, Ho, Nd, Pr, Pm, Sc, Sm, Sn, Re, Rh, Tb, Te, Yb and Yt	Ac, Ag, Be, Br, Cd, Co, Cr, Cu, Hf, Hg, Ir, In, Nb, Ni, Os, Pb, Pd, Po, Pt, Ra, Ru, Sb, Ta, Th, Ti, Tl and W
[Wood, 1974]		

The toxic level of PTEs has often been defined by a regulation, but definition of the regulation value is not a simple issue in practice. Different nations have adopted different, more or less arbitrarily defined levels - each reflecting their own local circumstances [Alloway, 1995] (see also table 2). Often these thresholds reflect a pragmatic compromise between what is technically desirable and what is achievable given local conditions in Europe.

No specific health guidelines for heavy metals associated with suspended or bed sediments have been established by the U.S. Environmental Protection Agency. This lack of national guidelines based on concise scientific criteria causes difficulty when it comes to evaluating the environmental effects of heavy metals in sediments [Garbarino et al., 1995].

According to a report published by the European Commission [Environment Directorate-General, 2002], the presence of heavy metals in modern society is a matter of ever-growing concern to politicians/authorities and the public in the European Union (EU) because heavy metals entering the techosphere (i.e. technology, economy and connected matters) will sooner or later be discharged to the environment or end up in waste, and they are not attractive for recycling. Few turnover analyses are available for each heavy metal in the EU member states [Environment Directorate-General, 2002]. It is therefore difficult to give a uniform description for all heavy metals. The European Freshwater Program has suggested an action plan at the EU level in order to protect human health and the environment from acute and chronic metal pollution [European Freshwater Program, 2000].

Table 2. Guideline levels for the recognition of metal contamination

Category	Soil (mg kg^{-1})			Drinking water (mg l^{-1})		
Nation	[1] Switzerland	[1] Germany	[1] Netherlands	[2] European Union	[3] USA	[4] World Health Organization
Situation	Guideline	Limit	Caution	Guideline	Limit	Guideline
As		20	30	0.01	0.01	0.01
Cd	0.8	3	1	0.005	0.005	0.003
Cr	75	100	250	0.05	0.01	0.05
Cu	50	100	100	2.0	1.3	2.0
Hg	0.8	2	2	0.001	0.002	[5] 0.001
Mn	—	—	—	—	0.002	0.4
Mo	5	5	100	—	—	0.07
Ni	50	50	100	0.02	—	0.02
Pb	50	100	150	0.01	0.0	0.01
Sb	—	—	—	0.005	0.01	0.02
Se	—	10	—	0.01	0.05	0.01
Sn	—	50	50	—	—	—
Ti	1	1	—	—	—	—
U	—	5	—	—	—	0.015
V	—	50	—	—	0.0	—
Zn	200	300	500	—	5.0	—

Note: [1]reviewed by [Haigh, 1995]; [2]Council Directive 98/83/EC; [3]based on [U.S. EPA, 1996]; [4][World Health Organization, 2006]; [5]total mercury (inorganic plus organic); and a dash represents no value listed.

APPROACH TO LAKE POLLUTION MODEL

The following factors must be taken into consideration to determine the sedimentation of materials in lakes [Hakanson and Jansson, 2002]: lake shape and volume, sediment discharge, sediment characteristics, delta development, turbidity currents, and hydrodynamical flow pattern; these factors vary in time and space. There seems to be no standard method available which describes all these complicated processes in an adequate manner [Hakanson and Jansson, 2002]. Mathematical models may not become the natural system (i.e. real world), but they can simplify and generalize real events by removing the incidental detail or noise [Wimpenny, 1999]. A representation of some phenomenon of the real world helps us gain better understanding, and the simulations using mathematical formulas allow administrators to predict positive or negative changes within the natural system due to management actions.

Equations for Mass Balance

The mathematical technique of differential equations is applied to solve problems where two or more substances are mixed together [Blanchard, Devaney and Hall, 1998]; this technique can be employed in order to simulate the mixing of pollutants in a lake [Hoggard, 1997; Aguirre and Tully, 1999]. The fundamental concept is based on the mass balance in a lake [Hakanson and Jansson, 2002].

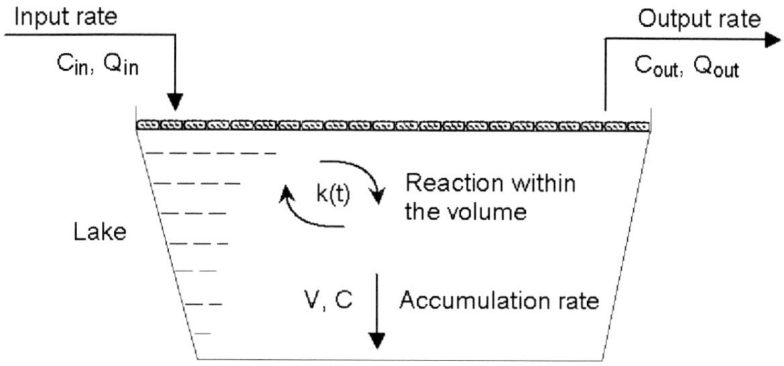

Figure 4. Conceptual diagram for well-mixed flow model in a lake.

First, the basic equation for mass per time is defined by subtracting the output rate from the input rate, and then variables in this subsection are listed as follows: C = pollutant concentration (= mass per unit volume); Q = the flow rate through the lake (= volume per unit time); V = the lake volume; M = the pollutant mass in the lake (= C·V); t = unit time, and k(t) = the reaction rate coefficient of the pollutant – k(t) = 0 where there is no reaction and k(t) = constant in 1st order reactions. The rate of dM/dt is defined by multiplying the concentration C by the volumetric flow rate·Q, then the mixture equation becomes [Hoggard, 1997; Aguirre and Tully, 1999];

$$dM/dt = C_{in}(t)\cdot Q_{in}(t) - C_{out}(t)\cdot Q_{out}(t)$$

The k value determines the order of chemical reactions (0th and 1st order reactions in this case). Thus another item is added to the above-mentioned equation, and it becomes [Hoggard, 1997; Aguirre and Tully, 1999];

$$dM/dt = C_{in}(t)\cdot Q_{in}(t) - C_{out}(t)\cdot Q_{out}(t) - k(t)\cdot C(t)\cdot V, \text{ or}$$

$$V\cdot dC(t)/dt = C_{in}(t)\cdot Q_{in}(t) - C_{out}(t)\cdot Q_{out}(t) - k(t)\cdot C(t)\cdot V$$

This equation gives a mathematical model for the pollutant concentration in a lake because it describes the amount of change between the pollutant input and the output at time t; that is, the accumulation of a pollutant is found by subtracting the output rate and the reaction rate from the input rate (cf. figure 4).

Assumptions are made to easily apply the above-mentioned equation, and these assumptions are summarized [Aguirre and Tully, 1999]: (i) the volume of the lake remains constant, (ii) the flow rate remains constant, (iii) the reaction rate remains constant, and (iv) the lake is well mixed. Lakes do not usually fluctuate in volume over a short period of time, so the 1st assumption seems plausible. In assuming that the volume is constant, the flow rate must also remain constant; therefore, the 2nd assumption is also plausible. The 3rd assumption is made to make the model easier to use but limited to only 0th and 1st order reactions. Assuming the lake is well mixed means that the concentration of the pollutant inside the lake is the same as the concentration of the outflow. This assumption is also to make the model easier to manage and is also limited to its use. The obtained equation is modified with these assumptions to become;

$$V \cdot dC(t)/dt = C_{in}(t) \cdot Q - C(t) \cdot Q - k \cdot C(t) \cdot V$$

Dividing each side by the volume V;

$$dC(t)/dt = C_{in}(t) \cdot (Q/V) - C(t) \cdot (Q/V) - k \cdot C(t)$$

Dividing the volume by the flow rate gives units of time; i.e. the term V/Q stands for residence time. If this residence time is represented by the symbol θ, the equation becomes;

$$dC(t)/dt = C_{in}(t) \cdot (1/\theta) - C(t) \cdot (1/\theta) - k \cdot C(t)$$

It is considered that there are three types of input function. These functions allow the pollution equation to be more accurate [Aguirre and Tully, 1999], and they are illustrated in figure 5.

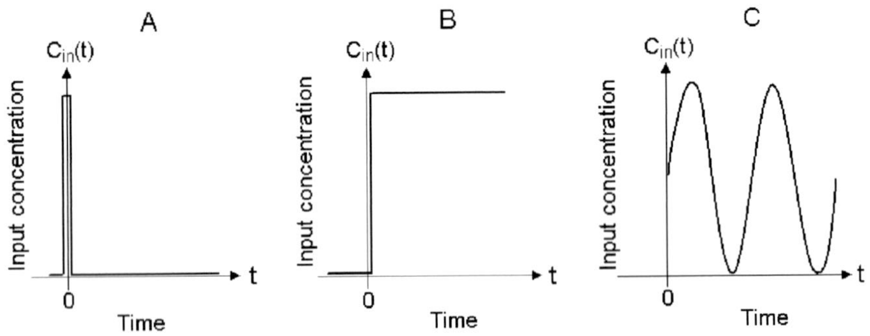

Figure 5. Different input functions. (A) the impulse input function has a spike, (B) the step input function is zero until the initial time and then increases, and (C) the sinusoidal input function varies periodically.

The above-mentioned three types of input function are summarized as follows [Hoggard, 1997; Aguirre and Tully, 1999]: (i) *impulse input of pollutant* – a load of pollutant is suddenly dumped into an otherwise clean lake. C_{in} is zero, so the flow of clean water into the lake gradually flushes the pollutant. In general, impulse functions could be made with the spike at any point, but the spike is always at the initial time in this model. An impulse input function is applied to represent the sudden introduction of a contaminant into the lake (see figure 5A); (ii) *step input of pollutant* – this function is zero up until t = 0, and then increases to a nonzero constant afterwards. That is, it actually looks like a "step". In this case, the initial time would be when the pollution source (e.g. factory) began operation. Before that time, no contaminant entered the lake, and after that time the contaminant entered at a constant rate (see figure 5B); and (iii) *sinusoidal input of pollution* – this input is represented by some type of a sin function, and the sinusoidal function in this model is written as;

$$C_{in}(t) = C_i \cdot \{1 + \sin(\alpha \cdot 2\pi / T)\}$$

where α = the normalized amplitude ($0 \leq \alpha \leq 1$), T = the fluctuation period, C_i = the average input concentration ($C_i(1 + \alpha)$ at the sin function's maximum and $C_i(1 - \alpha)$ at the minimum). The sinusoidal input function represents situations in which the input concentration of pollutant varies periodically (see figure 5C). While it may seem unusual, there are actually many situations which this models well. For example, seasonal variations may affect contaminant levels, and a plant dumping waste may produce less output at night than during the day when it operates [Hoggard, 1997; Aguirre and Tully, 1999].

Model for Overall Mass Balance

The above-mentioned model has four basic assumptions: (i) the volume is constant, (ii) the flow rate is constant, (iii) the reaction rate is constant, and (iv) the pollutant and the lake water are well mixed. Furthermore, only three input scenarios are considered. As a matter of fact, fluxes to lakes are related to other fluxes such as inflow from point sources, atmospheric deposition, background fluxes and so on; many of these fluxes could be related to different properties of the catchment areas [Lindstrom and Hakanson, 2001]. It is essential to contemplate a set of interwoven sub-models in order to simulate lake ecosystems that are large and complex, and this model structure is illustrated in figure 6.

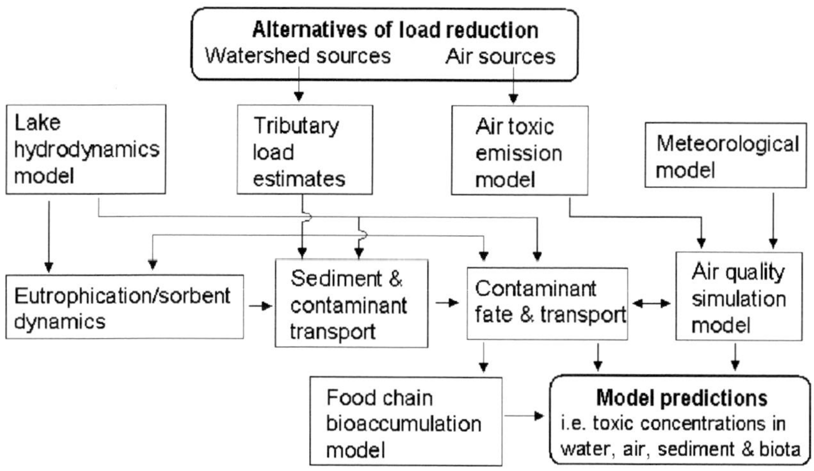

Figure 6. Model for overall mass balance (redrawn from [Blumberg et al., 2000; U.S. EPA, 2006]).

The modeling technique based on overall mass balance can represent a different key facet of the lake ecosystem [U.S. EPA, 2006]: (i) *hydrodynamics* – predicting water movements necessary to describe the 3-dimensional transport of dissolved and particulate constituents in the water column; (ii) *sediment (particle) transport* – describes the resuspension, transport, and deposition of particulate materials including sorbent phases necessary to describe the movement of particle-associated contaminants; (iii) *eutrophication/sorbent dynamics* – describes the production, respiration, grazing, and decomposition of planktonic biomass within the lake; (iv) *contaminant transport and fate* – describes contaminant partitioning between dissolved and sorbed phases, transfer between media (air, water, sediment), and biogeochemical transformations; and (v) *food web bioaccumulation* – simulated contaminant accumulation from water and sediments to predator fish via direct exposure and trophic transfer through benthic and pelagic food webs.

The model considers contaminant inter-actions with air, water and sediments, taking into account internal processes that may add or subtract mass. When modeling the behavior of a non-reacting dissolved substance such as chloride, in a simple system like a river flowing in one direction, the amount leaving the system should equal the amount entering. For reactive chemicals (e.g. PCBs) engaged in complex systems, a model must have many more parameters in order to provide the desired results. These parameters include quantitative estimates of the mass of contaminants that enter and leave the system, predictions concerning concentrations of pollutants for points in time and space, and the means to determine how much the inputs of a chemical must be reduced to reach a given concentration in the water, sediments or biota [Blumberg et al., 2000; U.S. EPA, 2006]. This information can be used to design optimal remedial strategies and to provide evidence that the economic investment of regulatory strategies will produce a definitive environmental benefit [Richardson et al., 1999].

Thus, the overall mass balance model takes raw data collected through field monitoring and inputs it into mathematical models to determine how concentrations change in relation to loadings from the atmosphere and tributaries. However, this approach has a large number of parameters and requires extensive data input; hence, a user would need to evaluate and input many different data points to accurately parameterize and sufficiently calibrate the model.

Case Study of a Subarctic Lake

A case study will help the reader to follow this chapter aimed at "understanding the relation between lake quality and pollution load". As most Arctic/subarctic lakes are located in remote places, these lakes generally seem not to suffer from anthropogenic impacts because heavy metals usually occur at levels below 1 mg l^{-1} in Arctic freshwater, similar to unpolluted areas outside the Arctic [Nilsson, 1997]. A survey of 250 wetlands in the Russian Arctic indicates three areas of heavy metal pollution – the Kola Peninsula (the northern part of the Russian territory in Europe), the Vorkuta vicinity (north Komi) and the Norilsk vicinity (central Siberia); metal levels are highly variable from lake to lake, but profile data show that metal levels are generally elevated in the upper parts of the lake sediment [Nilsson, 1997].

There is discussion as to whether this trend is due to natural (diagenic) processes or a result of anthropogenic metal input over time. The following parameters were determined in Staroe lake located on the Kola Peninsula in order to consider the cause of high metal concentrations in the upper sediment: (i) clarification of human impacts near the lake, (ii) analysis of bulk deposition as a factor of the atmospheric pathway, (iii) data collection of background levels as a factor of natural processes, (iv) water analysis as a factor of the present lake quality and (v) analysis of lake sediment as a factor of historical metal loads.

Study Area

Staroe lake (picture B in figure 7) is located on the east side of the Kola Peninsula, and it lies at approximately 68°02'N and 33°11E. The area encompasses about 2.8 km^2 and the mean depth is 5.8 m. The climate is harsh in this area – the period of snow cover is 7.5 months from October to mid-May, the mean temperature is -1.7°C and the mean wind speed is 5.9 km s^{-1}. The diurnal amounts of atmospheric bulk deposition averaged 2.2 mm (23.6 mm at the maximum) in 2004. The wind direction in this region is highly dependent on the movement of low pressure: the winds often blow from the ocean during the summer when the continent is warmer than the ocean, and they blow in the opposite direction from the continent during the winter [Mäkinen 1994].

The Kola Peninsula is geologically known all over the world for its large mineral deposits [Nilsson, 1997]. Surface mines are typically enlarged until either the mineral deposit is exhausted or the cost of removing larger volumes of overburden makes the mining uneconomical, and this overburden (also known as waste) is the term used to describe material that lies above a valuable deposit. Olenegorsk mining company put iron ore mines into operation in 1955 and put an open pit (designated Bauman pit in this section) into operation in 1987. Three overburden piles in this open pit are characterized as follows: 105 ha total area, 2,500 m length and 500 m width in each pile (see picture A in figure 7). Furthermore, there is a Cu-Ni smelter complex (Severonikel in Monchegorsk city) at a distance of 25 km from the Olenegorsk mining company (cf. figure 7). This smelter complex (picture C in figure 7) has been known as a great emitter of heavy metals on the Kola Peninsula [Nilsson, 1997]. Although the Staroe lake is less subjected to mining wastewater in terms of the water pathway (cf. figure 7), there is a possibility that the lake quality may be influenced by the mining-related emissions and/or the smelting-related emissions through the atmospheric pathway.

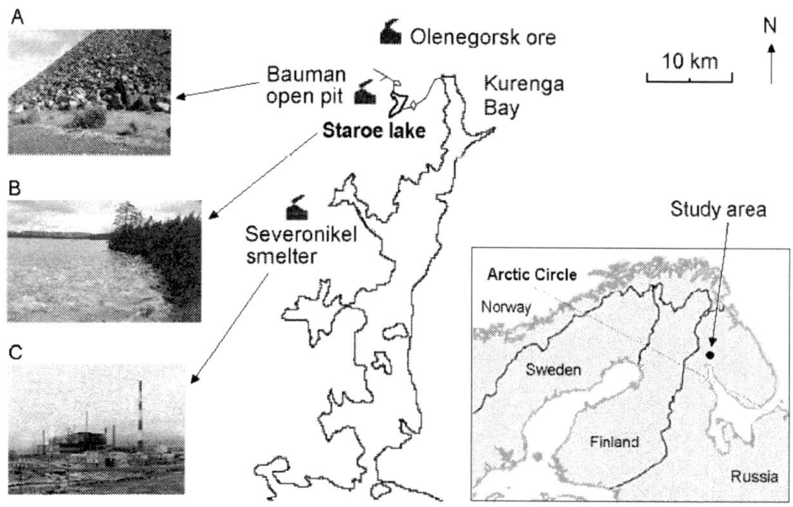

Figure 7. A map of the study area (68°02'N and 33°11E). Picture A – overburden pile in Bauman, picture B – Staroe lake, and picture C – Severonikel smelter complex in Monchegorsk.

Materials and Methods

Reference data were gathered through a literature research and our previous study: background levels on the Kola Peninsula [Vinogradov, 1962; Moiseenko, 1997; Dauvalter,

1999], and chemical properties of atmospheric deposition (snow precipitation) in the smelter region (Monchegorsk) and the background site (250 km to the south-west of the smelter) [Kikuchi and Gorbacheva, 2006]. Since the other data were measured, the applied materials and methods are briefly described in this section.

Water Sampling in Staroe Lake

Each 1,000 ml sample was collected at a surface point and a deep point (near the bottom layer) in July 2006. The collected samples were directly analyzed after filtration through a 0.45 μm membrane using the following techniques: (i) potentiometry for determining H^+, (ii) atomic absorption spectrophotometry for determining Al, Fe, Ca, Mg, K, Mn, Zn, Ni, Cu, Na and Sr, (iii) ion-exchange chromatography for determining SO_4^{2-}, Cl^- and NO_3^-, (iv) colorimetry for determining Si and P, and (v) the oxidability method using bichromate for C determination.

Bulk Deposition Near the Open Pit

Bulk deposition was collected using a set of 3 rainwater samplers. These samplers were placed in an open field halfway between Bauman open pit and Staroe lake, and the chosen field was away from vegetation influence and mining activities such as transport. Sampling was carried out every month from July to September 2006.

Each sampler was made of a polyethylene tube of 14.5 cm diameter having a few holes for ventilation. This tube was equipped with disposable polyethylene bags for sample collection and synthetic mesh for prevention of large solid particles. Three parallel samples from each plot were analyzed after filtration through filter paper with 2.0 μm pore size in the same way as the above-mentioned analysis; in addition, the filtered residue (insoluble particulates) was weighted to follow a mineralogical (composition) analysis.

Overburden in the Open Pit

The sampling (1 sample of ~1 kg per 25 ha) was carried out at 5 random points of the overburden piles. The collected samples were passed through different size sieves (2.00, 0.50, 0.25 and 0.10 mm). Each fraction was well mixed, and a 100 mg sub-sample in each fraction was used to perform mineralogical (composition) analysis using a microscope.

Bottom Sediment

Lake-bottom sediment was collected by an open-gravimetric column sampler (44 cm diameter) equipped with an automatic diaphragm. After collection, the sample columns were cut at a 1-cm interval for analysis, and they were then placed into a polyethylene container in the laboratory. The prepared samples were stored at 4°C until analysis.

Metal distribution has some patterns – (i) exchangeable, (ii) bound to carbonates, (iii) bound to metallic minerals, (iv) bound to organic matter and (v) residual or bound to crystalline minerals [Tiesser et al., 1979; Kersten and Förstner, 1987]. Considering these patterns and a preliminary analysis, the samples were digested with conc. HNO_3 after air drying in order to extract metals from bonding forms (iv) and (v). The samples thus prepared were analyzed by atomic absorption spectrometry for determining Na, Ca, Mg, Al, Fe, Zn, Ni, Cu, Mn and Sr, and colorimetry for determining P.

RESULTS

There are space limitations in this chapter; without special mention, all data are shown as the average values, which are considered to be representative. Standard deviation is abbreviated as SD, and *n* in brackets shows the number of measurements.

Water Quality

On the basis of the performed chemical analysis, there was no clear difference in water quality between the surface point and the deep point (near the bottom) in Staroe lake, and the pH value of 6.9 also showed no difference. As compared with Russian Standard SanPin 2.1.980-00, the contents of NO_3^- (~50 mg l^{-1}) and particulates (~2 mg l^{-1}) exceeded the standard levels (0.7 mg l^{-1} NO_3^- and 0.7 mg l^{-1} particulates); however, compared with the data collected from 460 lakes on the Kola Peninsula [Moiseenko, 1997], some contents exceeded the background levels. The measured values and the background levels are illustrated in figure 8. The contents of Fe, Ni, Cu, P, Si, Sr (surface point) and particulates were 3 to 6 times greater than the background levels, but the contents of Sr (deep point) and SO_4^{2-} were about 10 times greater than the background levels.

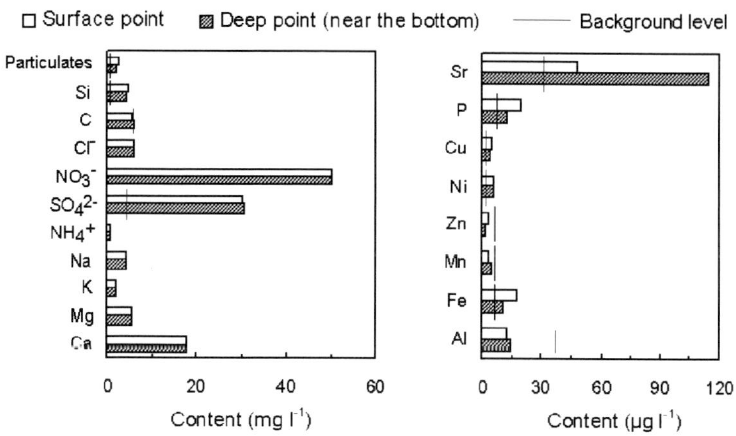

Figure 8. Water quality in Staroe lake (June 2006) and background levels [Moiseenko, 1997].

Atmospheric Deposition

Bulk deposition was measured from July to September 2006 halfway between Bauman open pit and Staroe lake (abbreviated as the bulk deposition near the study lake). The atmospheric deposition in the smelter region and the background levels are derived from a previous study and published data (see [Kikuchi and Gorbacheva, 2006]). These data are summarized in figure 9. The values shown in figure 9 are expressed as the monthly average to facilitate their comparison.

As compared with the background levels, the bulk deposition near the study lake contained great amounts of Mg, Ca and SO_4^{2-} (see figure 9). The Ni content was null at the background level, but the bulk deposition contained 0.10 mg m^{-2} Ni on a monthly average (see figure 9). According to the snow analysis in Monchegorsk [Jaffe et al., 1995], less than 1% of the emitted SO_2 is removed within a circumference of 20 km from the smelter complex; i.e. it implies long-range transport of SO_2 and a low rate of SO_2 oxidation in the atmosphere. Since not only SO_4^{2-} but also particles containing base cations (Mg and Ca) enter the lake through the atmospheric pathway, these particles contribute to the acid-neutralizing capacity of the lake.

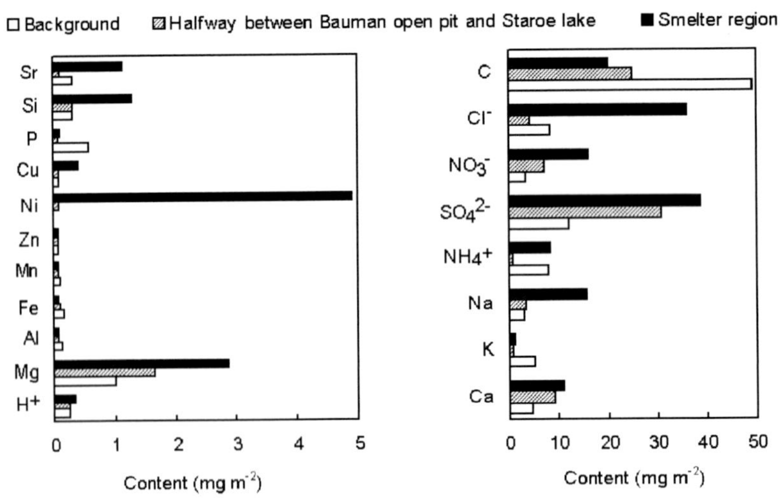

Figure 9. Atmospheric deposition on monthly average – background 250 km away from the smelter (bulk deposition, July to Sept. 2006), halfway between Bauman open pit and Staroe lake (bulk deposition, Jul to Sept. 2006), and the smelter region (snow deposition, Sept. 2004 to April 2005).

Overburden Quality

This study was performed to determine the percentage of different grain sizes and the main components contained in the overburden samples (cf. picture A in figure 7). The grain-size distribution obtained from the perfumed sieve analysis (n = 10) is summarized in table 3.

Table 3. Grain-size distribution of samples (rocks) in Bauman overburden piles

Size (mm)	> 2.00	2.00-0.50	0.50-0.25	0.25-0.10	0.10 >
Mean distribution (%)	53	26	6	9	6
Range (max. to min.) (%)	10 to 74	15 to 56	3 to 39	1 to 16	1 to 10

Fine-grained detritus are transported away from their original site by natural forces such as wind; there is hence a strong possibility that fine overburden detritus may deposit in Staroe lake near the Burman overburden piles. The mineral quality of this fine fraction (below 0.10 mm) was analyzed (n = 6), and the results show that the main component is quartz – an average content of 80%, with a maximum of 99% and a minimum of 60%.

Table 4. Quartz quality (wt %) contained in fine (< 0.10 mm) Bauman overburden

SiO_2	Al_2O_3	Fe_2O_3	FeO	Fe_2O_3T	CaO
65.60±0.47	16.20±0.19	1.02±0.13	2.97±0.05	4.35±0.14	3.17±0.08
MgO	Na_2O	K_2O	P_2O_5	TiO_2	
1.00±0.07	4.20±0.13	3.60±0.12	0.25±0.02	0.62±0.03	

The fine fraction of the Bauman samples was sent to mineral testing laboratories for analysis of quartz quality, and the results are summarized in table 4. It follows from table 4 that SiO_2 and Al_2O_3 account for 82% of the total content.

Sediment Profile

As stated in the introduction, the input of heavy metals to a lake can be basically divided into natural sources (e.g. weathering) and human sources. First, natural background levels for metals in sediment rocks are presented in table 5, and these levels are based on an intensive field study of sediment rocks located on the Kola Peninsula [Vinogradov, 1962]. Although metal values in lake sediments are variable and dependent on the bedrock structure, these mean values are basically similar to those of sediment rocks [Hakanson and Jansson, 2002].

Table 5. Representative composition (mg kg-1) of sediment rock on the Kola Peninsula

Cu	Ni	Zn	Sr	Mn	Na	Mg	K	Ca	Fe	Al
57	95	80	450	670	6,600	13,400	22,800	25,300	333,000	104,500

As mentioned in the section entitled "Basic information", it is essential to establish a background level in order to qualify the contamination degree and compare the specific metals which have different concentrations in lake sediments; the measured sediment profiles and background levels based on a lake study (n = 141 on the Kola Peninsula) [Dauvalter, 1999] are therefore illustrated in figure 10.

The sedimentation rate in a northern background lake is about 1mm per year [Norton et al., 1992]; it implies that the sediment collected at 20 cm depth was formed about 200 years ago. Initial increases in Cu, Ni, Sr, Mn Ca and Fe are observed at a depth of 6 cm, approximately corresponding to 1950. Na, Mg, K, and Al are variable and do not show a clear tendency to increase; these elements are abundant in natural surroundings (table 5).

Considering the sediment profile shown in figure 10 and the history of industrial movement on the Kola Peninsula, it is concluded that the lake sediment at 0-6 cm depth was formed during the rapid industrial development on the Kola Peninsula and the sediment below 6 cm depth was formed under natural background conditions.

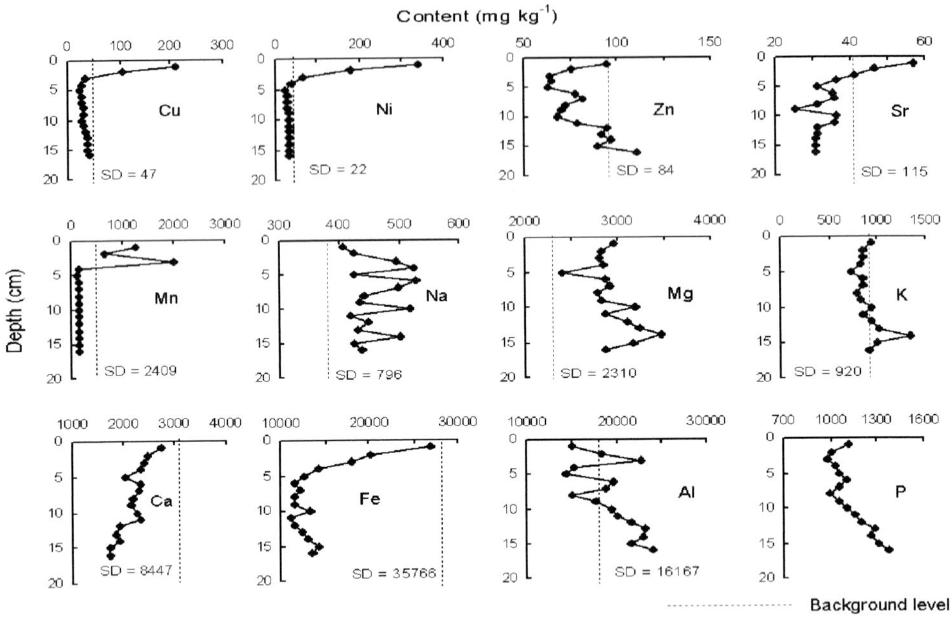

Figure 10. Background levels of lakes on the Kola Peninsula (redrawn from [Dauvalter, 1999]), and sediment profile determined in Staroe lake (July 2006). No data for P background level.

CONSIDERATION

The sediment columns (16 cm height) were cut at 1-cm intervals and 12 elements were analyzed at each interval (figure. 10); that is, a large volume of data was gathered. Principal component analysis (see the section entitled "Basic information") is applied to process these data. As before, principal component analysis (PCA) can reduce multidimensional data sets to lower dimensions for analysis.

Correlation Matrix

The correlation coefficients between elements are computed from the quantitative data obtained at each interval of the sediment column, and they are expressed in a matrix (table 6).

As seen in table 6, there are strong correlations between the metals (n = 16) described by positive correlation coefficients at the significant level of $p < 0.05$: (i) Zn-K, Mn-Ca and Mn-Sr for 0.6; (ii) P-Zn for 0.7; (iii) Ca-Mg, Ca-Cu and Ca-Ni for 0.8; and (iv) Mg-K, Sr-Ca, Sr-Cu, Cu-Ni and Sr-Ni for 0.9.

It follows from table 5 that Cu, Sr and Ni are poor in the sediment rocks, but these 3 metals have a co-occurrence pattern in the sediment profile. Table 1 shows that Cu and Ni have moderate or high toxicity in the hydrologic environment, so the potential adverse effect of these two elements on the bottom sediment seems to be considerable. Concerning the correlation between Zn and P, one of the factors controlling sorption of divalent metal ions (e.g. Zn^{2+}) is the existence of organic matter (as a complex agent) and a proton exchanger associated with the dissociation of carboxylic or phenolic groups in sediments [Fujioshi et al., 1997]: therefore, the Zn distribution pattern could reflect the organic bound P pattern.

Table 6. Correlation coefficients of metals in the lake sediment (0-16 cm depth)

	K	Na	Ca	Mg	Fe	Al	Cu	Ni	Zn	Mn	Sr	P
K												
Na	0.2											
Ca	0.3	0.1										
Mg	0.9	0.3	0.1									
Fe	0.4	-0.3	0.8	0.1								
Al	0.5	0.3	-0.1	0.5	0.5							
Cu	0.4	-0.5	0.8	0.1	0.1	-0.3						
Ni	0.4	-0.5	0.8	0.1	0.1	-0.3	0.9					
Zn	0.6	-0.4	0.2	0.5	0.5	0.2	0.5	0.5				
Mn	0.2	-0.1	0.6	0.0	0.0	0.3	0.5	0.5	0.0			
Sr	0.5	-0.2	0.9	0.2	0.2	-0.1	0.9	0.9	0.4	0.6		
P	0.5	-0.2	-0.2	0.5	0.5	0.2	0.1	0.1	0.7	-0.4	0.0	

Eigenvalues and Cumulative

Eingenvalues and variances by factor were calculated from the correlation coefficients shown in table 6, and the obtained values are summarized in table 7. As stated in the introduction, there are mainly five sources of metals; hence, it is difficult to suppose that one factor dominantly affects the sediment quality. As seen in table 7, components 1 and 2 account for about 70% of the total variance: component 1 is the most important and component 2 is the second-most important in lake management.

Table 7. Eigenvalues and cumulative of factors in the sediment (0-16 cm depth)

Component	Eigenvalue	Total variance (%)	Cumulative (%)
Factor 1	4.89	40.8	40.8
Factor 2	2.98	25.9	66.7
Factor 3	1.89	15.8	81.5

Component Loadings

Component loadings in PCA are the correlation coefficients between the variables and factors. According to one rule in confirmatory analysis, loadings are generally 0.7 or higher to confirm that independent variables are represented by a particular factor [Garson, 1998].

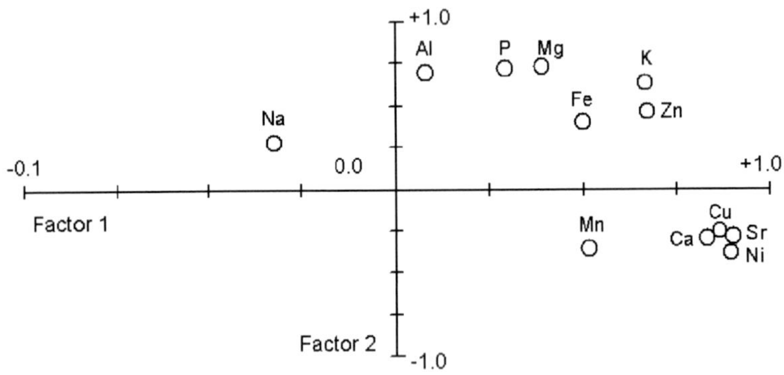

Figure 11. Loading plots of principal component analysis for elements in lake sediment (0-16 cm depth).

It follows from figure 11 that factor 1 chiefly consists of Ca, Cu, Sr and Ni; and factor 2 chiefly consists of Al, P and Mg in light of loading value > 0.7. Considering the abundance of metals in nature (P in figure 9, Al and Mg in table 5), factor 2 seems to be dependent upon natural sources such as bedrock weathering and the forest ecosystem. By contrast, Cu, Sr and Ni in factor 2 originate from anthropogenic sources, and there is a possibility that Ca may originate from both anthropogenic (cf. table 4 and figure 9) and natural (cf. table 5) sources.

Future Trends

The principal component (PC) score can be calculated for the standardized variable. The calculated PC scores indicate that the element group has shifted from component 2 (i.e. natural sources) to component 1 (i.e. anthropogenic sources) as a function of the time passage. If this trend continues without countermeasures, Staroe lake may lose its original characteristics in the near future.

SUMMARY AND FINDINGS

Lakes contain almost all of the fresh surface water on the planet. The water in rivers and streams makes up less than 1% of the volume in lakes. Furthermore, it takes many years to replenish lakes owing to the relatively small amount of precipitation that falls on lakes and the small amount of stream water that runs directly into lakes. On average, lake replenishment takes 100 years. This fact indicates that lakes require special protection from contamination.

The models (i.e. scenario) using simple mass balance will allow administrators to predict changes within the natural lake system due to management actions; however, simple models generally require some assumptions. The overall mass balance model has a large number of parameters and requires extensive data input; hence, a user would need to evaluate and input many different data points to accurately parameterize and sufficiently calibrate the model.

The presented case study of a subarctic lake shows that all factors (i.e. contaminants) are not always dominant over the quality change in the lake and sediment; that is, there are prime factors responsible for influencing the lake/sediment quality. Therefore, the combination of such key factors with a simple model may be useful and practical in daily lake management.

ACKNOWLEDGEMENT

The authors are grateful to Ms. R. M. Galimzyanova (senior researcher) of Geological Institute of Kola Science Center (Russian Academy of Science) for assisting field work and mineralogical analyses, Ms. E. O. Kiselyova (topic engineer) and Ms. G. N. Andreeva (topic engineer) of Institute of North Industrial Ecology Problems for supporting chemical analyses, Centro de Estudos de Recursos Naturais, Ambiente e Sociedade for offering data processing facilities, and Ms. C. Lentfer for English review.

REFERENCES

Alloway, B.J. (ed.). (1995). *Heavy Metals in Soils,* 2nd edition. London (UK): Blackie Academic and Professional.

Aguirre, J. and Tully, D. (1999). *Lake pollution model.* Trony (NY): Rensselaer Polytechnic Institute

Baumgartner, D.J. (1996). Surface water pollution. In I. Pepper, C. Gerba and M. Brusseau (Eds.), *Pollution Science* (pp. 188-209). San Diego (CA): Academic Press.

Blanchard, P., Devaney, P.L. and Hall, G.R. (1998) *Differential equations.* Pacific Grove (CA):Cole Publishing Co.

Blumberg, K., Botts, L., Brown, T.H., Holsen, T.M. and Jonson, A. (2000). *Atmospheric deposition of toxics to the Great Lakes – integrating science and policy.* Chicago (IL): Delta Institute.

Ceballos, B.S.O, Konig, A. and Oliveira, J.F. (1998) Dam eutrophication: A simplified technique for a fast diagnosis of environmental degradation. *Water Research* 32 (11), 3477-3483.

Cohen, Y. (1998). *Dry deposition.* Los Angeles (CA): Multimedia Envirosoft Corp.

Dunteman, G.H. (1989). *Principal component analysis*. Thousand Oak (CA): Saga Publications.

Dauvalter, V. (1999). Appropriateness of Sedimentation in Water Objects of European Subarctic - Nature Protection Aspects of Problem (PhD thesis). 1999. Apatity (Russia): Kola Science Centre (in Russian).

Environment Canada (1998). The Grasshopper effect and tracking hazardous air pollutants. *Science and the Environment Bulletin* 4/5, 2-3.

Environment Directorate-General (2002). *Heavy metals in waste* (ref. No. Env.E.3/Etu2000/0058). Brusseles (Belguim): Europeam Commission.

European Freshwater Program (2000). *Toxic waste storage sites in EU countries*. Brussels (Belgium): WWF-European Policy Office.

Forstner, U. and Wittman, G.T.W. (1979). *Metal pollution in the aquatic environment*. Berlin: Springer.

Fujioshi, R., Gomei, T. and Sawamura, S. (1997). Sorptive behavior of Zn (II) on a lake sediment by sequential extraction-radiotracer technique. *Talanta* 44, 1055-1061.

Garbarino, J.R., Hayes, H.C. Roth, D.A. and Antweiler, R.C. (1995). Heavy metals in the Mississippi River. *U.S. Geological Survey Circular* (No.1133). Reston (VG): U.S Geological Survey.

Garson, D. (1998). Quantitative research in public administration - *Factor analysis*. Raleigh (NC): North Carolina State University.

Haigh, M.J. (1996). Soil contamination and land reclamation – notes from the coal fields of Europe. *Envis Newsletter* 1 (3), 1-10.

Hakanson, L. (1982). Lake bottom dynamics and morphometry – the dynamic ration. *Water Resource Research* 18, 1444-1450.

Hakanson, L. and Jansson, M. (2002). *Principles of lake sedimentology*. Caldwell (NJ): Blackburn Press.

Hoggard, J. (1997). Lake pollution module - using differential equations to model lake pollution. Blacksburg (VA): Virginia Polytechnic Institute.

Jaffe, D., Cerundolo, B., Rickers, J., Stolzberg, R. and Baklanov, A. (1995). Deposition of sulfate and heavy metals on the Kola Peninsula. *The Science of the Total Environment* 160/161, 127-134.

Jolliffe I.T. (2002). *Principal Component Analysis* (Springer Series in Statistics), 2nd ed., New York (NY): Springer.

Kersten, M. and Förstner, U. (1987). Effect of sample pretreatment on the reliability of solid speciation data of heavy metals - Implications for the study of early diagenetic processes. *Marine Chemistry* 22, 294-312.

Kikuchi, R. and Gorbacheva, T. (2006). Vegetation recovery after environmental damage by metallurgic industry in the Arctic region: Transformation of soil chemistry in restored land. In: C.V. Loeffe (ed.), *Conservation and Recycling of Resources* (pp. 93-118). Hauppauge (NY): Nova Science Publishers.

Korhola, A., Sorvari, S., Rautio, M., Appleby, P.G., Dearing, J.A., Hu, Y., Rose, N., Lami, A. and Cameron, N.G. (2002). A multi-proxy analysis of climate impacts on the recent development of subarctic Lake Sannajärvi in Finnish Lapland. *Journal of Paleolimnology* 28 (1), 59–77.

Korhola, A. and Weckström, J. (2005). Paleolimnological studies in arctic Fennoscandia and the Kola Peninsula (Russia). In: R. Pienitz, M.S.V. Douglas and J.P. Smol (eds.), *Long-*

Term Environmental Change in Arctic and Antarctic Lakes (pp. 381–418). Dordrecht (Netherlands): Springer.

Krzanowski, W.J. (1988). *Principles of Multivariate Analysis*. Oxford (Uk): Oxford University Press.

Lambarkis, N, Antonakos, A. and Panagopoulos, G. (2004) The use of multicomponent statistical analysis in hydrogeological environmental research. *Water Research* 38 (7), 1862-1872.

Lee, G.F. (1970). *Factors affecting the transfer of materials between water and sediments*, Eutrophication Information Program (Literature Review No. 1), Madison (WI): Water Resources Center.

Lindstrom, M. and Hakanson, L. (2001). A model to calculate heavy metal load to lakes dominated by urban runoff and diffuse inflow. *Ecological Modelling* 137, 1-21.

Mäkinen, A. (1994). Biomonitoring of atmospheric deposition in the Kola Peninsula and Finnish Lapland based on the chemical analysis of mosses (report No. 4). Helsinki (Finland): Ministry of Environment (Environmental Policy Department).

Marsh, P. and Hey, M. (1989). The flooding hydrology of Mackenzie Delta lakes near Inuvik, N.W.T., Canada. *Arctic* 42, 41–49.

Moiseenko, T.I. (1997). Theoretical grounds for regulation of anthropogenic loads on Subarctic water reservoirs. Apatity (Russia): Kola Science Centre.

Nilsson, A. (1997). *Arctic pollution issues – a state of the Arctic environment report*. Oslo (Norway): Arctic Monitoring and Assessment Program.

Norton, S.A., Bienert, R.W.J., Binford, M.W. and Kahl, J.S. (1992). Stratigraphy of total metals in RIPLA sediment cores. *Journal of Paleolimnology* 7, 191-214.

Praus, P. (2005). Water quality assessment using SVD-based principal component analysis of hydrological data. *Water SA* 31(4), 417-422.

Reghunath, R, Murthy, T.R.S. and Raghavan, B.R. (2002). The utility of multivariate statistical techniques in hydrochemical studies - an example from Karnataka, India. *Water Research* 36 (10), 2437-2442.

Richardson, W.L.; Endicott, D.D. and Kreis, R.G. (1999). *Managing Toxic Substances in the Great Lakes - the Green Bay mass balance study* (draft report). Grosse Ile (MI): U.S. Environmental Protection Agency.

Smith, I.R. (1975). *Turbulence in lakes and rivers*. Ambleside (United Kingdom): Freshwater Biological Association.

Simons, T.J. (1980). *Circulation models of lakes and inland seas*. Ottawa (Canada): Department of Fisheries and Oceans.

Sternberg, R.W. and Larsen, L.H. (1975). Threshold of sediment movement by open ocean waves. *Deep Sea Research* 22, 299-309.

Thorpe, T. (2002). Windows and scum lines. *Water line* 6 (1), 3.

Tiesser A., Campbell, P.G.C. and Bisson, M. (1979). Sequential extraction procedure for the speciation of particulate trace metals. *Analytical Chemistry* 51, 844-851.

Turekian, K.K. and Wedepohl, K.H. (1961). Distribution of the elements in some major units of the Earth's crust. *Geological Society of America Bulletin* 72, 175-191.

U.S. EPA - U.S. Environmental Protection Agency (1996). *National Primary Drinking Water Regulation* (EPA 811 F95 002C). Washington D.C. (WA): Office of Ground Water and Drinking Water.

U.S. EPA - U.S. Environmental Protection Agency (2006). *Lakes Michigan Mass Balance*. Chicago (IL): Great Lakes National Program Office.

U.S. EPA - U.S. Environmental Protection Agency (2007). *Great Lakes monitoring - atmospheric deposition of toxic pollutants*. Chicago (IL): Great Lakes National Program Office.

Vega, M., Pardo, R., Barrado, E. and Deban, L. (1998). Assessment of seasonal and polluting effects on the quality of river water by exploratory data analysis. *Water Research* 32 (12), 3581-3592.

Vinogradov, A.P. (1962) Mean content of chemical elements in major types of rocks. *Geochemistry* 7. 555-571 (in Russian).

Welch, H.E., Legault, J.A. and Bergmann, M.A. (1987). Effects of snow and ice on the annual cycles of heat and light in Saqvaqjuac lakes. *Canadian Journal of Fisheries and Aquatic Sciences* 44,1451–1461.

Wimpenny, J. (1999). Modelling in microbiology. In: Bell, C.R., Brylinsky, M. and Johnson-Green, M (eds.), *proceedings of the 8th International Symposium on Microbial Ecology* (pp. 917-923), Halifax (Canada): Atlantic Canada Society for Microbial Ecology.

Woo, M.-K. (1980). Hydrology of a small lake in the Canadian High Arctic. *Arctic and Alpine Research* 12, 227–235.

Woo, M.K., Heron, R. and Steer, P. (1981). Catchment hydrology of a High Arctic lake. *Cold Regions Science and Technology* 5, 29–41.

Woo, M.K. and Xia, Z.J. (1995). Suprapermafrost groundwater seepage in gravelly terrain, Resolute, NWT, Canada. *Permafrost and Periglacial Processes* 6, 57–72.

Wood, J.M. (1974). Biological cycles for toxic elements in the environment. *Science* 183, 1049-1052.

World Health Organization (2006). *Guideline for drinking-water quality* (vol. 1). Geneva (Switzerland) : WHO Press.

Wunderlin, D.A., Diaz, M.P., Ame, M.V., Pesce, S.F., Hued A.C. and Bistoni M.A. (2001) Pattern recognition techniques for the evaluation of spatial and temporal variations in water quality - a case study: Suquía river basin (Córdoba-Argentina). *Water Research* 35 (12), 2881-2894.

In: Lake Pollution Research Progress
Editors: F. R. Miranda and L. M. Bernard

ISBN: 978-1-60692-106-7

Chapter 5

Aquatic Pollutant Assessment Across Multiple Scales

Clint D. McCullough*

Centre for Ecosystem Management and Centre of Excellence for Sustainable Mine Lakes, Edith Cowan University, 100 Joondalup Drive, Perth, WA 6027, Australia

Abstract

Aquatic pollutant testing using biological assays is useful for ranking the toxicity of different chemicals and other stressors, for determining acceptable concentrations in receiving systems and for elucidating cause and effect relationships in the environment. This 'ecotoxicological' testing approach supplants previous approaches that indirectly estimated toxicity using chemical and physical surrogate measurements alone.

Nevertheless, many published aquatic pollution studies are restricted to examining the effects of a single toxicant on only a single species. Moreover, laboratory-based ecotoxicity tests often intrinsically suffer from a number of limitations due to their small-scale. For example, a major criticism of single-species bioassays is their failure to

* *Clint D. McCullough: e-mail: c.mccullough@ecu.edu.au*

integrate and link toxicants (and other associated abiotic components) with higher scales of biological and ecological complexity (predation, competition, etc.). Many researchers have suggested that single-species toxicant testing has become so widely entrenched that it has hindered the development and greater use of testing at more ecologically-relevant scales. An improvement to single-species laboratory tests are microcosm and mesocosm studies using more complex and relevant measures to aquatic biotic communities.

Nevertheless, mesocosms still do not entirely simulate the ecosystem they come from, rather they mirror its general properties. As a result, there is increasing interest in correlating pollution measures from field surveys with measures of aquatic biotic community structure to determine a toxicant's scale of effect. However, field assessments, although extremely useful in determining site-specific impacts, may be limited by lack of experimental controls, too few or poorly-positioned regional reference sites and by confounding effects from impacts unrelated to the disturbance of concern.

A hegemony on the evolution of ecotoxicological science and practice is that the primary application of ecotoxicological data is regulatory. As ecological systems do not have a single characteristic of scale, "validation" of single-species toxicity assessments by higher ecological level assessments remains the highest standard for aquatic pollution studies. As such, multi-scale assessments at all of these scales are now being recognised as providing the highest reliability for environmental protection of lake ecosystems.

INTRODUCTION

Toxicity assessment of chemicals and other stressors is useful for ranking their toxicity to aquatic biota, for determining their acceptable concentrations in receiving systems and for elucidating cause and effect relationships in the environment [34]. A fundamental premise of toxicity assessment (toxicology) is that any chemical occurring at a high enough concentration may cause toxicity. Indeed, this original analysis can be traced back to Paracelsus, often recognised as the father of Toxicology [49].

> “Poison is in everything, and no thing is without poison. The dose makes it either a poison or a remedy.”
>
> Paracelsus (1492–1541)

Assessing toxicity to the environment is often formalised through practice of Ecological Toxicology or ‘Ecotoxicology’. Ecotoxicology is the science of investigating the effects of toxicants on, and their relationship with, the ecology of a receiving ecosystem. The ecosystem in question may be either the atmosphere, the soil or, as will be the focus of this discussion, an aquatic system such as a lake. Aquatic ecotoxicology, although a relatively new science, compares well with the advancement of other branches of ecotoxicology such as those of air and soil pollutant impacts [27]. As such, ecotoxicology provides an objective and mensurative basis for making decisions about the likely impact of a chemical or physical change on the ecosystem of a given receiving environment [34]. The essence of ecotoxicology is that risk evaluation and regulation of toxic discharges is incomplete unless biological organisms are also used as indicators of the presence of toxic effects [124]. Biological assessment of water quality, and ecotoxicological testing as a subset of this, arose from the recognition that measurement of physico-chemical variables alone does not allow for an assessment of the endpoint for most waterbody monitoring, that is, the suitability of such waters for sustaining biological communities [32; 111].

Biological methods of assessing pollution in aquatic environments have the capacity to integrate effects through continuous exposure and because they measure directly the level of change at which a particular stressor becomes toxic. This approach supplants earlier efforts at indirectly estimating toxicity, using chemical and physical surrogate measurements alone [5; 76; 77]. Comprehensive and effective assessment and management of water quality relies on integrating biological approaches with the more traditional chemical and physical-based approaches, where chemical data provide explanatory variables for trends observed for biota ("cause for consequence") [33].

Single-Species Laboratory Testing

Ecotoxicology is most typically used to determine at what concentration a detrimental effect will become apparent for a given toxicant. In this regard many ecotoxicological tests occur as single-species laboratory bioassays in the screening of complex effluents to waterways, or in cases where there is little knowledge of the toxicity and mechanism of toxicity of a discharged substance [21; 34]. Therefore, ecotoxicological testing can also identify the absence of toxic effects with regards to the environmental safety of a discharge [124]. By definition there is no disturbance (pollution) to a biological community without an observable change in the structure/function or other definable characteristics of that community [100]. The use of toxicity testing involving biological organisms (bioassays) is therefore of prime importance in anticipating the likelihood of environmental impacts [114].

Nonetheless, many studies are restricted to examining the effects of a single toxicant on only a single species [114]. Acute (short-term) studies have also dominated the ecotoxicological literature, although there is an increasing focus on chronic (long-term) and sub-lethal (e.g., population growth rate) assays [88; 4; 13]. Data from single-species tests are typically derived using established standard protocols, where tests are maintained in a relatively simple form, and at small scale under highly controlled ambient conditions. Such procedures enable tests to be more easily reproduced within and amongst different laboratories, and in a given laboratory at different times [25]. These features, together with use of standard test species (e.g., rainbow trout, *Oncorhynchus mykiss*) also enable comparisons of quantitative results from different studies and toxicants. Repeatability (precision) for chronic ecotoxicity tests may be quite high [3], and mode of toxicity information can typically be readily inferred from one toxicant to another.

Despite these advantages, laboratory-based ecotoxicity tests suffer from a number of limitations. These may include inappropriate test species selection, inherent confounding of toxicity responses in the experimental design, inappropriate test species selection, and a failure to accommodate all modes of toxicant exposure to an organism. As a result, there has been some published criticism that responses in laboratory-scale studies are not representative of responses in natural ecosystems [98; 83; 18; 22].

Figure 1. Counting *Hydra* abundances in a single-species, laboratory-based bioassay testing a mine water discharge. (Photo: Clint McCullough).

It is often not recognised by researchers that different species may respond differently in their tolerance to different pollutants [23]. However, species chosen for testing are generally those that are most easily maintained in captivity [83; 109]. This explicit selection represents a major bias toward the relatively few robust taxa occurring in an ecosystem that are amenable to laboratory testing. More often than not, the few taxa that are amenable to laboratory testing do not meet other important criteria relating their ecological significance to key ecosystem processes or to their toxicant sensitivity being representative of other receiving system taxa [6]. Laboratory-based, single-species tests have been found to best predict the field effects of strong toxicants upon *common* species [10; 81]. However, this is likely to be because taxa selected for single-species ecotoxicological testing are typically common species [18]. By definition, these species are unlikely to be of as much conservation significance as rare species which may already be constrained in abundance and distribution by narrower environmental tolerances. Yet these latter species may be the least protected by single-species testing [28; 29; 89]. Similarly, although laboratory bioassay tests are used to predict ecological effects on natural populations of species, laboratory test data are often based upon captive stocks; the gene pool of which may differ from wild stocks, yielding erroneous toxicology conclusions [113; 92].

The extent to which differing races/locales of the same species may differ in their response to toxicants is an important consideration for ecotoxicological science. This is especially so when environmental conditions from where test populations are sourced may be appreciably different [128]. Thus, a relative ranking of toxicant toxicity risk will depend very much upon which strain(s) are tested [58]. Testing for comparability of response of laboratory-maintained stock and wild stock is therefore also a good way of achieving high laboratory quality control [58; 122] as concerns have been voiced for asexually reproducing laboratory populations of test species in general [58]. It is possible that a clone with a slight selective advantage under specific laboratory conditions may successfully exclude other genotypes, resulting in less variability of test results within labs because of reduced genetic diversity but increased variability of test results between labs culturing populations of separate genotypes [113].

Experimental design limitations may also lead to confounding of toxicity responses. For example, test organism deaths during the period of culture and/or maintenance may remove weaker test individuals, thereby potentially underestimating toxicity for that population when tested [17; 23]. Alternatively, stress associated with the culture and testing procedures may reduce genetic variability of a test species population and thereby render individuals more susceptible to the stress of the test toxicant. This increased susceptibility may in turn lead to overestimation of toxicity compared with the unstressed populations in the actual toxicant receiving populations [58]. Repercussions expected from this altered gene pool following transfer from a natural population include predictions of increased magnitude of the effect concentration of a toxicant and at the same time reduced variation of their response [7]. Measures of toxicity obtained from a subsample of a natural population with reduced variability are therefore likely underestimate the variability of the response of the more genetically diverse natural populations, although by how much is difficult to quantify [87; 97] (Figure 2).

However, if a population receiving elevated toxicants is more resistant to the given toxicant than natural, reference populations, then this may be taken as direct evidence that the

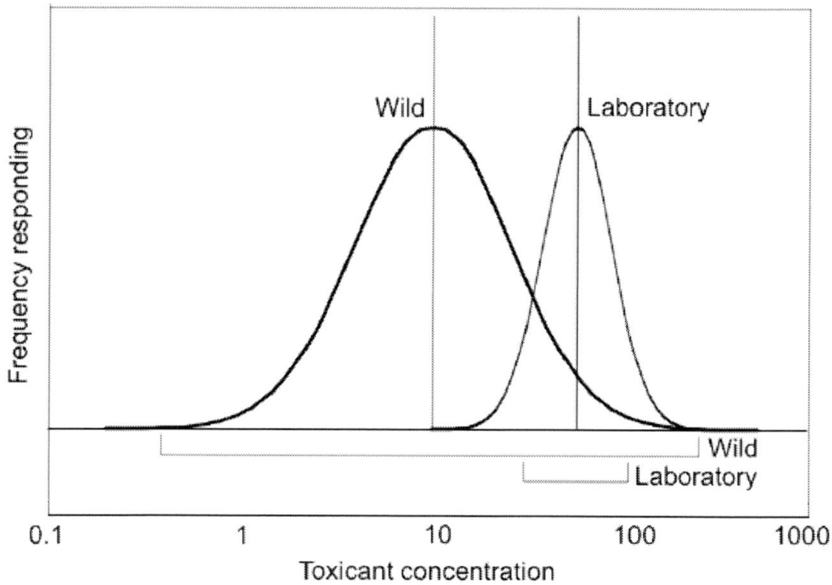

Figure 2. Expected response to a captive population of a test species relative to its original wild population (after Baird [7]).

toxicant concentration has been sufficient to elicit biological community changes [92]. Furthermore, this effect on the population of one species in the receiving system strongly implies that other species are likely to be affected by the toxicant in similar fashion [87]. The presence of tolerant forms may therefore be used as an indicator that a community is being subjected to an environmental stress [61; 24]. Furthermore, the ecological consequence of favouring obligatory pollution-tolerant genotypes may result in species being rendered less fit in other areas which are necessary for their success [70; 122]. Because of this physiological "cost", resistant strains have typically been found only in areas of sufficient toxicant concentration. The level of resistance to a toxicant has also been found to be directly related to the concentration of the toxicant to which the population is exposed [58].

Short-term, single-species laboratory tests for pollution may also fail to address issues of community complexity, as they have no way of incorporating factors such as the role of species interactions including predation and competition, or other population processes such as immigration and emigration [109]. Consequently, when there is differing sensitivity of test species to different toxicants, reliance on too few test species without regard to the test species' relevance to higher scales of ecological complexity may lead to a lack of agreement with field responses [93].

The extent to which one species can act as a proxy and provide predictive information on the effect of a toxicant to the receiving ecosystem is clearly a critical issue [88]. However, the ability of single-species assays to accurately predict the impact on an ecosystem, even when the species chosen for assay are indigenous to the receiving waters, has frequently been called into question both empirically [21; 99; 50] and philosophically [94]. Nonetheless, practicality and convenience remain the dominant influences on the choice of ecotoxicological testing. This is why single-species tests still predominate and are likely to do so for some time [42; 43].

Single-chemical laboratory toxicity tests also continue to play a role in pollution assessment for derivation of water quality standards or guidelines. In these assessments, the direct toxicity assessment (DTA) approach, using chronic (or sub-chronic) tests on appropriate species and using local dilution waters, is now regarded as the preferred, more realistic methodology for predicting effects on ecosystems [4]. DTA has been shown to result in stronger agreement between field validations and laboratory studies (see review by deVlaming and Norberg-King [50]).

There have, however, been few field validations of the conclusions of laboratory-based, single-species bioassays [44]. Nevertheless, reviews of studies comparing predictions of laboratory-based, single-species bioassays to field studies indicate good agreement when toxicity is high, such that only a single study of either laboratory or field approach may be required to characterise the pollutant's risk [10; 54; 51]. Nonetheless, there has also been criticism that responses of laboratory studies are not representative of responses in natural ecosystems where local differences in species composition and community complexity are integrated [18; 22]. Thus, studies which have compared predicted "ecosystem" responses (generally using community characteristics as a surrogate for ecosystem responses) by way of single-species bioassays have reached mixed conclusions.

In summary, the major criticism of single-species bioassays is their failure to integrate and link toxicants (and other associated abiotic components) with higher scales of biological and ecological complexity (predation, competition, etc.) [122; 125]. Nevertheless, the direct cause and effect nature and repeatability of a single-species test is also very important in maintaining the credibility of conclusions with scientists, regulators and lay-persons. Indeed, across various test types, measurement repeatability (precision) for chronic bioassays may be quite high [3]. Furthermore, more sensitive responses to stress have been predicted with smaller organisms that have shorter generation times [109]. Consequently, although there is more of a trend toward multi-species and community scale experimentation, single-species tests, often on non-indigenous species, remain the most frequent ecotoxicological approach [20; 88; 32; 34].

MICROCOSMS/MESOCOSMS

Compared with laboratory approaches, proponents of community-scale testing have argued that single-species toxicant testing has become so widely entrenched that it has hindered the development and greater use of community-scale testing [83]. Thus, despite continuing development of ecologically-relevant, population-based test endpoints in the laboratory, such as reproduction, growing concerns have been for a *complete* risk analysis of a pollutant requiring studies at the scales of populations, communities and finally ecosystems themselves [12].

Shortcomings of single-species studies have therefore been increasingly addressed through use of various multi-species tests, which may range in scale from laboratory studies using mixed-flask cultures, to field studies using algae and macroinvertebrates on artificial substrates, microcosm and mesocosm communities, and model streams and lake enclosures . Such mesocosm methods may be employed to complement existing single-species data as part of a multi-scale risk assessment.

There has been a growing interest amongst the ecotoxicological research community in the use of these larger scale (and consequently, higher scale of biological organisation) studies for pollution assessments. A compromise between the two extremes of spatial and organismal scales of assessment that both single-species tests and field studies represent, is intermediate scale microcosm and mesocosm systems. The concept of mesocosms was an extension of early experiments at the scale of farm ponds [78; 79]. The attractiveness of experimental mesocosms is that they retain a strong element of environmental realism and applicability, whilst permitting laboratory-like replication and manipulations to be performed. The greater realism of mesocosm environments is achieved through their capacity to include higher scale components of ecosystem responses such as micro/macroinvertebrate communities as well as basic ecosystem interactions.

A useful definition of the term "mesocosm" is provided by Odum [98], describing it as a middle-sized experimental environment falling between a laboratory-based (microcosms) and a full field-scale (macrocosm) multi-species study. Suitable enclosures representative of receiving water communities of micro- and macroinvertebrates have been variously described as artificial "ecosystems" and "multispecies toxicity tests". Mesocosm experiments may use artificial enclosures to achieve independent units, or they may make use of naturally isolated water bodies such as pools or backwaters. The most frequently cited advantages of these studies over single-species bioassays, are improved environmental realism and greater predictive abilities [88; 126; 31].

Figure 3. Large mesocosms may be labour-intensive and expensive to establish. (Photo: Clint McCullough).

Pollutant exposure durations in mesocosm pollutant exposure experiments are also frequently greater than those in single-species laboratory tests [50; 126]. The predominant reason for greater exposure times in mesocosm experiments is a function of their more expensive setup costs and cheaper ongoing costs than for the smaller-scale but more intensive single-species laboratory tests [98].

However, disadvantages to using mesocosms remain. Mesocosms typically exclude predators which may structure some communities e.g., zooplankton, macroinvertebrates [110]. For example, fish are generally excluded from the enclosures in natural water bodies and populations of test organisms are assessed by conventional field sampling techniques [114]. Replication is also often poor while variability amongst the communities of replicate mesocosms is often very high [85; 110].

It has also long been proposed by some that there is a direct link between ecosystem complexity and its stability [41; 55; 119]. This proposal suggests that, for an ecosystem already near a threshold level of tolerance to a stressor, the loss of even a small portion its species diversity may reduce both the resistance and resilience of the ecosystem to further perturbation [48]. Therefore, the loss of any taxa from complex ecosystems has been interpreted as having an effect of reducing the stability of the system to perturbations and stressors [119]. Consequently, deletion of a single species from a natural community may cause a simplification of the biological system and that this may then lead to species losses, possibly including losses of significant species such as keystone, flagship or economically important taxa [45; 101; 46]. Suitability of environmental conditions for the continued existence of a species can, therefore, only strictly be evaluated by assessing an impact at a community scale. Testing of more than one community type within mesocosms is also therefore recommended.

Ecological communities of organisms that are potentially useful for assessing water quality in artificial (experimental) and natural settings include the following.

Although single-species toxicity testing datasets used to derive guideline values are typically under-represented in aquatic insects [66], macroinvertebrates are a popular biological group chosen to assess effects of aquatic pollution at the community scale. Internationally, analysis of benthic macroinvertebrate communities has been the foremost tool for biological assessment of aquatic ecosystems due to the availability of good taxonomy and extensive literature of pollutant effects [111; 4]. Macroinvertebrates are also often the most speciose community in aquatic ecosystems [65]. Their typical characteristics of relatively long life cycles, ecological relevance, low motility and comparatively simple identification are good qualities for assessing the impacts of pollutants [105; 108; 106].

Periphyton is often the greatest source of primary production for communities of both littoral and pelagic habitats [65] and may represent the major source of energy to secondary and higher trophic scales [86]. Along with macroinvertebrates, periphyton have historically received the most attention in community-scale assessments of environmental quality [42; 43]. Periphytic diatoms (family Bacillariophyceae) have been especially suitable for the biomonitoring of aquatic ecosystems [52; 104]. Diatoms occur ubiquitously in high numbers and diversity, are generally relatively sensitive to changes in water chemistry, are easily collected, analysed and preserved and can be readily identified to species scale [104]. Good preservation in sediments also enhances the usefulness of these algae in both temporal and historical studies such as restoration and palaeolimnology [96; 38]. The distributions and associated water chemistry of diatom taxa are cosmopolitan and well documented and there is typically good information available on their environmental requirements. Furthermore, diatoms often have narrow ranges of tolerance to pH, nutrients and salinity, which have often been widely studied and defined [52; 104]. Nevertheless, as for microinvertebrates, the application of diatoms may be limited because of their small size and requirements for preparation prior to sorting and expert identification [74; 104; 75].

Figure 4. Macroinvertebrates are a diverse and easily identified biotic community of lake littoral margins (Photo: Chris Humphrey).

Sampling of diatom communities on natural rather than artificial substrates has been recommended by some authors, although many have realised the limited statistical power that the intrinsic variability of these methods allows for in experimental studies. However, artificial substrates are generally unrepresentative analogues for naturally occurring diatom communities. This may occur through a "founder effect" with differing inter-replicate seeding potential for the algae themselves, or through differences at higher trophic scales such as zooplankton grazers or their fish predators [2].

Figure 5. Artificial substrates, such as this glass slide-based 'periphytometer' allow easy quantification of periphyton abundances. (Photo: Clint McCullough).

Another group of biota suitable for water quality assessment is the aquatic microinvertebrates. Conventionally, microinvertebrates are invertebrates less than 250 μm in body length that share many of the desired characteristics of macroinvertebrates, but with shorter life cycles and thus faster community responsiveness to environmental change [123]. Commonly encountered microinvertebrates are zooplankters, either littoral or pelagic. Microinvertebrates are important components of aquatic ecosystems, grazing on detritus, bacteria and phytoplankton and often forming an important link between lower organisational levels of energy (primary producers) and those of higher trophic scales such as the numerous fish species [57]. A significant disadvantage of this group for water quality assessment is their smaller size which may make their enumeration and taxonomic identification difficult and consequently limit their use to more specialised applications.

As primary producers, unicellular algae phytoplankton are also often used as indicators of water quality because of their high sensitivity to environmental change and short generation time. Indeed, van Dam *et al.* [121] considered phytoplankton studies as potentially the most promising indicators of shallow lake and wetland degradation. Consequently, phytoplankton are often employed in single-species pollutant assay tests. Phytoplankton are also useful indicators of high nutrient conditions due to their ability to reproduce rapidly under ideal conditions. They provide fundamental information on an important trophic scale and act as an interface between the water chemistry and a significant component of the aquatic food web. In primarily autotrophic-based communities, phytoplankton are of great importance to ecosystem functioning [116].

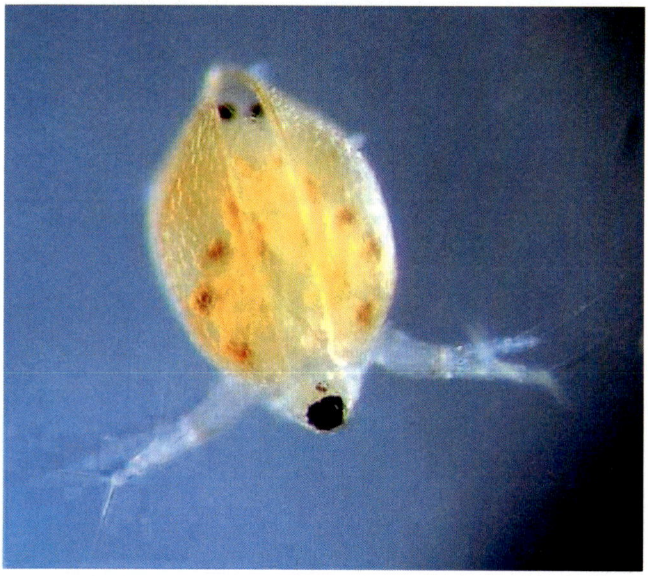

Figure 6. Microinvertebrates, such as this cladoceran, are small and may be hard to identify but they form a crucial trophic linkage in most lake food webs. (Photo: Clint McCullough).

Water body chlorophyll *a*, *b* and *c* concentrations can be used as proxies for phytoplankton biomass and general composition. The concentrations of these different photosynthetic pigments from various receiving conditions of mine waters may, therefore, be expected to act as proxies for validating the (single-species) laboratory *Chlorella* sp. bioassay results in an ecosystem-scale field setting [72].

However, higher-level endpoint criteria for manipulative ecotoxicity experiments are still unclear [26]. Moreover, there is a general consensus amongst researchers that the results of community and ecosystem studies are often complex, highly variable, and therefore, difficult to interpret [42; 68]. Natural spatial and temporal variability of communities may render detection of effects of stressors difficult at community scale in all but extreme cases [97]. For example, there may be difficulty in ascribing the change seen in a community to the toxicant in question when different concentrations and different community endpoints are simultaneously assessed [42]. Thus, historically, one of the most challenging tasks for ecologists is determining whether or not a stressor is detrimentally affecting the biological communities of a receiving aquatic ecosystem [51]. Mesocosms do not entirely simulate the ecosystem they come from, rather they mirror only the general properties that characterise that system [120].

The responses of biological communities to a toxicant often lends a different, more informative insight into the expected effects of perturbation at the scale of an ecosystem than single-species data. However, the scale and multivariate nature of their data also requires a different statistical approach. Many ecological applications of multivariate techniques are readily extended to ecotoxicological field studies. As a consequence, caveats and considerations for ecological study design, such as optimum scale of taxonomic resolution, type of community summary and replication, must also be provided for during the design phase of experiments involving communities. Low amount of replication is often associated with both mesocosm and field studies due to the relatively high intrinsic costs of construction and establishment [90]. Low replication and high intra-treatment variability frequently encountered in mesocosm studies may lead to low reliability of community data caused by artificial enclosures diverging for reasons unrelated to that of the dosed toxicant, such as confounding by unconstrained variables and founder effects [60]. Hence, replication and associated variability must be fundamental in consideration of study design [90]. Nevertheless, the more environmentally realistic scale of data gives field community studies more power to predict expected ecological effects than single-species studies [4]. Consequently, mesocosm studies increasingly contribute to aquatic pollutant studies by a variety of approaches which are particularly relevant for the management of large and unique ecological systems [112].

Field Studies

As discussed, estimates of toxicant risk to ecosystems are generally extrapolated from single-species bioassays without substantial validation of the accuracy of the specific response of the toxicant in the field [67]. This lack of field validation represents a limitation of both the real and perceived value of these toxicological data. Field validation of single-species studies of aquatic pollution has, therefore, been found to be a useful approach in assessing the ultimate effects of stressors and also in determining the confidence in the prediction of simpler ecotoxicological models [51; 122]. Since pollutant studies are typically aimed at determining effects of a stressor on higher scales of organisation, it is logical that analysis and interpretation should be performed at the same scale [90].

Manipulative field exposure experiments are typically of similar cost to mesocosm tests with high initial but lower ongoing costs. The reason for greater exposure times in field studies is by reason of their observational nature, where existing exposures are assessed retrospectively. These greater exposure durations of the large-scale experiments would be expected to increase the sensitivity of these studies. Further differences in sensitivity between single-species toxicity testing and the community assessments of the mesocosm and field studies may also be expected to be due to the absence or, at best, greatly reduced contribution of the single-species bioassay test species themselves to the mesocosm communities (i.e., possible lack of relevance of a single-species test species).

Observational field exposure tests, by their nature, typically involve testing over long durations regardless of their initial setup expenses. However, the exposure of entire ecosystems in field studies means that these studies are frequently reactive rather than proactive. Similarly to mesocosm studies, the reliability of laboratory studies to predict effects in the field is also often inconsistent. Some comparisons have found underestimations of effects whilst others have found overestimations [37]. To this end, it highly desirable to validate single-species laboratory tests by studying effects in a functioning community [114]. While laboratory bioassays may be useful as initial screening tools, more comprehensive studies must form the basis of ecosystem management [109]. Consequently, it may not be sufficient to simply detect a change in a sensitive or "early detection" indicator, because such a change cannot easily be linked to prediction of a change at the population, community or ecosystem scale in the field. Instead, responses must be sought in the field from suitable surrogates for these higher scales of organisation and complexity [70] using, for example:

species richness, community composition or structure [9],

patterns of abundance and distribution of species of high conservation value or ecological significance [4],

physical, chemical or biological processes e.g., production:respiration ratios, primary production, energy flow pathways [14; 47; 15; 11; 84].

Many studies have used results from field surveys showing correlations of pollutants with measures of biotic community structure to determine a toxicant's scale of effect [63; 1; 115]. Nevertheless, one of the great difficulties in evaluating ecosystem-scale risks is replicating treatments adequately enough to account for high inter-treatment variability. Every water body serving as a replicate is also different in many more ways than just the treatment in question e.g., physical morphometrics, micro-climatic patterns, water retention time etc. [109].

Such field assessments, although extremely useful in determining site-specific impacts, are still frequently limited through a lack of proper experimental control including too few or poorly-positioned reference sites and confounding effects from impacts unrelated to the disturbance of concern [66]. Consequently, the scale of biological organisation producing the most ecosystem-relevant dataset (e.g., biological communities) conversely also provides the least reliable database for regulators [83; 91]. Nonetheless, although there is a need to trade-off between test simplicity, costs and environmental relevance; relevance may still warrant greatest consideration in many pollutant studies and broader risk assessments [39]. Furthermore, in addition to having more ecological relevance, evidence also suggests that pollutant studies that address biotic effects at higher-scales of ecological organisation are at least as powerful in detecting the biological effects of pollutants as are single-species approaches [19; 90]. This is especially so when multivariate analyses are applied to these community data [39].

Multiple Scales of Evidence Studies

As discussed, the failure of smaller-scale, single-species bioassay designs to address all the complexities of actual receiving ecosystems has long been recognised [20; 99; 50]. Large-scale system behaviour cannot generally be predicted from individual sub-units [83] and, indeed, in many situations it may be that laboratory-based ecotoxicological testing measures the *wrong* variables *more precisely* than less repeatable and replicable (but more environmentally realistic) field-based experiments [26]. For example, although population abundance and growth may be measured, this measurement is made in the absence of many factors that limit abundance such as direct predation or competition and consequent resource limitation. Ecological field studies are sometimes initiated when laboratory tests have indicated the existence of a potential risk [79]. An integrated assessment approach has been especially recommended for situations where the effect of the toxicant is subtle [43]. The integration of different scales of organisation in this manner may provide complementary information and, ultimately, a better understanding of both the scales at which a stressor is likely to affect a community and in what manner this stress will reveal itself [43].

Ecotoxicological assessment of aquatic pollution is already, and needs to be more widely recognised, as a multi-disciplinary subject [8]. As such, there is always scope for criticisms of particular individual discipline's methods; all have different advantages and disadvantages, and consequently there is a need for multiple approaches to an ecosystem-scale toxicity assessment [80]. Such a more holistic ecotoxicological approach would ideally employ laboratory and field-based approaches as complementary methods. What is important is that their limitations and context are realised and accounted for. Ideally, ecotoxicological assessments should include endpoints from many different scales ranging from cellular and physiological processes at the individual level through to ecosystem changes such as functional feeding group relationships and food web structure [12] (Figure 7).

Although numerous endeavours have been made, there still appears to be no parsimonious way of marrying data from very different scales of study together into a single holistic trigger value derivation, without significant loss of information specific to each scale. An example of such an holistic assessment method, the "weight-of-evidence" approach (also know as "multiple lines of evidence" or a "meta-analysis" approach) [40; 53] seeks concordance between controlled experimental findings and actual field results [117; 71]. The weight-of-evidence approach has been recommended where different types of site-specific data provide partial information on different aspects of a stressor's action at those sites [95]. The weight-of-evidence approach is achieved at its simplest through such testing at a variety of ecological scales [83]. The essence of this multi-scalar process is that, where necessary, toxicant information from water physico-chemistry, single-species tests, multi-species tests, and ecosystem processes (e.g., changes in trophic relationships as indicated by functional feeding groups) are all considered in the ecotoxicological assessment [83]. As such, a multi-scalar type weight-of-evidence approach has proven to be a reliable complex risk assessment strategy [127].

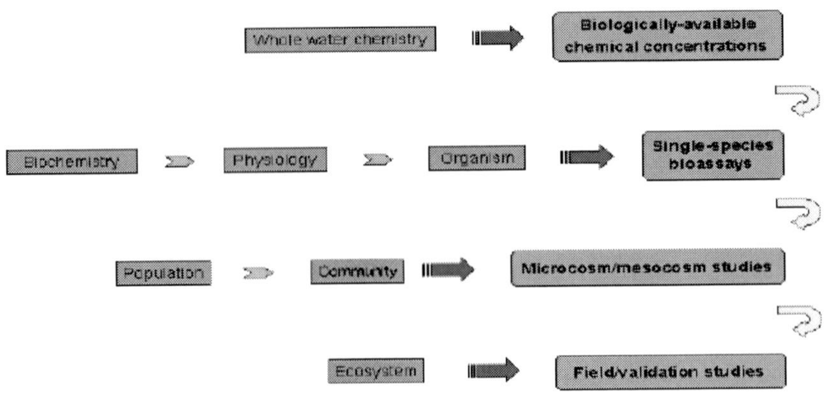

Figure 7. Increasing environmental relevance at increasing scale of complexity of pollution assessment typically undertaken and their relationship to each other in order.

CONCLUSION

As a result of the continuing evolution of "ecotoxicology", the emphasis on extrapolation of laboratory results to field expectations is considered an increasingly important requirement [35; 69]. A large part of the change in emphasis is, although toxicological data derived from laboratory bioassays may provide repeatable results, they may still not provide an accurate assessment of actual receiving system ecotoxicity.

Inaccurate pollutant assessments may either fail to sufficiently protect receiving ecosystems or, conversely, may contribute to overly conservative guidelines which unnecessarily restrict industry more than is required for environmental protection [35; 30]. Consequently, a hegemony on the evolution of ecotoxicological science and practice is that the primary application of pollutant toxicity data is regulatory [20]. Scientific results from the various forms of ecotoxicological testing available are frequently held in higher esteem by regulatory bodies when they are more precise and repeatable. This is the case even if these data may not be relevant or sufficiently comprehensive for the receiving system of concern.

Given that the two activities of science and policy are distinct [20], some authors suggest that such pollutant toxicology data should not provide guidelines alone. Instead ecotoxicological data are best incorporated into a risk assessment framework which accounts for the confidence in quality and relevance of the data to indicate whether environmental harm will occur, and if so, the acceptability of this harm to society [64]. Formal risk assessments provide an ideal framework for integrating ecotoxicological and other data [16] as economic, political and sociological considerations are also important in ecological risk-management [68], but are often not considered in risk-assessment [107]. As a result, single-species toxicity tests have been, are, and most likely also will remain, the mainstay of toxicological assessment [4]. Many of the criticisms of the validity of single-species, laboratory-test data to pollutants in a receiving environment therefore relate more to the *regulatory* application of the science within its limitations, than to the way the science itself is executed [20; 62].

Risk predictions of environmental impacts arising from pollutants in aquatic environments are likely to remain an inexact science for some time yet. A broader criticism of single-species testing is therefore that it is relatively rare for a single line of evidence to reach a definitive risk conclusion [56]. Indeed, some authors consider that single-species studies alone cannot be used to assess environmental toxicant risk e.g., Joern and Hoagland [73]. While extrapolation of results from small-scale studies to larger scales may be sufficient for generic or screening assessment, small-scale study data on their own are likely to be unreliable for water bodies with site-specific and high-reliability requirements for environmental protection [36; 59]. Furthermore, there is a danger that these smaller-scale studies can easily become the over-interpreted object of study, rather than the toxicological issue they are designed to address [30; 36]. Studies at smaller, proxy scales to that of the receiving system in question often encourage a "Type III error" of the original definition *sensu* Kimball [82], where an irrelevant question is answered. In this type of error, the questions "is there an impact on the proxy?" and "what mode of action on the proxy does the toxicant take?", etc. become the sole questions of analysis, with little regards to the context of scale or management application which initiated them. Given that ecological systems do not have a single characteristic of scale, the context and relevance of toxicant guidelines are now considered more reliably derived from studies at more appropriate, higher scales [30].

Nevertheless, although differences between assessment scales are expected, the alternative approach of employing more relevant field-scale data is still rarely used to derive water quality guidelines for aquatic pollutants [102]. Some authors have even questioned whether existing field study design can give managers enough information about the risks of potential pollutants on which to base management decisions [103]. Nevertheless, on the rare occasion that they have been completed, studies that have combined data from laboratory tests with data from field observations have often been found to provide the greatest information about a toxicant and its likelihood of ecosystem-scale effects [118]. This greater understanding of a toxicant's risk increases both the scientific validity of the water quality criteria derivation process, and also improves the confidence for managers and stakeholders. Consequently, "validation" (*sensu* Cairns [19]) of single-species toxicity assessments of this sort (or from tests of similar lower levels of complexity) by high level assessments, remains the highest standard for ecotoxicological trigger value derivation.

REFERENCES

[1] Agard, JBR; J Gobin and R Warwick (1993). Analysis of marine macrobenthic community structure in relation to pollution, natural oil seepage and seasonal disturbance in a tropical environment (Trinidad, West Indies). *Marine Ecology Progress Series*. 92: 233-243.

[2] Aloi, JE (1990). A critical review of recent freshwater periphyton field methods. Canadian *Journal of Fisheries and Aquatic Sciences*. 47: 656-670.

[3] Anderson, SL (1991). Letter to the Editor: Precision of short-term chronic toxicity tests in the real world. *Environmental Toxicology and Chemistry*. 10: 143-145.

[4] ANZECC/ARMCANZ (2000). Australian and New Zealand guidelines for fresh and marine water quality. National Water Quality Management Strategy Paper No 4. Canberra, Australian and New Zealand Environment and Conservation Council and Agriculture and Resource Management Council of Australia and New Zealand.

[5] Auer, CM; JV Nabholz and KP Baetcke (1990). Mode of action and the assessment of chemical hazards in the presence of limited data: use of structure activity relationships (SARS) under TSCA Section 5. *Environmental Health Perspectives*. 87: 183-197.

[6] Bacher, GJ; JC Chapman and RP Lim (1992). The impact of agriculture on inland temperate waters: field validation. *Australian Biologist*. 5: 196-202.

[7] Baird, DJ (1992). Predicting population response to pollutants; in praise of clones. A comment on Forbe and Depledge. *Functional Ecology*. 6: 616-617.

[8] Bartell, SM (1997). Charlatan or Sage-A dichotomy of views on ecological risk assessment. *Environmental Management*. 21: 822-824.

[9] Baskin, Y (1994). Ecosystem function of biodiversity. *Bioscience*. 44: 657-660.

[10] Birge, WJ; JA Black; TM Short and AG Westerman (1989). A comparative ecological and toxicological investigation of a secondary wastewater treatment plant effluent and its receiving stream. *Environmental Toxicology and Chemistry*. 8: 437-450.

[11] Boulton, AJ (2003). Parallels and contrasts in the effects of drought on stream macroinvertebrate assemblages. *Freshwater Biology*. 48: 1173-1185.

[12] Bradbury, SP (1995). Ecological risk assessment for chemical stressors: challenges in predictive ecotoxicological research. *Australasian Journal of Ecotoxicology*. 1: 3-9.

[13] Bunce, NJ and RBJ Remillard (2003). Haber's Rule: the search for quantitative relationships in toxicology. *Human and Ecological Risk Assessment*. 9: 1547-1559.

[14] Bunn, SE (1995). Biological monitoring of water quality in Australia: workshop summary and future directions. *Australian Journal of Ecology*. 20: 220-227.

[15] Bunn, SE and PM Davies (2000). Biological processes in running waters and their implications for the assessment of ecological integrity. *Hydrobiologia*. 422: 61-70.

[16] Burgman, MA (2005). Risks and decisions for conservation and environmental management. Cambridge, UK, Cambridge University Press.

[17] Cairns, J, Jr. (1983). Are single species toxicity tests alone adequate for estimating environmental hazard? *Hydrobiologia*. 100: 47-57.

[18] Cairns, J, Jr. (1986). The myth of the most sensitive species. *Bioscience*. 36: 670-672.

[19] Cairns, J, Jr. (1986). What is meant by validation of predictions based on laboratory toxicity tests? *Hydrobiologia*. 137: 271-278.

[20] Cairns, J, Jr. (1988). Should regulatory criteria and standards be based on multispecies evidence? *Environmental Professional*. 10: 157-165.

[21] Cairns, J, Jr. (1995). Future trends in ecotoxicology. Ecological toxicity testing: Scale, complexity and relevance. J. Cairns Jr and B. R. Niederlehner. Boca Raton, Florida, Lewis Publishing: 217-222.

[22] Cairns, J, Jr. and BR Niederlehner (1987). Problems associated with selecting the most sensitive species for toxicity testing. *Hydrobiologia*. 153: 87-94.

[23] Cairns, J, Jr. and JR Pratt (1986). On the relation between structural and functional analyses of ecosystems. *Environmental Toxicology and Chemistry*. 5: 785-786.

[24] Calow, P (1992). Can ecosystems be healthy? Critical consideration of concepts. *Journal of Aquatic Ecosystem Health*. 1: 1-5.

[25] Calow, P (1992). The three Rs of ecotoxicology. *Functional Ecology*. 6: 617-619.

[26] Calow, P (1994). Ecotoxicology; what are we trying to protect? *Environmental Toxicology and Chemistry*. 13: 1549.

[27] Calow, P (1995). Risk assessment: principles and practice in Europe. *Australasian Journal of Ecotoxicology*. 1: 11-13.

[28] Cao, Y and DD Williams (1999). Rare species are important in bioassessment (reply to the comment by Marchant). *Limnology and Oceanography*. 44: 1841–1842.

[29] Cao, Y; DD Williams and NE Williams (1999). How important are rare species in aquatic community ecology and bioassessment? *Limnology and Oceanography*. 43: 1403–1409.

[30] Carpenter, SR (1996). Microcosm experiments have limited relevance for community and ecosystem ecology. *Ecology*. 77: 677-680.

[31] Caux, P-Y and RA Kent (2001). Exploring future directions in environmental quality guideline development in Canada. *Australasian Journal of Ecotoxicology*. 7: 13-30.

[32] Chapman, JC (1995). The role of ecotoxicity testing in assessing water quality. Australian *Journal of Ecology*. 20: 20–27.

[33] Chapman, PM (1990). The Sediment Quality Triad approach to determining pollution-induced degradation. *Science of the Total Environment*. 98/98: 815-825.

[34] Chapman, PM (1995). Bioassay testing for Australia as part of water quality assessment programmes. *Australian Journal of Ecology*. 20: 7-19.

[35] Chapman, PM (1995). Ecotoxicology and pollution - key issues. *Marine Pollution Bulletin*. 31: 167-177.

[36] Chapman, PM (2002). Integrating toxicology and ecology: putting the "eco" back into ecotoxicology. *Marine Pollution Bulletin*. 44: 7-15.

[37] Chapman, PM; A Fairbrother and D Brown (1998). A critical evaluation of safety (uncertainty) factors for ecological risk assessment. *Environmental Toxicology and Chemistry*. 99–108.

[38] Clark, JS; T Hussey and PD Royall (1996). Presettlement analogs for Quarternary fire regimes in eastern North America. *Journal of Paleolimnology*. 16: 76-96.

[39] Clarke, KR (1999). Non-metric multivariate analysis of changes in community-level ecotoxicology. *Environmental Toxicology and Chemistry*. 18: 118-127.

[40] Clarke, KR and RM Warwick (2001). Change in marine communities: an approach to statistical analysis and interpretation. Plymouth, Plymouth Marine Laboratory.

[41] Clements, FE (1916). Plant succession: an analysis of the development of vegetation. Washington D.C., Carnegie Institution of Washington: 512.

[42] Clements, WH (1994). Editorial: Assessing contaminant effects at higher levels of biological organisation. *Environmental Toxicology and Chemistry*. 13: 357-359.

[43] Clements, WH and PM Kiffney (1994). Integrated laboratory and field approach for assessing impacts of heavy metals at the Arkansas River, Colorado. *Environmental Toxicology and Chemistry*. 13: 397-404.

[44] Connell, D; P Lam; B Richardson and R Wu (1999). Introduction to Ecotoxicology. Oxford, United Kingdom, Blackwell Science.

[45] Connell, JH (1978). Diversity in tropical rainforests and coral reefs. Science 199: 1302-1310.

[46] Davic, RD (2003). Linking keystone species and functional groups: a new operational definition of the keystone species concept. *Conservation Ecology*. 7: r11.

[47] Davies, PM (1997). Assessment of river health by the measurement of community metabolism. Canberra.

[48] de March, BGE (1988). Acute toxicity of binary mixtures of five cations (Cu^{2+}, Cd^{2+}, Zn^{2+}, Mg^{2+}, and K^{+}) to the freshwater amphipod *Gammarus lacustris* (Sars): alternative descriptive models. *Canadian Journal of Fisheries and Aquatic Sciences*. 45: 625-633.

[49] Deichmann, WB; D Henschler; B Holstedt and G Keil (1986). What is there that is not a poison? *Archives of Toxicology*. 58: 207-213.

[50] deVlaming, V and T Norberg-King (1997). A review of single species toxicity tests: are the tests reliable predictors of aquatic ecosystem community responses?, U.S. Environmental Protection Agency Office of Research and Development.

[51] Dickson, KL; WT Waller; JH Kennedy and LP Ammann (1992). Assessing the relationship between ambient toxicity and instream biological response. *Environmental Toxicology and Chemistry*. 11: 1307-1322.

[52] Dixit, SS; JP Smol and JC Kingston (1992). Diatoms: powerful indicators of environmental change. *Environmental Science and Technology*. 26: 23-33.

[53] Downes, BJ; LA Barmuta; PG Fairweather; DP Faith; MJ Keough; PS Lake; BD Mapstone and GP Quinn (2002). Monitoring ecological impacts: concepts and practice in flowing waters. Cambridge, Cambridge University Press.

[54] Eagleson, KW; DL Lenat; LW Ausley and FB Winborne (1990). Comparison of measured instream biological responses with response predicted using the *Ceriodaphnia dubia* chronic response toxicity test. *Environmental Toxicology and Chemistry*. 9: 1019-1028.

[55] Elton, CS (1958). The ecology of invasions by plants and animals. London, UK, Methuen.

[56] Fairbrother, A (2003). Lines of evidence in wildlife risk assessments. *Human and Ecological Risk Assessment.* 9: 1475-1491.

[57] Fernando, CH (1994). Zooplankton, fish and fisheries in tropical waters. *Hydrobiologia.* 272: 105-123.

[58] Forbes, VE and MH Depledge (1992). Predicting population response to pollutants: the significance of sex. *Functional Ecology*. 6: 376-381.

[59] Freckleton, RP (2004). The problems of prediction and scale in applied ecology: the example of fire as a management tool. *Journal of Applied Ecology.* 41: 599-603.

[60] Futuyma, DJ (1998). Evolutionary Biology. Sunderland, USA, Sinauer Associates, Inc.

[61] Grant, BF (1976). Endrin toxicity and distribution in freshwater: a review. *Bulletin of Environmental Contamination and Toxicology.* 15: 283-290.

[62] Gray, JS (1994). Science and the environment. *Marine Pollution Bulletin.* 28: 270-271.

[63] Gray, JS; KR Clarke; R Warwick and G Hobbs (1990). Detection of initial effects of pollution on marine benthos: an example from the Ekofisk and Eldfisk oilfields, North Sea. *Marine Ecology Progress Series.* 66: 285-299.

[64] Hart, BT; MA Burgman; M Grace; C Pollino; C Thomas and JA Webb (2006). Risk-based approaches to managing contaminants in catchments. *Human and Ecological Risk Assessment.* 12: 66-73.

[65] Havens, KE; LA Bull; GL Warren; TL Crisman; EJ Phlips and JP Smith (1996). Food web structure in a subtropical lake ecosystem. *Oikos.* 75: 20-32.

[66] Hickey, CW and E Pyle (2001). Derivation of water quality guideline values heavy metals using a risk-based methodology: a site specific approach for New Zealand. *Australasian Journal of Ecotoxicology.* 7: 137-156.

[67] Holdway, DA (1997). Truth and validation in ecological risk assessment. *Environmental Management.* 21: 816-819.

[68] Horwitz, P and P Nichols (2002). Located toxicology: the need for alternative methodologies to address toxicological significance. *Australasian Journal of Ecotoxicology.* 8: 45-50.

[69] Hose, GC and PJ Van den Brink (2004). Confirming the species-sensitivity distribution concept for Endosulfan using laboratory, mesocosm and field data. *Archives of Environmental Contamination and Toxicology*. 47: 511-520.

[70] Humphrey, CL; DP Faith and PL Dostine (1995). Baseline requirements for assessment of mining impact using biological monitoring. *Australian Journal of Ecology.* 20: 150–166.

[71] Humphrey, CL; L Thurtell; RWJ Pidgeon; RA van Dam and CM Finlayson (1999). A model for assessing the health of Kakadu's streams. *Australian Biologist.* 12: 33-42.

[72] Jeffrey, SW; SW Wright and M Zapata (1999). Recent advances in HPLC pigment analysis of phytoplankton. *Marine and Freshwater Research.* 50: 879-896.

[73] Joern, A and KD Hoagland (1996). In defense of whole-community bioassays for risk assessment. *Environmental Toxicology and Chemistry.* 15: 407-409.

[74] John, J (1993). The use of diatoms in monitoring the development of created wetlands at a sand-mining site in Western Australia. *Hydrobiologia.* 269/270: 427-436.

[75] John, J (2000). A guide to diatoms as indicators of urban stream health. Perth, Curtin University of Technology.

[76] Karr, JR and EW Chu (1997). Biological monitoring: essential foundation for ecological risk assessment. *Human and Ecological Risk Assessment.* 3: 993-1004.

[77] Karr, JR and EW Chu (1999). Restoring life in running waters: better biological monitoring. Washington, D.C., Island Press.

[78] Kedwards, TJ; SJ Maund and PF Chapman (1999). Community level analysis of ecotoxicological field studies: I. Biological monitoring. *Environmental Toxicology and Chemistry.* 18: 149-157.

[79] Kedwards, TJ; SJ Maund and PF Chapman (1999). Community level analysis of ecotoxicological field studies: II. Replicated-design studies. *Environmental Toxicology and Chemistry.* 18: 158-171.

[80] Kefford, BJ; PJ Papas; D Crowther and D Nugegoda (2002). Are salts toxicants. *Australasian Journal of Ecotoxicology.* 8: 63-68.

[81] Kefford, BJ; PJ Papas; L Metzeling and D Nugegoda (2004). Do laboratory salinity tolerances of freshwater animals correspond with their field salinity? *Environmental Pollution.* 129: 355-362.

[82] Kimball, AW (1957). Errors of the third kind in statistical consulting. *Journal of the American Statistician Association.* 52: 133-142.

[83] Kimball, KD and SA Levin (1985). Limitations of laboratory bioassays: the need for ecosystem-level testing. *Bioscience.* 35: 165-171.

[84] Kremen, C (2005). Managing ecosystem services: what do we need to know about their ecology? *Ecology Letters.* 8: 468-479.

[85] Lawrence, JR and MJ Hendry (1995). Mesocosms for subsurface research. *Water Quality Research Journal of Canada.* 30: 493-512.

[86] Lewis, W, M. Jr.; SK Hamilton; MA Rodríguez; J Saunders, F. III and MA Lasi (2001). Foodweb analysis of the Orinoco floodplain based on production estimates and stable isotope data. *Journal of the North American Benthological Society.* 20: 241-254.

[87] Luoma, SN (1977). Detection of trace contaminant effects in aquatic ecosystems. *Journal of Fisheries Research Board Canada.* 34: 436-439.

[88] Maltby, L and P Calow (1989). The application of bioassays in the resolution of environmental problems: past, present and future. *Hydrobiologia.* 188/189: 65-66.

[89] Marchant, R (1999). How important are rare species in aquatic community ecology and bioassessment: A comment on the conclusions of Cao et al. Limnology, *Oceanography.* 44: 1840–1841.

[90] Maund, SJ; PF Chapman; TJ Kedwards; L Tattersfield; P Matthiessen; R Warwick and E Smith (1999). Editorial: Application of multivariate statistics to ecotoxicological field studies. *Environmental Toxicology and Chemistry*. 18: 111-112.

[91] McArdle, BH; KJ Gaston and JH Lawton (1990). Variation in the size of animal populations: patterns, problems and artefacts. *Journal of Animal Ecology*. 59: 439-454.

[92] McCullough, CD; AC Hogan; CL Humphrey; RA van Dam and MM Douglas (in press). Failure of *Hydra* populations to develop tolerance, indicates absence of toxicity from a whole-effluent. Proceedings of the 30th Congress of the International Association of Theoretical and Applied Limnology Montreal, Canada.

[93] McPherson, CA and PM Chapman (2000). Copper effects on potential sediment test organisms: the importance of appropriate sensitivity. *Marine Pollution Bulletin*. 40: 656-665.

[94] Mentis, M (1988). Hypothetico-deductive and inductive approaches in ecology. *Functional Ecology*. 2: 5-14.

[95] Menzie, C; MH Henning; J Cura; K Finkelstein; J Gentile; J Maughan; D Mitchell; S Petron; B Potocki; S Svirsky and PA Tyler (1996). Special report of the Massachusetts Weight-of-Evidence Workgroup: a weight-of-evidence approach to evaluating ecological risks. *Human and Ecological Risk Assessment*. 2: 277-304.

[96] Millspaugh, SH and C Whitlock (1995). A 750 year fire history based upon lake sediment records in central Yellowstone National Park. *Holocene*. 5: 283-292.

[97] Millward, RN and A Grant (1995). Assessing the impact of copper on nematode communities from a chronically metal-enriched estuary using pollution-induced community tolerance. *Marine Pollution Bulletin*. 30: 701-706.

[98] Odum, EP (1984). The mesocosm. *Bioscience*. 35: 419-422.

[99] Parkhurst, BR (1995). Are single species toxicity test results valid indicators of effects to aquatic communities. Ecotoxicological testing: scale, complexity and relevance. J. Cairns Jr and B. R. Niederlehner. Boca Raton, Florida, Lewis Publishing: 105-121.

[100] Parliamentary Commissioner for the Environment. (2006). "Glossary." Retrieved July, 2006, from http://www.pce.govt.nz/reports/pce_reports_glossary.shtml#p.

[101] Pimm, SL (1979). Complexity and stability: another look at MacArthur's original hypothesis. Oikos 33: 351-357.

[102] Pollino, C and BT Hart (2005). Bayesian approaches can help make better sense of ecotoxicological information in risk assessments. *Australasian Journal of Ecotoxicology*. 11: 57-58.

[103] Power, M and LS McCarty (1997). Fallacies in ecological risk assessment practices. *Environmental Science and Technology*. 31: 370-375.

[104] Reid, MA; JC Tibby; Penny and PA Gell (1995). The use of diatoms to assess past and present water quality. *Australian Journal of Ecology*. 20: 57–64.

[105] Resh, VH and JK Jackson (1993). Rapid assessment approaches to biomonitoring using benthic macroinvertebrate. Freshwater biomonitoring and benthic macroinvertebrates. D. M. Rosenberg and V. H. Resh. New York, Capman and Hall: 195-233.

[106] Resh, VH; RH Norris and MT Barbour (1995). Design and implementation of rapid assessment approaches for water resource monitoring using benthic macroinvertebrates. *Australian Journal of Ecology*. 20: 108-121.

[107] Roelofs, W; WAJ Huijbregts; T Jager and AMJ Ragas (2003). Prediction of ecological no-effect concentrations for initial risk assessment: combining substance-specific data and database information. *Environmental Toxicology and Chemistry*. 22: 1387-1393.

[108] Rosenberg, DM and VH Resh, Eds. (1993). Freshwater biomonitoring and benthic macroinvertebrates. New York, Chapman and Hall.

[109] Schindler, DW (1987). Detecting ecosystem responses to anthropogenic stress. Canadian *Journal of Fisheries and Aquatic Sciences*. 44: 6–25.

[110] Schmidt, K; M Koski; J Engström-öst and A Atkinson (2002). Development of Baltic Sea zooplankton in the presence of a toxic cyanobacterium: a mesocosm approach. *Journal of Plankton Research*. 24: 979-992.

[111] Schofield, NJ and PE Davies (1996). Measuring the health of our rivers. Water May-June 1996: 39-43.

[112] Schrader-Frechette, KS and ED McCoy (1993). Method in ecology: strategies for conservation. London, UK, Cambridge University Press.

[113] Snyder, TP; KM Switzer and RE Keen (1991). Allozymic variability in toxicity-testing strains of *Ceriodaphnia dubia* and in natural populations of *Ceriodaphnia*. *Environmental Toxicology and Chemistry*. 10: 1045-1049.

[114] Sprague, JB (1990). Aquatic toxicology. Methods for fish biology. C. B. Schreck and P. B. Moyle. Bethesda, USA, American Fisheries Society: 491-528.

[115] Stark, JS (1998). Heavy metal pollution and macrobenthic assemblages in soft sediments in two Sydney estuaries, Australia. *Marine and Freshwater Research*. 49: 533-540.

[116] Stauber, JL (1995). Toxicity testing using marine and freshwater unicellular algae. *Australasian Journal of Ecotoxicology*. 1: 15-24.

[117] Suter II, GW (1996). Abuse of hypothesis testing statistics in ecological risk assessment. *Human and Ecological Risk Assessment*. 2: 331-347.

[118] Thompson, SA and GG Thompson (2004). Adequacy of rehabilitation monitoring practices in the Western Australian mining industry. *Ecological Management and Restoration*. 5: 30-31.

[119] Tilman, D (1999). The ecological consequences of changes in biodiversity: a search for general principles. *Ecology*. 18: 1455-1474.

[120] Tsirtsis, G and M Karydis (1997). Aquatic microcosms: a methodological approach for the quantification of eutrophication processes. *Environmental Monitoring and Assessment*. 48: 193-215.

[121] van Dam, RA; C Camilleri and CM Finlayson (1998). The potential of rapid assessment techniques as early warning indicators of wetland degradation: a review. *Environmental Toxicology and Water Quality*. 13: 297-312.

[122] van Dam, RA and JC Chapman (2001). Direct toxicity assessment (DTA) for water quality monitoring guidelines in Australia and New Zealand. *Australasian Journal of Ecotoxicology*. 7: 175-198.

[123] Van den Brink, PJ; J Hattink; F Bransen; E Van Donk and TCM Brock (2000). Impact of the fungicide carendazim in freshwater microcosms. II. Zooplankton, primary producers and final conclusions. *Aquatic Toxicology*. 48: 251-264.

[124] Wall, TM and RW Hanmer (1987). Biological testing to control toxic water pollutants. *Journal of the Water Pollution Control Federation*. 59: 7-12.

[125] Ward, JV and K Tockner (2001). Biodiversity: towards a unifying theme for river ecology. *Freshwater Biology.* 46: 807-819.

[126] Warne, MS (1998). Critical review of methods to derive water quality guidelines for toxicants and a proposal for a new framework. Canberra, Supervising Scientist.

[127] Wickwire, WT and C Menzie, A. (2003). New approaches in ecological risk assessment: expanding scales, increasing realism, and enhancing causal analysis. *Human and Ecological Risk Assessment.* 9: 1411-1414.

[128] Wu, R (1996). Editorial: Ecotoxicology: problems and challenges in Australasia. Australasian *Journal of Ecotoxicology.* 2: 1.

[129] Reviewed by Associate Professor Mark Lund, School of Natural Sciences, Edith Cowan University, Australia.

In: Lake Pollution Research Progress
Editors: F. R. Miranda and L. M. Bernard
ISBN: 978-1-60692-106-7

Chapter 6

FISH ASSEMBLAGE SUBJECTED TO STRONG ANTHROPOGENIC STRESS: THE CASE OF THE BARRA BONITA RESERVOIR, TIETÊ RIVER BASIN, SÃO PAULO, BRAZIL

M. L. Petesse[*]
Departamento de Ecologia, IB, Universidade Estadual Paulista – UNESP. CEP -13506-900 Rio Claro (SP) Brazil

ABSTRACT

The Barra Bonita reservoir (S=310 km^2; $\bar{z}$=10,1 m) is an ecosystem subjected to strong anthropogenic stress. It is located in the central part of the Tietê river basin - São Paulo State, characterized by the widespread urbanization, industrial development and intensive agricultural use of the soil. The Barra Bonita reservoir has high social-economic importance due to hydropower generation, navigation and fisheries. From a limnological point of view, it is classified as polymictic and eutrophic. The purpose of this study was

[*] M. L. Petesse: e-mails: mlpetesse@gmail.com

to characterize the actual fish assemblage in the reservoir and, by means of the species-abundance relationship, value its organization in relation to the disturbances caused by the anthropogenic impact. For the fish sampling we chose 24 sites located in three different habitats: reservoir shoreline, mouth of tributary and centre. Samples were taken in two periods of the year: dry season (winter, August-September 2003), and rainy season (summer-February 2004). Fish sampling was standardized by using 10 gillnets with mesh sizes ranging from 3 to 12 cm between opposite knots and funnel traps. At each point we also measured some morphological, physical-chemical and environmental variables. Multivariate analyses (three way-anova and ancova) were employed to, respectively, point out space and temporal variations in the Catch per Unit Effort in weight ($CPUE_W$) and to detect the importance of morphological, physical-chemical and environmental variables on Shannon species diversity (H' in number and weight). A total of 35 species, belonging to 14 families and 4 orders, were caught. Fish assemblage is composed of small body-sized species, with wide feed flexibility and high reproductive compensation. The superposition of the biological cycles of the fishes with the hydrological management of the reservoir, suggests that only those with multiple spawn and/or parental care have success. The $CPUE_W$ distribution shows significant statistical differences between seasons, zones and habitat, indicating the presence of transversal migration from the centre to the shoreline and mouth of the tributary habitats, especially in the rainy season. The variables that influenced the diversity of the fish assemblage were: depth, transparency and landscape for H'_N and H'_W; temperature (H'_N); conductibility and macrophytes (H'_W). The correlation of diversity with the depth was negative, showing that the tributary mouths and the shoreline habitats of the reservoir are the most explored by the ichthyofauna. The environmental variables selected (surrounding landscape and macrophytes beds) act respectively as external feeding support and shelter/nursery habitat, revealing the essential importance of the structural environmental complexity for the diversity maintenance in reservoirs.

INTRODUCTION

Reservoirs are the principal source of energy in Brazil. They started to be built with this purpose at the beginning of the second half of the XX century. A rough survey of the existing reservoirs, reveals that almost 600 can be classified as bigger, with dam height superior to 15 m or volume superior to 100,000 m^3 (Agostinho *et al.*, 2007). Approximately 90% of the entire Brazilian energy consumption is from hydroelectric origin. Almost 70% of it is generated in the high Paraná River basin (Petrere *et al.*, 2002) which is one of the most highly populated and industrial regions of the country. In this basin, at the present time, there are 146 reservoirs with a surface superior to 100 km^2 (Agostinho and Ferreira, 1999). The price paid for this policy was the transformation of the Paraná River and its principal tributaries (Grande, Tietê, Paranapanema and Iguaçu Rivers), into a succession of reservoirs with complete alteration of the original ecological characteristics. Reservoirs, in Brazil, are a reality which strongly marks the landscape and the economy of wide regions as they are considered "obligatory" for the need of development in the country. Despite this, with the exception of the most recent ones, reservoirs are insufficiently studied and basic information such as occlusion date, surface inundate, average depth and retention time are fragmented and inconsistent for many of them (Agostinho *et al.*, 2007). With regard to the fish, even more scarce is the biological information on species composition before and immediately after the occlusion of the dam, or on the effects of the introduction of alochthonous and exotic species.

This practice, widely used in the 1970's, was considered a measure of compensation for the negative impacts caused by the impoundment and a way to improve the fishing production.

The reservoir is an artificial ecosystem with cycles and dynamics strongly dependent on human action. In this way, the organisms are submitted to a continuous reorganization that does not allow a balanced evolution, endangering the productivity of the entire ecosystem (Tundisi *et al.*, 1999). Britski (1994) affirms that, the construction of a dam causes deep alterations in the environment and that the losses of biodiversity produce consequences, in the medium and long periods, that are still not well understood. In this context, Lowe-McConnell (1975) observed that in reservoirs the fish fauna is less diversified in comparison with that of the rivers from which they originated.

According to Castro and Arcifa (1987), the transformation of the river in reservoir alters the alimentary supplies of the riverine fish, which are accustomed to explore benthic fauna and flora. In the new environment they have restricted access to such food sources and the occurrence of an anoxic hypolimnion during part of the year restricts the ichthyofauna to the shoreline.

The nutrients accumulations determine the increase in the plankton production that can favor species with such diet. In the south of Brazil reservoirs, however, these species are rare with the exception of *Hypophthalmus edentatus* (Pimelodedae) that occurs in the Itaipu reservoir (Paraná state) feeding on zooplankton (Petrere, 1996).

Important changes also affect the success of reproduction. In this case, Okada *et al.* (1996) observed that in the reservoirs of the Paraná River basin, the highest catches (kg/ha) occur in reservoirs with greater fluvial reaches or with important tributaries.

Considering the complexity of the reservoir ecosystem, it is important to understand how the organisms respond to the disturbances and to promote studies scientifically appropriate, constantly updated and optimized through programs of biological monitoring. Only in this way we can improve the managing action to support aquatic life and help the ecosystem to maintain its functionality.

The management of the ecosystem and of the ichthyofauna in particular, is justified due to the fact that fish represent an important resource, not only from the natural-ecological point of view, but also from the social and economical perspective. Fishing is, for many poorer populations, the main source of protein and economical subsistence and it is also an important attraction for the tourist-recreational development of a region.

Area of Study: Barra Bonita Reservoir (Tietê River Basin)

Barra Bonita Reservoir is located in the Tietê River basin (20° 31' S; 48° 32' O), in the proximity of Barra Bonita and Igaraçu do Tietê cities (SP). The Tietê is the main River of the São Paulo State. It originates in the Serra do Mar Mountain, Municipal district of Salesopolis, and after 1050 km flows in the Paraná River. The river has been widely exploited for hydroelectric production and has seen its course transformed into a cascade of reservoirs. Barra Bonita reservoir is the first and the oldest of the six forming the Tietê River reservoirs cascade system. They are in order: Barra Bonita, Bariri, Ibitinga, Promissão, Nova Avanhandava e Três Irmãos.

Barra Bonita reservoir drains 44% of the 71988 Km^2 of the Tietê river basin. In this portion, the most intense industrial activity of the São Paulo State is located and 17 million people are concentrated in the metropolitan region of São Paulo capital city (Great São Paolo). The absence of adequate treatment of industrial and domestic sewage, causes significant problems on the water quality of the river (Barrella and Petrere, 2003). This especially affects the tract from the Great São Paulo to the proximities of the Barra Bonita reservoir where the river partially recovers and presents acceptable conditions of water quality (CETESB, 2001). Downstream of Barra Bonita dam, improvements in the water quality are verified, because the cascade of reservoirs work as decantation tanks supporting the auto purification of the water (Barbosa *et al.*, 1999; Petrere *et al.*, 2002).

Barra Bonita Reservoir was impounded in 1962 and starting hydropower generation in May 1963. The primary purpose of the reservoir is hydropower generation, but other uses are contemplated, especially navigation, recreational and small-scale professional fisheries, irrigation, urban and industrial provisions.

According to the limnological aspects, it is classified as polymictic and eutrophic (Tundisi and Matsumura-Tundisi, 1990; Barbosa *et al.*, 1999). Nevertheless, the fish production is higher than the downstream reservoirs, because important reproductive areas are still located in the two main tributaries (Tietê and Piracicaba rivers) (CESP, 1996; Okada *et al.* 2003).

The morphological characteristics of Barra Bonita reservoir are in table 1.

Table 1. Morphological characteristics of Barra Bonita reservoir (AES-Tietê, 2003)

Catchment area	32330 km²
Reservoir surface at maximum usable level of 451.5 m	310 km²
Dam elevation	454.5 m
Altimetry elevation of maximum maximorum	453.0 m
Altimetry elevation of usable maximum	451.5 m
Altimetry elevation of usable minimum	439.5 m
Shoreline development at the maximum usable level of 451.5 m	525 km
Average depth ($\bar{Z}$)at the maximum usable level of 451.5 m	10.1 m
Maximum depth	25 m
Dead volume	569×10^6 m³
Usable volume	2.566×10^6 m³
Total volume	3.622×10^6 m³
Average flow (period 1931/1992)	414 m³/s
Daily maximum observed flow (07/06/83)	4011 m³/s

Considering the surface and volume, Barra Bonita reservoir can be classified as a medium sized reservoir according to the category listed for Straskraba (1999) in table 2.

Table 2. Size category of reservoir

Category	Area (km^2)	Volume (m^3)
Large	10^4 - 10^6	10^{10} - 10^{11}
Medium	10^2 - 10^4	10^8 - 10^{10}
Small	1 – 10^2	10^6 - 10^8
Very small	<1	<10^6

The soil use in the surroundings of the reservoir is dominated by the intensive cultivation of reed sugar and pasture. Reminiscent areas with riparian vegetation are rare. Generally these are restricted to narrow and fragmented strips in correspondence to the tributary.

Other potential impairments for the water quality of the reservoir come from the industrial activity of the region such as that of textile and cellulose productions, food transformation, alcohol and sugar plants (CETESB, 2001).

In synthesis, Barra Bonita reservoir, due to its origin and for the characteristics of its catchment basin, represents a particular object of scientific study. It is an ecosystem submitted to high environmental stress and at the same time is of great economic and social interest due to the hydropower generation and for the presence of a relevant fishing community.

Our objective was to characterize the actual fish assemblage in the reservoir and, by means species-abundance relationship, value its adaptation in relation to the disturbances caused by the anthropogenic impact. To accomplish this, in the first instance, the morphological aspect and the hydraulic management of the reservoir will be analyzed and, in the second, will be approached the fish assemblage structure in terms of species richness, diversity and catch per unit effort in weight ($CPUE_W$). Trophic and reproductive aspects, based on bibliography recognition, will be used to show the species adaptation to the impoundment condition.

METHODS

Morphologic and Hydraulic Management of Barra Bonita Reservoir

In order to characterize the reservoir morphology, the sinuosity index (Marchetti, 1989) and the yearly maximum range (YMR, m) of water level (Cohen and Radomski, 1993) were calculated. The first express the relationship between the reservoir shoreline development and the circumference of a circle with the same reservoir area. Its mathematical form is:

$$SI = \frac{P}{2 * \sqrt{\Pi * A}}$$, where P – shoreline development and A – reservoir surface.

The index assumes values close to one when the morphology of water body is near to a circle and moves away from it when the shoreline shows high articulations.

The YMR is the difference between the yearly maximum and minimum level of the water in the reservoir. It is considered an index representative of the reservoir's hydraulic dynamics, being related to the emersion/submersion of the low deep areas particularly important for the biotic and abiotic processes (Cohen and Radomski, 1993).

In order to characterize the reservoir's hydraulic management, the retention time (RT, days), the rain regimen of the region, the reservoir water inflows and outflows were analyzed. The time series of flows and rains data were respectively provided by the agency of electric plant management AES- Tietê, (www.aestietê.com.br, 2003) and from the Sistema Informativo de Gerenciamento de Recursos Hídricos of São Paulo State (SIGRH: www.sigrh.sp.gov.br, 2003).

The Durbin-Watson Index was used in order to test the existence of autocorrelation in the regression between the retention time (response variate) and year (independent variate) (Chatterjee and Price,1991; Draper and Smith, 1981). The Durbin-Watson Index is given by the *d* statistic, defined as:

$$d = \frac{\sum_{2}^{n}(e_t - e_{t-1})^2}{\sum_{1}^{n} e_t^2}$$

where: e_t = residual in time t; e_{t-1} the residual in time t-1.

The relation between d and the correlation coefficient r is: $d \approx 2\ (r\text{-}1)$ showing that d can range between 0 e 4. Values of d close to 2 indicate absence of autocorrelation (ρ=0) (Chatterjee and Price, 1991).

Fish Assemblage

Observing the reservoir morphology, four zones can be recognized: (i) two fluvial (FL), in correspondence with its forming rivers (Tietê and Piracicaba), (ii) transition (TR) in its central part, where the transversal section increases and occurs the sedimentation of suspended particles, and (iii) lentic (LE) in the final portion near to the dam, where lacustrine characteristics prevails.

In each of these zones, stations were chosen in three different types of habitat: lateral, tributary mouths and centre. The first one is close to the reservoir shoreline and the sample stations have a mean depth of 4.5 m. The soil use in the surrounding landscape is characterized by intensive cultivation of reed sugar and pasture. The shoreline is deprived of riparian vegetation, while flooding and rooting macrophytes are common in this habitat. The extension of rooting macrophyte beds varies with the reservoir's water level, while the flooding macrophytes are principally concentrated in the littoral areas protected from the flow and wind. Beach areas without macrophytes are rare and located in the proximity of human villages.

The sample stations at the mouths of tributaries have a mean depth of 4 m. This habitat is characterized by the remains of riparian forest. Rooting and flooding macrophytes are also present, contributing to the creation of a more complex environment.

The stations in the "centre" habitat were located at the bottom of pelagic areas at a mean depth of 15 m. This habitat has never been investigated in the past studies on the fish fauna of Barra Bonita Reservoir.

Rocky areas are absent in the reservoir due to the geological and morphological nature of the area and to the sedimentary process that affects the reservoir storage capacity.

A total of 24 sample stations were established in accordance with figure 1: six in the lentic zone, six in the transition zone, six in the fluvial zone of the Piracicaba River, and six in the fluvial zone of the Tietê River.

The samples were collected twice a year: in the dry season (winter, August-September 2003) and in the rainy season (summer, February 2004). The sampling gears were standardized using 10 gillnets, with mesh ranging from 3 to 12 cm between opposite knots, and funnel traps. The latter were set in the tributaries (two in each sampling station) below

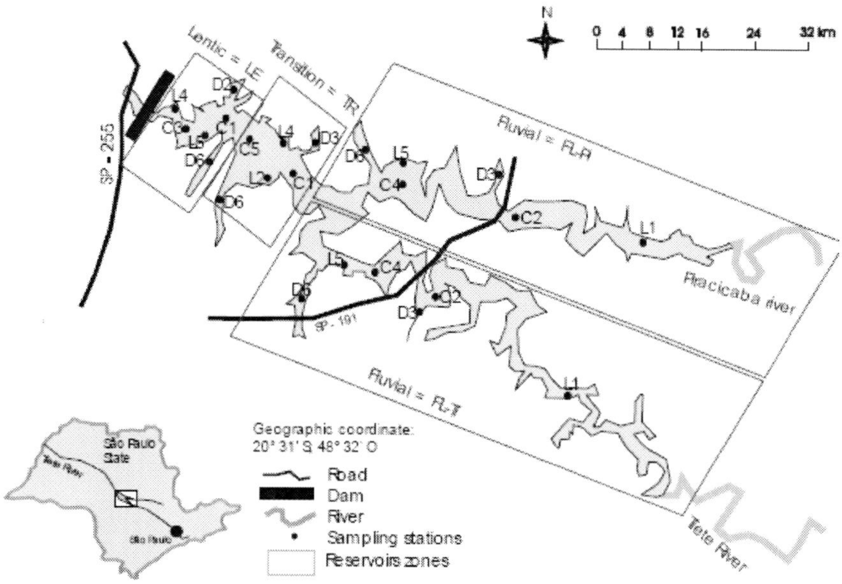

Figure 1. Barra Bonita Reservoir and sampling stations (•). Code zones: FL-PI= Fluvial Piracicaba. FL-TI= Fluvial Tietê, TR= Transition, LE=Lentic; code habitats: C= Centre; L= Shoreline; D = mouth of tributary. Station code= code zone+code habitat+station number (1-6).

floating macrophytes with the objective of integrating the list of species by capturing fish of a small size. Each gillnet was 20 m long with heights varying from 1.5 to 2.4 m. They were placed in the proximity of macrophytes beds parallel to the shoreline or inclined with variable angles depending on the flooding direction. The gear was deployed at the end of the afternoon and retrieved on the morning of the subsequent day, yielding a total fishing time of 18 hours.

At each sampling station, the following variables were measured: water depth (m), water temperature (°C), dissolved oxygen (mg/l), transparency (m), pH (unit of pH), and conductivity (µS/cm). Water depth and transparency were measured with a graduated rope and Secchi disk, respectively. Water quality variables were measured with a Horiba, while the dissolved oxygen was determined by Winkler´s method.

Some habitat traits, such as the surrounding landscape typology, bottom type, and the presence/absence of macrophytes and riparian forest, were also recorded and categorized according to table 3.

Table 3. Habitat variables category

	Category I	Category II	Category III
Typology of the surrounding landscape	condominium/ cultivated	pasture	forest
Riverbed type	muddy/ sand	gravel	Stone/gravel
Macrophytes	absent/rare	fragmented	present
Riparian Forest	absent/rare	fragmented	present

Considering that the number of sampled species is generally fewer than the actual (Agostinho *et al.*, 1997), the jackknife method was used with the objective of making a non vitiate estimate of the number of species present in the Barra Bonita reservoir. This is a nonparametric method based on the observed frequency of rare species in the community (Krebs, 1998). Its mathematical form is:

$$\hat{S} = s + \left(\frac{n-1}{n}\right) * k$$

where: Ŝ = Jackknife estimator of richness species;

s = Total number of species in n sample;
n = total number of samples;
k = number of unique species; (unique species is that which occurs only in a single sample)

The analysis was performed by the jackknife routine, available in Krebs (1998).
In order to calculate confidence limits, the jackknife variance was calculated as:

$$\text{var}\left(\hat{S}\right) = \left(\frac{n-1}{n}\right) * \left[\sum_{j=1}^{s}\left(j^2 f_j\right) - \frac{k^2}{n}\right],$$

where:

$Var(\hat{S})$ = Jackknife's variance of richness of species; f_j = number of sample with j unique species (j = 1,2,3.....,s); k = number of unique species; n = total number of samples.

The community structure of the fish assemblage was described by the diversity index (H'), Evenness (E) and Dominance (D).

In the first case, we used the Shannon index as:

$$H' = -\sum_{i=1}^{s} \frac{n_i}{N} * \log_2 \frac{n_i}{N}$$

where: s= species number in the sample; n_i= individuals number of i-esima specie; N= total number of individuals in the sample.

H' increase with the number of species and, in theory, can reach high values, but for biological community, seems not to exceed the 5 unit (Krebs, 1998). Fausch *et al.* (1990) argues that, although the community's structure is influenced by the number and weight of the individuals in the sample, it is not clear which is the best unit of measure for the diversity index calculation. In order to respond to this observation, we calculated the index using the number (H'_N) and the weight (H'_W) of individuals. The same was done for the evenness (E_N and E_W), obtained from:

$$E = H' / H'_{max}$$

where H' is the Shannon Index; $H'_{max} = \log_2(S)$, with S =species number in the sample.

The dominance (D) was calculated using the number of individuals (D_N) from the:

$$\hat{D} = \sum_{i=1}^{s} \frac{[n_i(n_i - 1)]}{[N(N-1)]}$$

where s= species number in the sample; n_i= individuals number of i-esima specie; N= total number of individuals in the sample.

Catch for Unit Effort (CPUE)

This is the ratio between the catches (C= individual number or weight) and the fishing effort (*f*):

CPUE= C/*f*

In this case, the effort was defined as gillnet total area (=358 m^2) for a mean fishing time of 18 h. The results were standardized to 1000 m^2 of catching area and for 24 h of fishing time. We utilized the capture per unit effort in weight ($CPUE_W$), which is considered more appropriate to show the importance and the contribution of the different species in the fish assemblage. The use of CPUE is based on the assumption that for a data quantity of effort, the catches are proportional to the total size of population at the sampling time (Ricker, 1975). This, however, cannot be considered an absolute index of abundance due to the variability of species catchability. In experimental fishing, the use of multiple fishing gear limits this variability and the CPUE can be considered an appropriate measure of fish abundance allowing the comparison between community, habitats or different periods (Tejerina-Garro *et al.*, 1998).

The Trophic Structure

This analysis is considered important as the changes in the environment or in the water quality may affect food availability and may cause changes in the fish assemblage (Araújo, 1998). The adopted trophic categories were (Lowe-McConnel, 1999; Hahn *et al.,* 1997; Celi-Fedatto-Abelha *et al.,* 2001): omnivorous, herbivorous, iliophagous, detritivorous, invertivores, insenctivorous, planktivorous and carnivorous. The attribution of trophic category to each fish species was based on a bibliographical survey taking into account the dominant items of the diet of each species.

The Reproductive Structure

This is a very important aspect of the dynamics and success of the species living in a highly disturbed environment such as a reservoir. We considered the following aspects: type of spawning, reproductive period, reproductive strategy (i.e. migration and parental care) and species resilience. The information results from a bibliographical survey (www.fishbase.org, 2004; Nakatani *et al,* 2001; Santos, 1980; Vazzoler and Menezes, 1992; Vazzoler *et al,* 1997;

Braga and Gennari, 1991; Braga, 1997, 1999; Gennari and Braga, 1996). In the case of resilience, defined as the species capacity to double its population in time, the informative source was the archive Fishbase (www.fishbase.org, 2004), based on the International Centre for Living Aquatic Resource Management (ICLARM - Manila, Philippines). According to Musick (1999), four resilience categories are considered: very low (population doubles in more than 14 years); low (population doubles in 4.5-14 years); medium (population doubles in 1.4-4.4 years); and high (population doubles before 13 months). These categories are based on population parameters such as: r_m (intrinsic rate of population growth, $year^{-1}$), K (von Bertalanffy growth coefficient, $year^{-1}$), t_{max} (maximum age, years), t_m (age at first maturity, years) and fecundity (number of eggs). The species classification in relation to its reproductive habits follows Vazzoler and Menezes (1992) that distinguish: migratory species, non-migratory species without parental care, and non-migratory species with parental care.

According to Nakatani *et al.* (2001), the reproductive potential of a species depends on: 1. spawning success; 2. eggs fertility rate; 3. structural sex balance in the reproductive stock. In the case of the reservoir, the success of reproduction strongly depends on the hydraulic management of the reservoir. For this the hydraulic cycle of the reservoir will be crossed with the reproductive cycle of each species in order to show the critical periods for the recruitment and evaluate the impoundment impact on the species.

Dajoz Constancy Index

The list of species collected in this work was compared with those of Torloni *et al.* (1993), Castro (1997), Barrella and Petrere (2003), Freitas (1999) and Smith (2004) through the Dajoz constancy index (Dajoz, 1973) that allow us to make an analyses of the ichthyofauna historical transformation. The sampling method was the same for all the authors and involved similar effort, with the exception of Torloni *et al.* (1993), who only considered the species caught by professional fishermen. The constancy index is the percentage ratio between the number of samples in which a species is present and the total of samples. It is defined as: constant species, the ones present in more than 50% of the samples; accessory species, those present between 25% and 50% of the samples; accidental species, those present in less than 25% of the samples.

Statistical Analysis

Cluster Analysis

In order to evaluate the similarity between the sampling stations in relationship to species distribution, a cluster analysis was performed. The original matrix of presence/absence, for the total sample, was formed by 35 species and 24 sampling stations. The clustering method was the UPGMA (Unweighted Pair Group Method Average), and the Jaccard index was the measure of distance selected. This index, excludes the problem of double absence due to the non homogeneous spatial distribution of the species (Valentin, 2000). To evaluate the distortion caused by the clustering method, the cophenetic coefficient was calculated by the correlation of original and cophenetic matrices. A value higher than 0.75 is considered a good representation of the original data as suggested by McGarigal *et al.* (2000).

Analyses of Variance

The temporal (season) and space (zones and habitats) variability of the H' (in number and weight) and of $CPUE_W$ were evaluated through a three-way ANOVA. The analysis of residuals was employed to verify the assumption of normality and homocedasticity of variances. If F was significant at p=0.05 or p=0.01, the HSD Tukey multiple comparison test *a posteriori* was employed.

Analyses of Covariance (ANCOVA)

ANCOVA is a statistical method that combines regression analyses with analyses of variance (Camussi *et al.*, 1986). This analysis was employed in an exploratory form with the aim of verifying which categorical factors (landscape, type of substrata, macrophyte and riparian forest) and continuous variables (depth, water temperature, pH, transparency, conductibility and oxygen) were related to diversity (dependent variable) in number (H'_N) and weight (H'_W). For the model definition, a "*backward*" procedure was adopted, that is: in the first instance, all the variables and factors were put in the model and soon afterwards, the variables and factors that were not significant at the level of probability of p=0.15, were excluded in order to define the minimal model able to explain their effects on the dependent variable (H'_N and H'_W). The analysis of residuals was employed to verify the assumption of normality and homogeneity of variance.

RESULTS

Morphological Aspects

The sinuosity index of Barra Bonita reservoir is 8.4. This value is far from the ideal of one indicating a very complex morphology justified by the presence of numerous lateral tributaries and bays. These increase the tendency for the nutrient accumulations, especially in areas far from the general water circulation, promoting the trophic degradation. The high vulnerability of Barra Bonita reservoir is also related to its localization in the cascade reservoir system. It is the first of the cascade and, in this context, Braga *et al.* (1998), relate that Barra Bonita reservoir keeps at least 50% of the entire nitrogen and phosphorous load produced in the upper part of its catchment basin.

The reservoir's yearly maximum range (YMR) lies between 1.5 m (1976) and 9.6 m (1968) with a mean of 5.2 m (figure 2). During the analyzed period (1969-2002), the YMR was higher than 7 m in 1968, 1969, 1973, 1980, 1999, and 2000. Jackson and Marmulla (2005) relate that a range of 2.5-4 m can be considered reasonable, but most of the cases regarding Barra Bonita reservoir, are over this range. This can impair reproduction and recruitment of fish as well as the access to alimentary resources and shelter for the emersion of shallow areas (Rodriguez and Lewis, 1994).

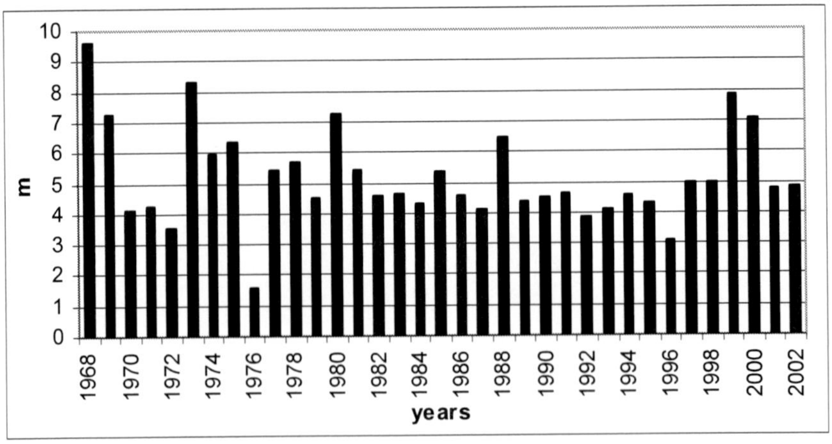

Figure 2. Yearly maximum range (YMR, m) of Barra Bonita Reservoir.

The Hydraulic Management

The retention time (RT) of Barra Bonita reservoir was obtained by the analyses of the historical daily series of the reservoir's water level from 1968 to 2002. The results show that most of the water retention time (RT) values are between 40 and 80 days with a maximum of 131 (1971) and a minimum of 25 days (1983); the mean for the period is 69 days. The tendency of RT to decline along the time is shown in figure 3. The linear regression equation (RT vs. years) shows a significant negative relationship between the two variables, being Pearson's correlation coefficient r = - 0.65 (n =33; $p<0,01$). The Durbin-Watson index, whose value is close to 2 (d = 1.94, $p> 0.05$), indicates the absence of autocorrelation, validating the analysis. Note that the reservoir is losing on average 1.22 days/year in its retention time, which seems quite high.

In order to characterize the rain regimen of the region, the mean months rain data in the stations of Laranjal Paulista (Tietê river), Santa Teresinha (Piracicaba river) e Barra Bonita (dam), were analyzed. The results show the presence of two periods in the year: a dry and a rainy one. The former, with a mean of 55 mm/month of rain, extends from April to September; the latter, with a mean of 170 mm/mouth of rain, extends from October to March. The rain regimen is similar in all the examined stations and the descriptive statistics (table 4-6) show that the period of more intense rains is concentrated from October to March with peaks in December and January.

The analyses of river inflows (m^3/s) show a period of minimum flow from July to September and a maximum from December to March (figure 4). The descriptive statistics of the river inflows (period 1968-2002) show that the minimum flow occurs in August with 196

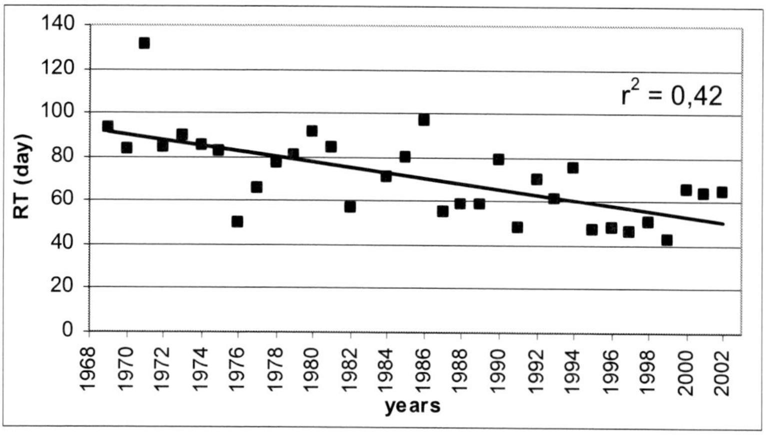

Figure 3. Relationship between RT (retention time) and year from 1969 to 2002. The year of 1983 is not included because it is an outlier. RT = 2486.1 – 1.22 year, n =33, r = - 0.65 (r^2 = 0.42) ($p<0{,}01$).

Table 4. Descriptive statistics of pluviometer station (mm/month): Laranjal Paulista – Tietê River

	N. Values	Mean	Median	Minimum	Maximum	Variance
JAN	35	203.8	187.5	75.1	425.5	7646.8
FEB	35	175.6	142.6	28.1	434.4	11917.1
MAR	35	136.5	117.5	32.5	451.5	7817.2
APR	35	67.1	57.5	0.2	269.4	2590.1
MAY	35	67.7	46.1	0.2	272.6	3729.0
JUN	35	56.0	38.9	0	249.1	3814.2
JUL	35	34.1	19.0	0	162.2	1570.6
AUG	35	33.8	14.2	0	169.0	1781.2
SEP	35	67.2	56.4	2	276.8	3404.6
OCT	35	120.4	114.9	14.1	231.1	3683.9
NOV	35	113.4	99.0	44.8	329.4	3268.0
DEC	35	194.5	181.2	53	519.7	8631.2

Table 5. Descriptive statistics of pluviometer station (mm/month): Santa Terezinha – Piracicaba River

	N. Values	Mean	Median	Minimum	Maximum	Variance
JAN	35	234.1	206.2	62.1	495.5	10781.3
FEB	34	195.3	157.9	60.8	515.2	13045.8
MAR	33	177.4	140.9	15.8	442.5	13241.6
APR	34	71.5	56.0	0	214.4	3397.2
MAY	34	63.1	48.9	0	337.4	5412.0
JUN	34	48.3	28.7	0	185.7	2806.7
JUL	33	29.9	11.8	0	119.0	1525.0
AUG	33	34.3	12.0	0	137.3	1654.7
SEP	33	67.2	53.2	0	215.9	2998.1
OCT	33	133.5	138.1	12.6	290.7	4470.2
NOV	34	135.0	125.5	34.2	315.6	4155.7
DEC	34	220.4	211.0	90.5	454.6	7787.3

Table 6. Descriptive statistics of pluviometer station (mm/month): Barra Bonita – Dam

	N. Values	Mean	Median	Minimum	Maximum	Variance
JAN	22	221.2	205.0	82.8	404.3	7995.7
FEB	22	160.0	148.4	16.4	381.2	9125.8
MAR	22	145.4	132.8	26.4	408.9	8613.6
APR	22	73.8	68.9	2.7	172.5	2293.0
MAY	22	79.7	71.5	6.2	232.4	3405.5
JUN	22	61.7	57.8	0.0	194.5	2951.8
JUL	22	37.2	24.6	0.0	126.0	1202.2
AUG	22	37.4	10.8	0.0	138.7	2226.7
SEP	22	71.4	65.2	10.9	189.7	2130.6
OCT	22	117.7	95.3	28.1	260.1	3812.3
NOV	22	148.5	143.1	38.5	362.8	5439.4
DEC	22	233.5	221.7	105.5	431.0	9487.0

m^3/s and the maximum in February with 721.6 m^3/s (table 7) according to the rain regimen. This is confirmed by the positive and highly significant correlation ($r = 0.55$; $p < 0.01$) between river inflow and rain in the Barra Bonita station (figure 5).

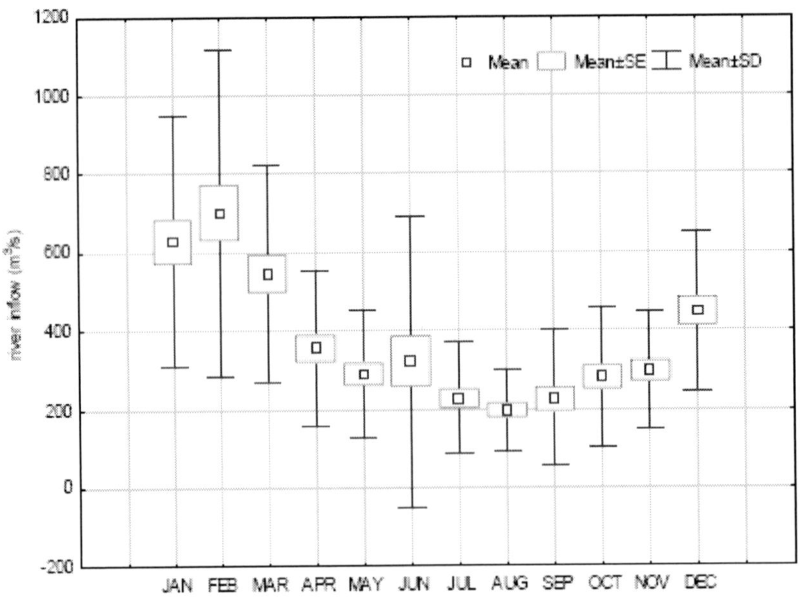

Figure 4. River inflow to Barra Bonita Reservoir. Historical series: 1968-2002.

The effect of hydraulic management on the river flow can be seen from the analyses of dam discharges. In this case, the maximum values can be observed in January and February according to the summer main rains events. Contrary to the river inflow, from April to November the discharges are, in general, homogeneous (figure 6) being artificially controlled and maintained in the proximity of 300 m^3/s. The minimum discharge observed, in this case, is in August with 263 m^3/s (table 8).

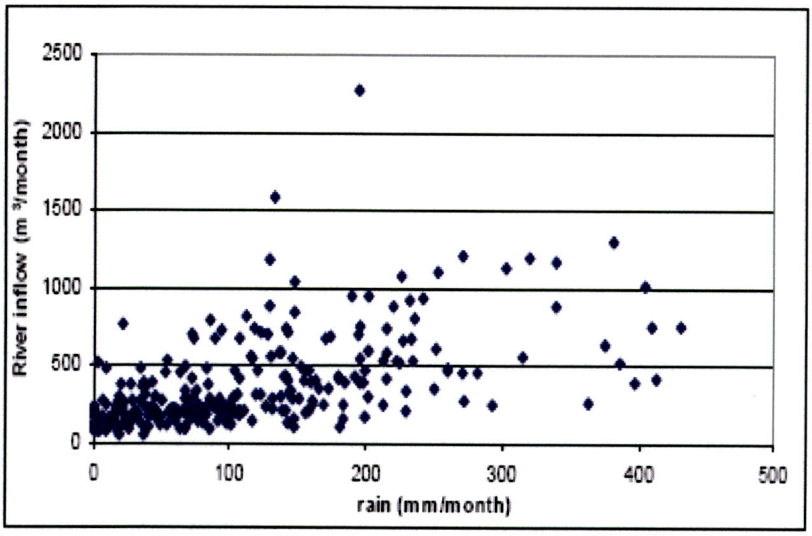

Figure 5. Correlation between river inflow and rainy. Mean month of series 1968-2002.

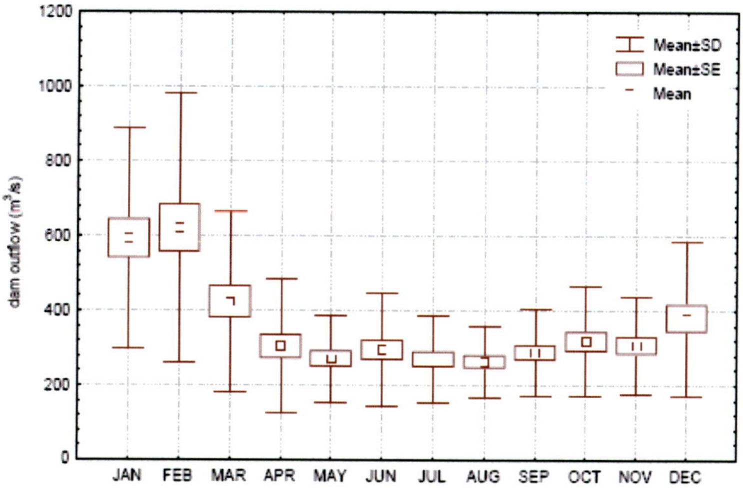

Figure 6. Barra Bonita reservoir outflow. Historical series: 1968-2002, without 1983 year. (outlier).

These results clearly show that the regulation of the dam discharge is not only conditioned by the electric production, but also by the need to maintain the water level of the reservoir in a specific range depending by the rain regimen (Kennedy, 1999). Thus, years with a greater rainfall (such as 1976 and 1983) produced an increase in the monthly discharges. On the other hand, particularly dry years such as 1969, 1978, 1979, 1986 and 2000 result in a decrease of reservoir water level and strong reduction of discharges (figure 7).

Table 7. Descriptive statistics of river inflow (m^3/s, per month) to Barra Bonita reservoir. Historical series1968-2002.

	N. Values	Mean	Median	Minimum	Maximum	Variance
JAN	34	646.4	561.9	123.0	1341.0	92104.5
FEB	34	721.6	668.0	87.0	1872.0	164826.7
MAR	35	545.5	477.0	119.0	1193.0	76000.9
APR	35	354.9	316.0	100.1	1032.0	40116.5
MAY	35	287.8	252.0	61.0	920.0	27094.6
JUN	35	319.8	207.0	82.0	2269.0	139191.6
JUL	35	227.4	196.6	61.0	760.0	20100.1
AUG	35	196.0	186.0	58.0	551.0	11056.2
SEP	35	225.3	200.0	53.0	941.0	30198.5
OCT	35	279.5	196.0	91.0	795.0	32263.1
NOV	35	296.8	273.0	65.0	738.0	22806.8
DEC	35	446.8	411.0	127.0	1106.0	41234.3

Table 8. Descriptive statistics of river outflow (m^3/s, per month) from Barra Bonita reservoir. Historical series: 1968-2002; in bracket data without 1983 year (outlier)

	N. Values	Mean	Median	Minimum	Maximum	Variance
JAN	34	610.6 (592.9)	520.5	140.0	1285.0	95479.5
FEB	34	648.9 (620.7)	584.0	72.0	1696.0	153727.6
MAR	35	434.6 (422.3)	378.0	93.0	960.0	62434.7
APR	35	317.5 (303.6)	283.0	53.0	960.0	37985.1
MAY	35	287.7 (270.5)	268.0	38.0	872.0 (525.0)	23759.6
JUN	35	356.1 (295.1)	270.0	29.0	2430.0 (757.0)	152517.2
JUL	35	283.2 (269.4)	270.0	65.0	751.0 (682.0)	19382.0
AUG	35	272.7 (263.5)	273.0	82.0	587.0 (561.0)	11738.4
SEP	35	309.1 (287.8)	292.0	77.0	1034.0 (732.0)	28746.8
OCT	35	333.7 (317.7)	314.0	50.0	879.0 (742.0)	29783.4
NOV	35	321.5 (307.3)	278.0	66.0	805.0 (633.0)	23734.1
DEC	35	392.5 (380.6)	350.0	109.0	1168.0	46509.3

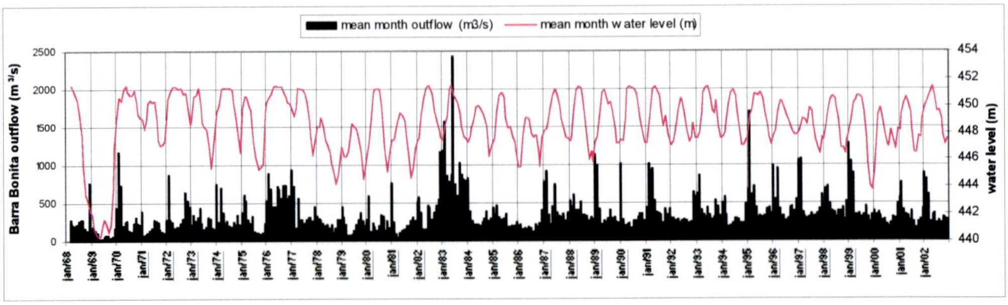

Figure 7. Mean outflow (m^3/s per month) and mean water level variation of Barra Bonita Reservoir. Historical series: 1968-2002 (data source: AES-Tietê).

The hydraulic cycle of the reservoir is shown in figure 8. The reservoir flooding phase ranges from December to May and the emptying phase from June to November. It is also possible to observe that the flooding phase ends approximately when the mean water reservoir level of 450 m is reached and a new phase starts again at the mean level of 446.8 m.

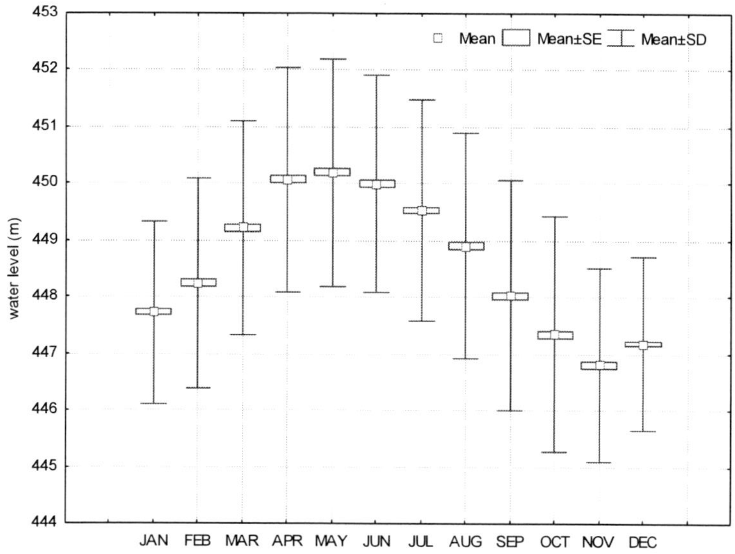

Figure 8. Monthly mean water level of Barra Bonita reservoir, from 1969 to 2002.

Analyzing the daily level fluctuations (figure 9), it can be observed that the period from December to May (flooding period), contrary to the steady nature of the emptying period, is characterized by unpredictable micro-pulses (short-time high and low water levels fluctuations), caused by the frequency and intensity of summer rains.

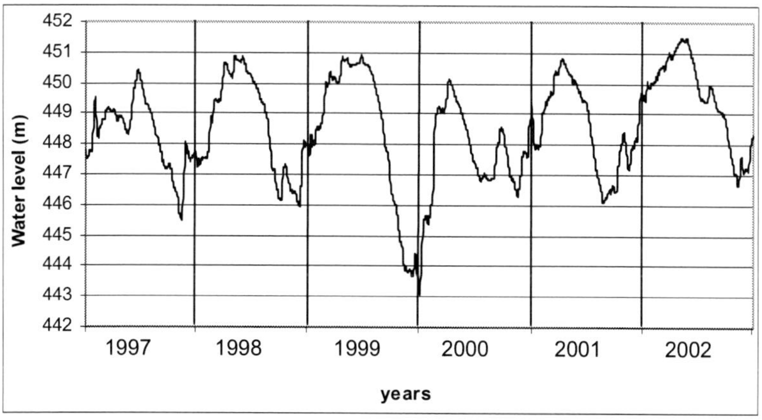

Figure 9. Daily water level (m) of Barra Bonita Reservoir from January 1997 to December 2002.

Species Assemblage

In the Barra Bonita reservoir 6252 individuals were collected for a total weight of 453.1 kg. The total number of species observed was 35 belonging to four orders and 14 families (table 9). The taxonomic classifications follow Reis *et al.* (2003) and Britski *et al.* (1999).

Table 9. Fish species collected in the Barra Bonita Reservoir in the dry and rainy seasons (*= introduced species)

Family/species	Dry	Rainy
Anostomidae		
Leporinus lacustris	X	X
Leporinus obtusidens	X	X
Schizodon intermedius		X
Schizodon fasciatus	X	X
Schizodon nasutus	X	X
Acestrorhynchidae		
Acestrorhynchus lacustris	X	X
Characidae		
Astyanax altiparanae	X	X
Astyanax fasciatus	X	X
Astyanax schubarti	X	
Moenkhausia intermedia		X
Salminus hilarii		X
Triportheus paranensis	X	X
Hyphessobrycon eques		X
Piaractus mesopotamicus		X
Serrasalmus maculatus		X
Serrasalmus spilopleura	X	X
Metynnis maculatus *		X
Curimatidae		
Cyphocharax modestus	X	X
Cyphocharax nagelii	X	X
Steindachnerina insculpta	X	X
Erythrinidae		
Hoplias malabaricus	X	X
Parodontidae		
Apareidon piracicabae	X	X
Prochilodontidae		
Prochilodus lineatus	X	X
Callichthyidae		
Hoplosternum littorale	X	X
Loricariidae		
Liposarcus anisitsi*	X	X
Hypostomus ancistroides	X	X
Pimelodidae		
Pimelodus maculatus	X	X
Heptapteridae		
Pimelodella sp.		X

Table 9. (Continued)

Family/species	Dry	Rainy
Rhamdia quelen	X	X
Gymnotidae		
Gymnotus carapo	X	X
Cichlidae		
Crenicichla haroldoi	X	X
Geophagus brasiliensis	X	X
Satanoperca jurupari *	X	X
Oreochromys niloticus *	X	X
Scianidae		
Plagioscion squamosissimus *	X	X

The order with the largest number of species was Characiformes (65.7%), followed by Siluriformes (17.1%), Perciformes (14.3%) and Gymnotiformes (2.9%). The predominance of Characiformes is common in the Neotropical ichthyofauna (Britski, 1994). The more abundant family was Characidae (Characiformes) with 11 species.

In the dry sampling season (August-September 2003), 27 species were captured for a total of 2140 individuals and 171.9 kg. No capture was registered with the funnel trap.

In the rainy season the species catches rise to 34 for a total of 4112 individuals and 281.2 kg including the funnel trap catch. With this gear, three species were sampled: *H. eques*, *M. maculatus* and *P. mesopotamicus*. Most of these (99%) belong to the first species comprising very tiny forms, with mean weight of 1.2 g/individual. Also in the case of *P. mesopotamicus* and *M. maculates*, the individual belonged to juvenile forms.

The total number of species in each sampling station range between two and 27. In the fluvial zones of the Tietê and Piracicaba Rivers, the minimum and the maximum richness were respectively observed (figure 10a). Considering the reservoir habitats, the minor number of species was observed in the centre habitat where the total number of species ranges between two and 10 (figure 10 b).

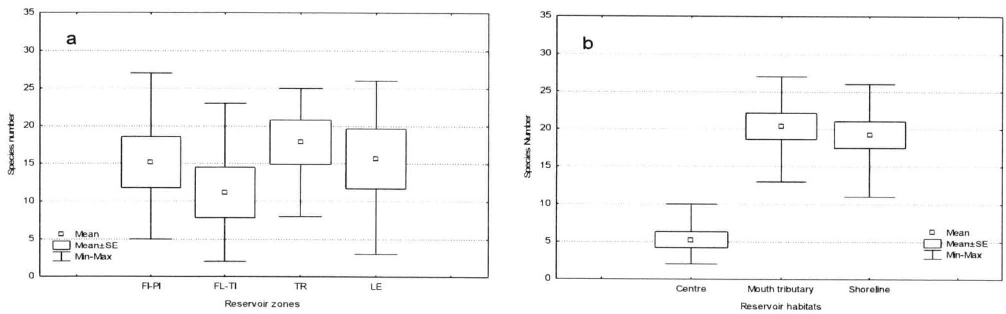

Figure 10. Species number for reservoir zones (a) and habitats (b) – total species, N=35.

In general, 16 species were present in more than 50% of the stations showing a wide distribution (figure 11). Species with a limited distribution were: *Pimelodella sp.*, *M. maculatus* and *A. schubarti*. These were collected in only one habitat: the fluvial zone of the Piracicaba River. Other such as: *S. intermedius*, *S. maculatus*, *H. eques*, *S. fasciatus*, *A.*

fasciatus, *S. hilarii* and *P. mesopotamicus*, were caught at the mouth of tributary associated to the presence of flooding macrophytes. The specimens of *S. hilarii* and *P. mesopotamicus* belong to juvenile forms, indicating the importance of this habitat as a nursery for these species. The introduced species *L. anisitsi,* on the other hand, was found in all the sampling stations, showing a high potential of adaptability and diffusion of the specie in this environment.

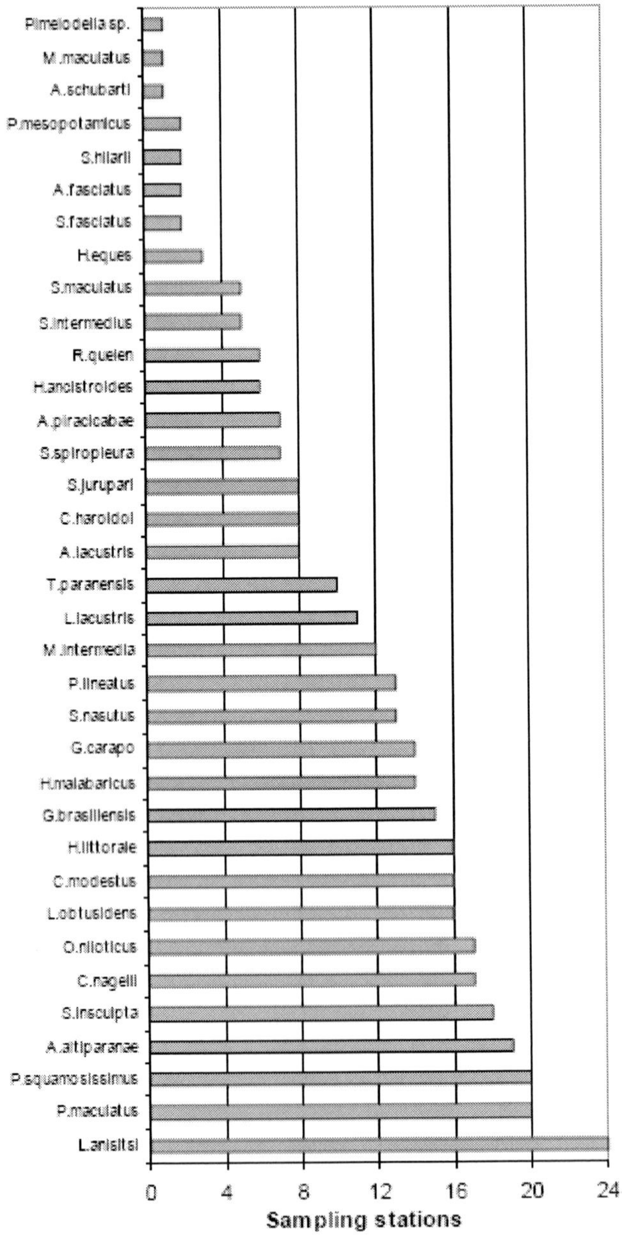

Figure 11. Species distributions.

The cluster analyses reveal the similarity among the sampling stations based on the presence/absence of the species. It distinguishes three groups of stations: 1. stations with predominant lotic characteristics; 2. stations with remarkably lentic characteristics and 3. pelagic stations (figure 12).

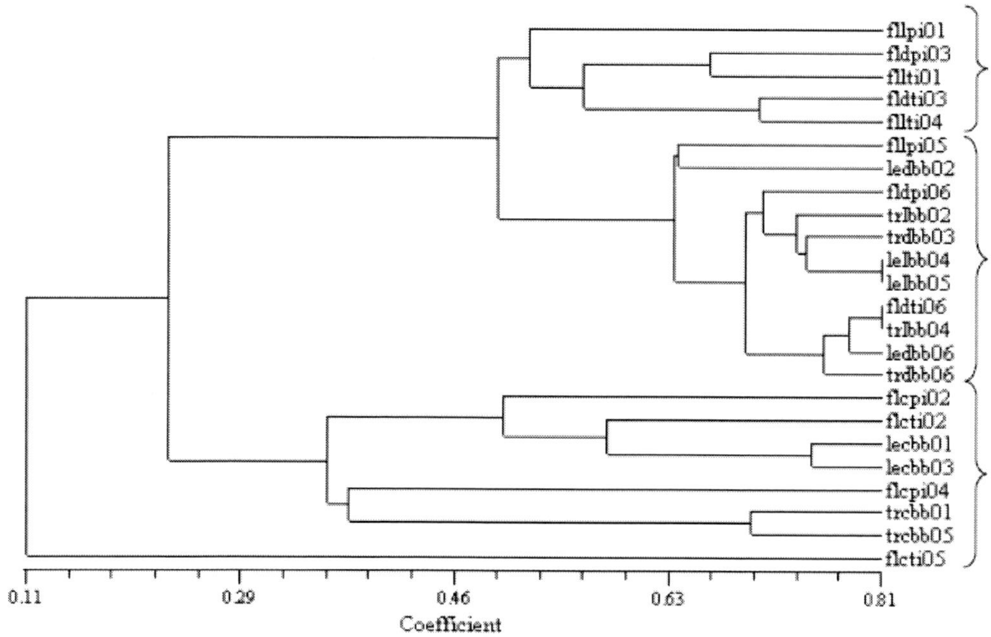

Figure 12. Cluster analyses of species distribution (presence/absence) in the sampling stations. Clustering method: UPGMA; distance measure: Jaccard; cophenetic correlation coefficient, r_c=0.91). Station code in figure 1.

The first group includes the stations of Piracicaba and Tietê Rivers with the exclusion of the centre stations and of the principal tributaries (FL-Pi-D-06 =Turvo River and FL-Ti-D-06= Capivara River). In this group, 23 species were registered and those with wider distribution were: *H. littorale, S. insculpta, C. modestus, L. anisitsi, A. altiparanae, P. lineatus, G. carapo, C. nagelii, L. obtusidens* and *O. niloticus*. The captures occurred at a mean depth of 3 m and 7 m in the dry and rainy seasons respectively.

The second cluster, groups the stations of the reservoir body (TR and LE zones) and those of Piracicaba River with lentic characteristics. The total number of species found in this group is 33 and the most frequent are: *S. insculpta, A. altiparanae, H. littorale, G. brasiliensis, C. nagelii, P. squamosissimus, C. modestus, L. anisitsi, P. maculatus, S. nasutus, H. malabaricus, M. intermedia, L. obtusidens, O. niloticus, L. lacustris, G. carapo, P. lineatus, C. haroldoi* and *S. jurupari*. The captures occurred at a mean depth of 3.5 and 7 m in the dry and rainy seasons respectively.

The third, groups the centre stations corresponding to the demersal habitat. The most frequent species in this group were: *L. anisitsi, P. maculatus* and *P. squamosissimus*. In this case, the captures occurred at a mean depth of 13 m (dry) and 28 m (rainy).

The cophenetic correlation coefficient is r_c = 0.91, indicate that the result is a good representation of the original similarity data matrix.

The most abundant species, in the total sample, were: *S. insculpta* (22.7%), *A. altiparanae* (10.4%), *H. littorale* (9.6%) and *C. nagelii* (7.5%). These represented almost 50% of the total abundance sampled. In relation to weight, 50% of the total frequency observed was displayed by *P. maculatus* (14.4%), *H. littorale* (13.4%), *S. insculpta* (11.2%) and *L. anistsi* (10.7%).

Among the 35 species collected, five had been introduced: four from other Brazilian hydrographic basins (*S. jurupari* and *P. squamosissimus* - Amazonian basin, *M. maculatus* and *L. anisitsi* - other basins) and one from Africa (*O. niloticus*), as reported by Smith *et al.* (2002).

Most of the introductions were undertaken by CESP (Energy Company of São Paulo State) in order to compensate the impoundment impact, which caused the loss of the great migratory species such as *Salminus maxillosus*, *Pseudoplatistoma corruscans* and *Zungaro zungaro,* and improve the fishing production of the reservoir. In the decade of the 1970's, the species introduced were: *Triportheus a. angulatus* (= *T. signatus*) from the Parnaíba river basin, *Oreochromis hornorum* from Africa, *Plagioscion squamosissimus*, *Astronotus ocellatus* and *Cichla ocellaris* all from Amazonian basin (Torloni *et al.,* 1993a). Some of these have not appeared in the professional or experimental fisheries since 1989, indicating the probable failure of the species adaptation process in the Barra Bonita reservoir. The current program of fish stocking, undertaken by the Aquaculture and Hydrobiology Station of Promissão, at the present only promote the stocking of "native species" (www.aestiete.com.br, 2003).

Jackknife Estimate

The number of expected species, calculated with the Jackknife method for the total sample of the reservoir, varied between 34 and 41. The observed number of species (35) fell into this confidence interval, thus we think that the fishing effort employed sufficiently represents the fish assemblage in the Barra Bonita Reservoir.

Analyzing the data for habitat, it can be verified that the largest expected number of species is in the shoreline habitat with 36 species and the smallest in the centre with 17 (table 10).

Table 10. Jackknife estimate for the habitat species richness

	Observed species number (N)	Species number estimate (N)	Standard deviations	Confidence interval (95%)
Centre (C)	14	17.5	1.87	13.1-21.9
Mouth of tributary (D)	32	35.5	1.87	31.1-39.9
Shoreline (L)	32	36.4	1.84	32-40.7

Diversity Index (H'_N) and Evenness (E_N)

The diversity index in number (H'_N) in the dry season (table 11) varied between a minimum of 0.49 bits/individual and a maximum of 3.49 bits/individual; the mean of the period was 2.28 bits/individual. In the rainy season (table 11) the minimum was 0.81

bits/individual and the maximum 3.88 bits/individual; the mean of the period was 2.46 bits/individual

Evenness values in number (E_N) range between 0.5 and 1 (figure 13).

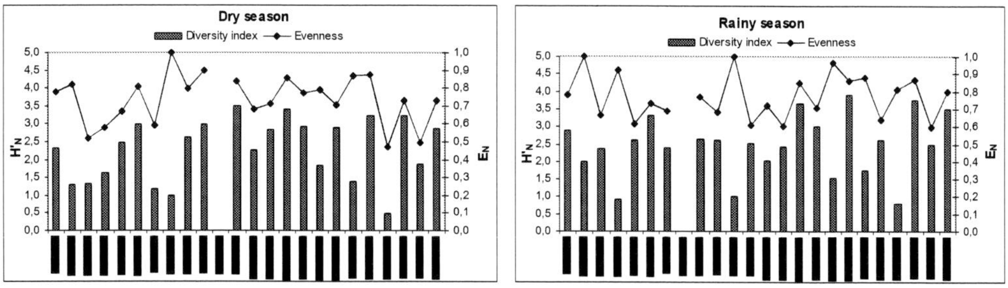

Figure 13. Diversity index in number (H'_N) and E_N in the dry and rainy seasons.

Table 11. Descriptive statistics of Shannon diversity index in number (H'_N) in the two sampling seasons

	Mean	Median	Minimum	Maximum	Int.Var.	Variance
DRY	2.28	2.47	0.49	3.49	3.02	0.76
RAINY	2.46	2.53	0.81	3.88	3.07	0.75

The analysis of variance (three-way) for the diversity index in number (H'_N), detects significant differences only among habitats (table 12) and the HSD Tukey test shows that the centre habitat differs from the others ($p<0.01$) (table 13). Two stations were excluded from the analyses of data because no fish were caught.

Table 12. Analysis of variance: Shannon diversity index (H'_N) per year season, zone and habitat

Depended variable: H'_N ; N= 46, multiple R = 0.872, R^2= 0.761

Source of variations	SQ	gl	MQ	F	p
YEAR SEASON	0.341	1	0.341	0.932	0.345
ZONE	2.886	3	0.962	2.631	0.075
HABITAT	19.428	2	9.714	26.567	0.000
YEAR SEASON*ZONE	0.278	3	0.093	0.254	0.858
YEAR SEASON*HABITAT	0.135	2	0.068	0.185	0.833
ZONE*HABITAT	1.571	6	0.262	0.716	0.641
YEAR SEASON*ZONE*HABITAT	1.079	6	0.180	0.492	0.807
ERROR	8.044	22	0.366		

Table 13. HSD Tukey probability test applied to detect difference in H'_N values among habitats (C=Centre; D=Mouth of tributary e L=Shoreline)

	C	D	L
C	1.000		
D	0.000	1.000	
L	0.000	0.236	1.000

Diversity Index (H'$_W$) and Evenness (E_W)

The diversity index in weight, in the dry season (table 14) ranged between a minimum of 0.11 (bits/individual) and a maximum of 3.72 (bits/individual); the mean for the period was 2.25 bits/individual. In the rainy season (table 14) the minimum was 0.43 bits/individual and the maximum 3.97 bits/individual; the mean for the period was 2.45 bits/individual

The evenness in weight (E_W), range between 0.1 and 0.9 (figure 14)

Table 14. Descriptive statistics of Shannon diversity index in weight (H'$_w$) in the two sampling seasons

	Mean	Median	Minimum	Maximum	Int.Var.
DRY	2.25	2.76	0.11	3.72	1.47
RAINY	2.45	2.84	0.43	3.97	1.27

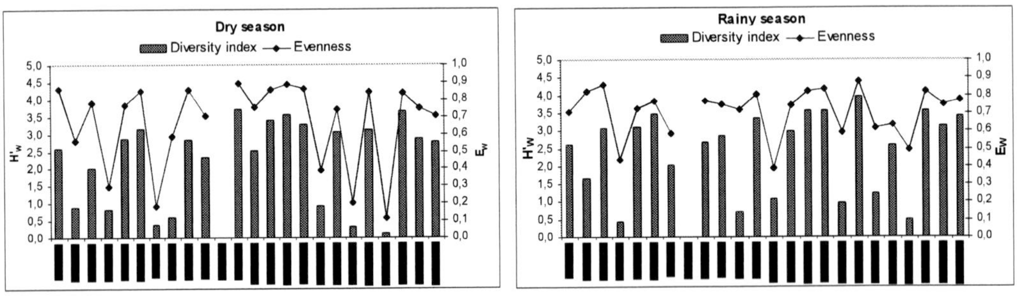

Figure 14. Diversity index in weight (H'$_W$) and E_W in the dry and rainy seasons.

In the case of H'$_W$, the analysis of variance (three-way) (table 15) shows significant differences between zones and habitats revealing the effect of unit of measure in the fish diversity distribution.

Table 15. Analysis of variance: Shannon diversity index (H'$_W$) per year season, zone and habitat
Depended variable: H'$_W$; N= 46, multiple R = 0.938, R^2= 0.879

Source of variations	SQ	gl	MQ	F	p
YEAR SEASON	0.449	1	0.449	1.345	0.259
ZONE	3.808	3	1.269	3.807	0.024
HABITAT	41.495	2	20.748	62.224	0.000
YEAR SEASON*ZONE	0.376	3	0.125	0.376	0.771
YEAR SEASON*HABITAT	0.104	2	0.052	0.156	0.856
ZONE*HABITAT	3.513	6	0.586	1.756	0.155
YEAR SEASON*ZONE*HABITAT	1.920	6	0.320	0.959	0.475
ERROR	7.336	22	0.333		

The result of the HSD Tukey comparison test shows that inside the four zones, the centre habitat differs from the others (p <0.05), with the exception of the Tietê river zone, where the centre habitat differs only from the mouth of tributary (table 16).

Table 16. HSD Tukey Probability test applied to detect differences in H'_W values among zone (1= Fluvial Piracicaba; 2= Fluvial Tietê; 3= Transition Barra Bonita e 4= Lentic Barra Bonita) and habitat (C=Centre; D=Mouth of tributary e L=Shoreline)

zone	habitat	C	D	L	C	D	L	C	D	L	C	D	L
1	C	1.000											
1	D	0.001	1.000										
1	L	0.002	1.000	1.000									
2	C	1.000	0.002	0.004	1.000								
2	D	0.000	1.000	0.998	0.001	1.000							
2	L	0.411	0.275	0.491	0.314	0.093	1.000						
3	C	0.994	0.013	0.034	0.931	0.003	0.964	1.000					
3	D	0.000	0.918	0.739	0.000	0.997	0.008	0.000	1.000				
3	L	0.000	0.998	0.967	0.000	1.000	0.035	0.001	1.000	1.000			
4	C	0.995	0.000	0.000	1.000	0.000	0.051	0.605	0.000	0.000	1.000		
4	D	0.001	1.000	1.000	0.001	1.000	0.234	0.010	0.944	0.999	0.000	1.000	
4	L	0.000	0.998	0.969	0.000	1.000	0.037	0.001	1.000	1.000	0.000	0.999	1.000

Dominance

The dominance index reveals that the values range between a minimum of zero and a maximum of 0.8. The maximum value occurs in the centre habitats in the two sampling periods (figure 15), revealing that few species are adapted to exploit the resources in this habitat.

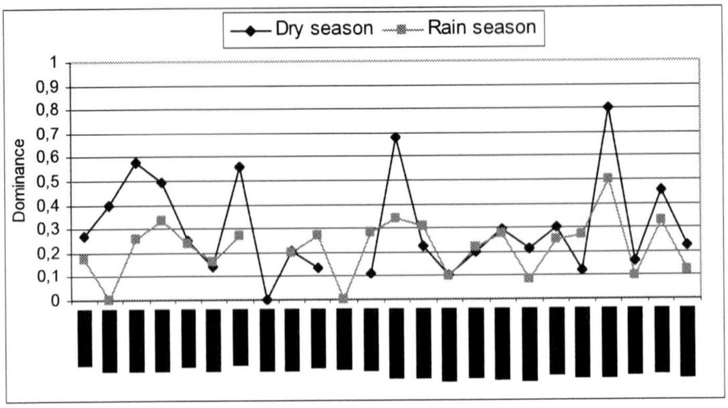

Figure 15. Dominance index in the two sampling seasons. See figure 1 for sampling stations codes.

The analysis of variance (three-way) do not show significant differences between seasons, zones and habitats.

Catch Per Unit Effort (CPUE)

The CPUE in weight ($CPUE_W$) was higher in the rainy season especially at the mouth of tributary stations (figure 16). The maximum value of this period was 103.16 kg/1000m^2/day. In the dry season the maximum value observed was 92.67 kg/1000m^2/day (table 17).

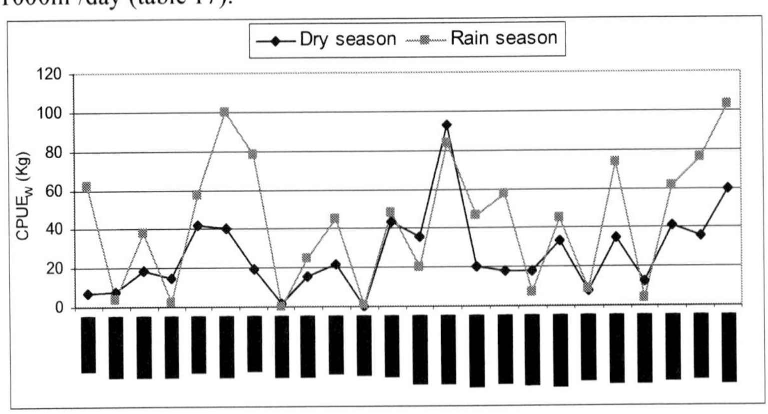

Figure 16. $CPUE_W$ (kg/m^2/day) in the two sampling periods: dry and rainy seasons. See figure 1 for sampling stations codes.

Table 17. Descriptive statistics of $CPUE_W$.

$CPUE_W$	Mean	Median	Minimum	Maximum	Int.Var.	Variance	Std. Dev.
Dry	26.68	19.71	0.00	92.67	92.67	426.17	20.64
Rainy	43.54	45.69	0.00	103.16	103.16	1070.19	32.71
Mean total sample	35.11	33.91	0.48	88.26	87.77	607.21	24.64

Data were $\sqrt{x+0,5}$ transformed to meet the assumption of normality and homocedasticity of variance. The result of the analysis of variance (table 18) show significant differences among sampling periods, zones and habitat. Also the interaction periods*habitats was significant, suggesting that the period of sampling influenced the $CPUE_W$ distributions in the reservoir habitats.

Table 18. Analyses of variance per year season, zone and habitat.
Depended variable: $CPUE_W$; N= 48, multiple R= 0.919, R^2= 0.844

Source of variations	SQ	gl	MQ	F	p
YEAR SEASON	15.300	1	15.300	7.524	0.011
ZONE	25.747	3	8.582	4.220	0.016
HABITAT	171.028	2	85.514	42.052	0.000
YEAR SEASON*ZONE	1.627	3	0.542	0.267	0.849
YEAR SEASON*HABITAT	26.855	2	13.428	6.603	0.005
ZONE*HABITAT	18.652	6	3.109	1.592	0.211
YEAR SEASON*ZONE*HABITAT	4.595	6	0.766	0.377	0.887
ERROR	48.805	24	2.034		

The distribution of the $CPUE_W$ in the two sampling periods (figure 17 and 18) displays the higher values during the rainy season. The lower values in the dry season are in the fluvial zones (Piracicaba and Tietê Rivers) (table 19), indicating that the community is constituted mainly of small body-sized fishes with lighter weight.

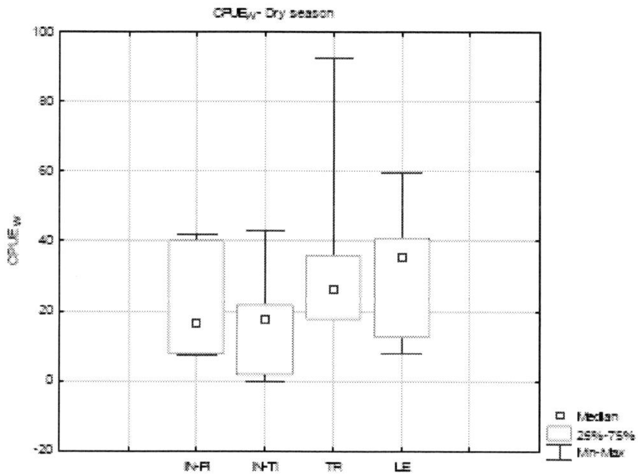

Figure 17. $CPUE_W$ distribution values for reservoir zones in the dry season (FL-Pi=Fluvial Piracicaba; FL-Ti=Fluvial Tietê; TR= Transition; LE=Lentic).

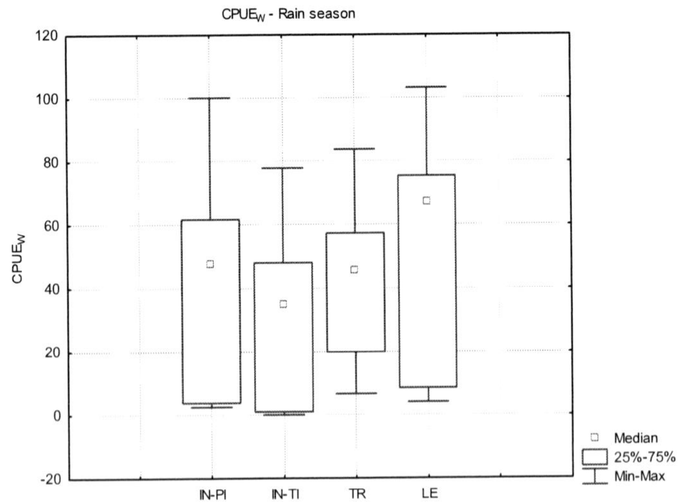

Figure 18. CPUE$_W$ distribution values for reservoir zones in the rainy season (FL-Pi=Fluvial Piracicaba; FL-Ti=Fluvial Tietê; TR= Transition; LE=Lentic).

Table 19. Descriptive statistics of CPUE$_W$ for reservoir zones

Dry	N. cases	Mean	Median	Minimum	Maximum	Variance
FL-PI	6	21.71	16.62	7.27	41.77	238.96
FL-TI	6	16.96	17.64	0.00	43.20	248.55
TR	6	36.15	26.40	17.79	92.67	828.60
LE	6	31.89	35.05	8.07	59.64	361.17
Rainy						
FL-PI	6	43.94	47.69	2.55	100.15	1402.55
FL-TI	6	32.78	34.82	0.00	78.00	915.03
TR	6	43.15	45.69	6.61	83.84	749.95
LE	6	54.30	67.37	4.02	103.16	1576.92

The same considerations can be made for the habitats (figures 19 and 20). In this case, in the rainy season, the mouth of tributary and lateral habitats shows a relevant increase in CPUE$_W$ (table 20), suggesting a transversal migration of the fish from the centre to the shoreline and mouth of tributary habitats.

The species most representative in weight (figure 21) in the dry season were: *P. maculatus, H. littorale* and *L. anisitsi.* In the rainy season the explosion of iliophagous species such as *S. insculpta,* can be observed. Can be noticed also the considerable increase of *P. maculatus, H. littorale, L. anisitsi* and *P. squamosissimus.* Finally, was registered the presence of *M. intermedia* that was captured only in this season and the disappearance of herbivorous such as *A. piracicabae.*

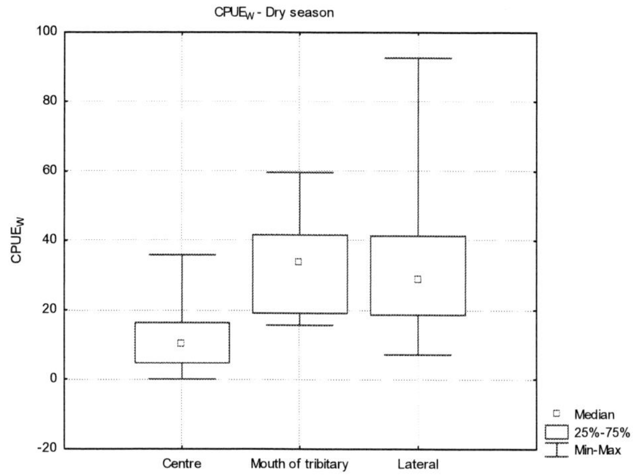

Figure 19. CPUE$_W$ values for habitats in the dry season.

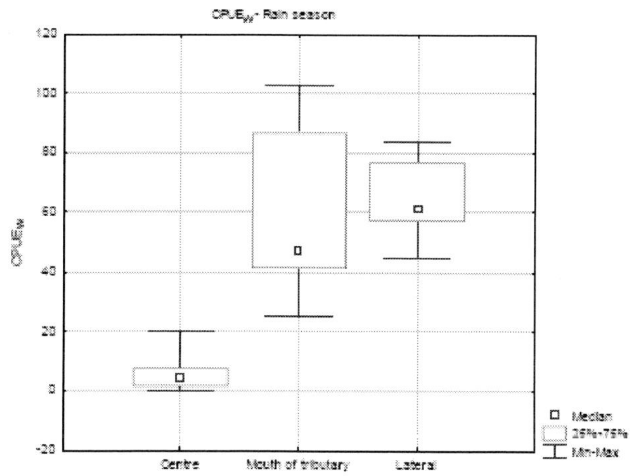

Figure 20. CPUE$_W$ values for habitats in the rainy season.

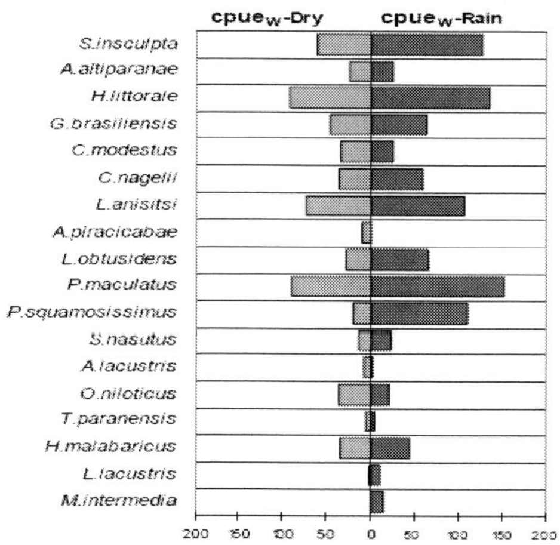

Figure 21. CPUE$_W$ more important species in the dry and rainy seasons.

Table 20. Descriptive statistics of $CPUE_W$ for habitat in the two sampling periods

Dry	N. cases	Mean	Median	Minimum	Maximum	Variance
Centre	8	12.34	10.38	0.00	35.80	127.34
Mouth of tributary	8	33.06	33.76	15.72	59.64	221.11
Lateral	8	34.65	28.62	7.27	92.67	697.67
Rainy						
Centre	8	5.77	3.91	0.00	19.80	39.74
Mouth of tributary	8	59.87	47.25	24.77	103.16	845.90
Lateral	8	65.00	61.64	44.87	83.84	168.90

ANALYSES OF COVARIANCE

The results of the co-variance analyses for the diversity in number (H'_N) select the factor landscape and the co-variables: depth, water temperature and transparency. The regression coefficients at the probability level of p=0.15 are in table 21.

Table 21. Regression coefficients of factor and co-variables selected in the minimum model of exploratory ANCOVA, *backward* with p = 0.15
Depended variable: Diversity in number (H'_N); N= 46, multiple R = 0.802, R^2= 0.643

	Level	Regression coefficients
Constant		1.696
Landscape	1	-0.709
Landscape	2	0.108
Landscape	3	0.101
Depth		-0.107
Temperature		0.044
Transparency		0.529

The minimum model of the covariance analyses is presented in table 22.

Table 22. Results of co-variance analyses, ANCOVA minimum model (*backward,* p = 0.15). Dependent variable: (H'_N)

Source of variation	SQ	gl	MQ	F	p
Landscape	2.094	3	0.698	2.272	0.095
Depth	17.524	1	17.524	57.025	0.000
Temperature	0.982	1	0.982	3.195	0.082
Transparency	1.395	1	1.395	4.539	0.039
Residual	11.985	39	0.307		

In the case of the diversity in weight (H'_W) the covariance analyses selects the followed factors: landscape and macrophytes and the co-variables: depth, transparency and conductibility. The regression coefficients at the probability level of p=0.15 are in table 23.

Table 23. Regression coefficients of factors and co-variables selected in the minimum model of exploratory ANCOVA, *backward* with p = 0.15
Depended variable: Diversity in weight (H'_W); N= 46, multiple R = 0.896, R^2= 0.803

	Level	Regression coefficients
Constant		3.494
Landscape	1	-0.988
Landscape	2	0.152
Landscape	3	0.101
Macrophytes	1	-0.493
Macrophytes	2	0.260
Macrophytes	3	0.095
Depth		-0.123
Transparency		0.466
Conductivity		-0.003

The minimum model for H'_W is presented in table 24.

Table 24. Results of co-variance analyses, ANCOVA minimum model (*backward,* p = 0.15). Dependent variable: H'_W

Source of variation	SQ	gl	MQ	F	P
Landscape	3.315	3	1.105	3.321	0.030
Macrophytes	2.076	3	0.692	2.080	0.120
Depth	11.245	1	11.245	33.806	0.000
Transparency	0.967	1	0.967	2.906	0.097
Conductivity	1.544	1	1.544	4.640	0.038
Residual	11.975	36	0.333		

The selected variables and factor have a great importance in limnology. According to Barbanti *et al.* (1993) the nutrients are used with more effectiveness in few deeps habitats allowing a larger biological production. In complex systems, like reservoirs, the depth varies with the hydropower generation and the level fluctuations have short (daily) and long (annual) periods. The first are related to the slight daily level fluctuations (few centimeters) in general with little importance for the biotic organism, while the second are related to the reservoir flooding and empting phases. In this case the level variation is higher (meters) and the accessibility to food resources, shelter and reproduction areas can be affected, limiting the potential productivity of the biotic system (Cohen and Radomski, 1993). In the case of the Barra Bonita reservoir, the morphology of the surrounding landscape is slightly declining thus the mouth of tributary and shoreline habitats are especially sensitive to the emersion risk.

The temperature influences the kinetics of chemical and biochemical reactions in the water as well as the physics property of the water (i.e. density), controlling the dynamics of oxygen and nutrient transfers in the water body (Marchetti, 1989).

Transparency is an indirect measure of the trophic state, being related with the chlorophyll production (Marchetti, 1989). The Organization for Economic Co-operation and Development (OECD) proposed a probabilistic model for the trophic classification based on the mean annual transparency value. In the case of Barra Bonita reservoir, the model reveals that it has a 55% of probability to be ipereutrophic and 40% to be eutrophic. This classification is confirmed by Freitas (1999) that defines the reservoir as ipereutrophic and for Barbosa *et al.* (1999) that observed the tendency to a progressive increase of eutrophy in the cascade Tietê reservoir system.

Conductibility, especially inform on the alteration in the water mineral composition. Its variability is associated to the quality of the water inflow and can reveal sources of water pollution (Marchetti, 1989; Esteves, 1998). The data analyses for Barra Bonita reservoir shows that the high values found in the dry season at the fluvial zones of Piracicaba and Tietê Rivers, can be related to the organic and inorganic inputs from the urban and industrial areas in its upper basins. In the transition zone a certain variability can be observed in the two periods sampled, while in the lentic zone a stabilization of the values occurs due to the biologic purification process and sedimentation of inorganic and organic particles.

In the case of the factor landscape, the territories around Barra Bonita reservoir are strongly affected by the widespread cultivation of the reed sugar. This type of "industrial" cultivation limits the structural heterogeneity of the surrounding environment decreasing the potential input of allochthonous food resources to the aquatic system.

The macrophytes, on the other hand, have wide distribution in the reservoir and, for this reason, can be considered the most important element of structural complexity in the system.

Trophic Structure

The trophic structure of the fish assemblage of Barra Bonita reservoir considered eight trophic categories as presented in table 25.

The most heterogeneous and abundant category is that of the omnivorous that groups 14 species, mostly belonging to the families of Characidae, Anostomidae, Cichlidae and Pimelodidae. They are species without a specialized diet but that displays a preferential alimentary tendency (such as insectivorous, herbivorous, detritivorous) in relationship to the availability of the food resources in the environment.

The second category (herbivores) is represented by two species (*P. mesopotamicus* e *H. ancistroides*) that feed mainly on superior vegetable remains or epiphytic algae associated to the bottom.

The third and fourth categories include species that feed respectively on fine sediments (Curimatidae and Prochilodontidae families) and grossly detritus (Loricaridae family). Following Hahn *et al.* (1997), we considered these as separate categories due to the different functional roles acted in the food chain. Some authors (Lowe-McConnell, 1999; Bowen, 1983), consider the fish with these types of diets, as the most specialized and the adopted separation justifies their different functions in the ecosystem.

The species belonging to the category of the invertivores have a diet primarily composed of microcrustaceos and secondarily of gastropods and insects. This specialization is typical of fish that prefer habitats without muddy sediment. In the Barra Bonita reservoir this group is represented by the family of Callichthyidae with the only one species *H.littorale*.

The insectivore category is quite wide, with five species all small in size that look for food all around their habitat, exploring: bottom (G. *carapo, Crenicichla sp., Pimelodella sp.*), roots of flooding macrophytes (*H. eques*) and the surface (*T. paranensis*).

In the planktivorous category, only one species was included: *M. intermedia* that feeds mainly on zooplankton. *O. niloticus* is also considered a planktivorous species, but when the food resources are abundant its diet is more heterogeneous and, for this reason, it was put in the omnivorous category (Nakatani *et al.*, 2001).

Table 25. Trophic categories. Number in parentheses after the species name corresponds to the bibliography reference

Trophic category	Species	Authors
Omnivorous	S.jurupari (1), A. fasciatus (2), M.maculatus (3), S. intermedius (4), S.fasciatus (5), S.nasutus (6)	1: Fishbase (2004); Smith et al. (2003); 2: Lopes (1997); 3: Smith et al. (2003); Nomura (1984); 4: Celi-Fedatto-Abelha et al. (2001); 5: Fishbase (2004); 6: Nomura (1984)
	R. quelen (7), P. maculatus (8), A.altiparanae (9)	7: Nakatani (2001); Nomura (1984); Smith et al. (2003); Hahn et al. (1997); 8: Hahn et al (1997); Agostinho et al. (1997); Lima (2000); Araújo (1998) 9: Costa and Braga (1993)
	G. brasiliensis (10)	10: Lopes (1997); Araújo (1998)
	A.schubarti (11)	11: Costa and Braga (1993); Lopes (1997)
	L.obtusidens (12), L.lacustris (13)	12: Hahn et al. (1997); Nomura (1984); 13: Hahn et al. (1997); Fishbase (2004); Smith et al. (2003)
	O.niloticus (14)	14: Nakatani (2001);
Herbivorous	P.mesopotamicus (15); H.ancistroides (22)	15: Hahn et al. (1997); Fishbase (2004); 22: Hahn et al. (1997); Nomura (1984)
Iliophagous	P.lineatus (16), S.insculpta (17), C.modestus (18), C. nagelii (19); A.piracicabae (20)	16: Hahn et al. (1997); 17: Hahn et al. (1997); 18: Hahn et al. (1997); Nomura and Taveira (1979); 19: Hahn et al. (1997); 20: Hahn et al. (1997);
Detritivorous	L.anisitsi (21),	21: Hahn et al. (1997);
Invertivorous	H.littorale (23)	23: Hahn et al. (1997); Signorini (1999)
Insectivorous	G.carapo (24), Crenicichla sp. (25)., Pimelodella sp.(26), H.eques (27), T. paranensis (28)	24: Carneiro (1998); Hahn et al. (1997); 25: Hahn et al. (1997); 26: Hahn et al. (1997); 27: Nomura (1984); 28: Nakatani (2001);
Planktivorous	M. intermedia (29)	29: Costa and Braga (1993)
Carnivorous	P.squamosissimus (30), A.lacustris (31), S.spilopleura (32), S. maculatus (33), S.hilarii (34), H.malabaricus (35)	30: Braga (1998); Smith et al (2003); 31: Hahn et al. (1997); 32: Hahn et al. (1997); Nomura (1984); 33: Fishbase (2004); 34: Nomura (1984); 35: Hahn et al. (1997);

Finally, in the carnivorous category, the species that feed on meat or on organic matter of animal origin were included. These feeding on scales and parts of fins of other fish, such as *S. spilopleura*, are considered, for some authors, as parasites (Hahn *et al.*, 1997, Sazima and Pombal, 1988) while for others this is a strategy of high specialization (Lowe-McConnel, 1999).

The trophic structure of the fish assemblage of Barra Bonita Reservoir (figure 22) shows that the widest category is that of the omnivorous including 40% (14) of the species. The second is that of the carnivorous with 17% (6), followed by the iliophagous and insectivorous, both with 14% (5). The detritivorous, invertivores, and planktivorous categories, which are known as species with highly specialized diets, are the least represented.

The analysis of frequencies in number and weight of trophic structure show a different result (figure 23). The more important category in terms of abundance are those of iliophagous (38%) and omnivorous (28%), that added together, form 66% of the total community. Among the other categories the invertivores represent almost 10% of the

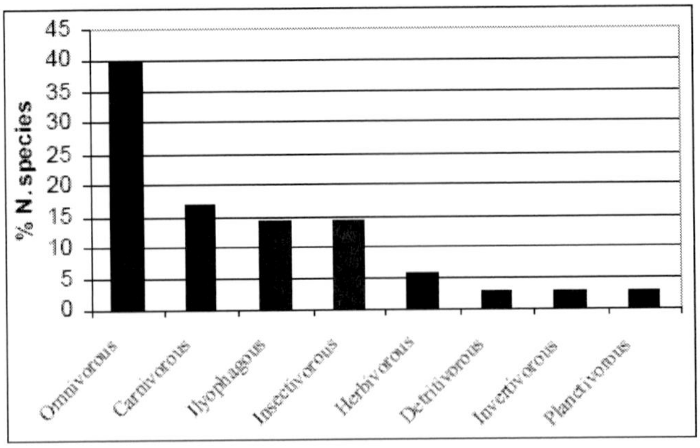

Figure 22. Trophic structure of fish assemblage of Barra Bonita reservoir (literature source).

community abundance. Considering that this category is represented by only one species (*H. littorale*), its contribution can be considered relevant. The carnivorous category, on the other hand, represents only 8.2%.

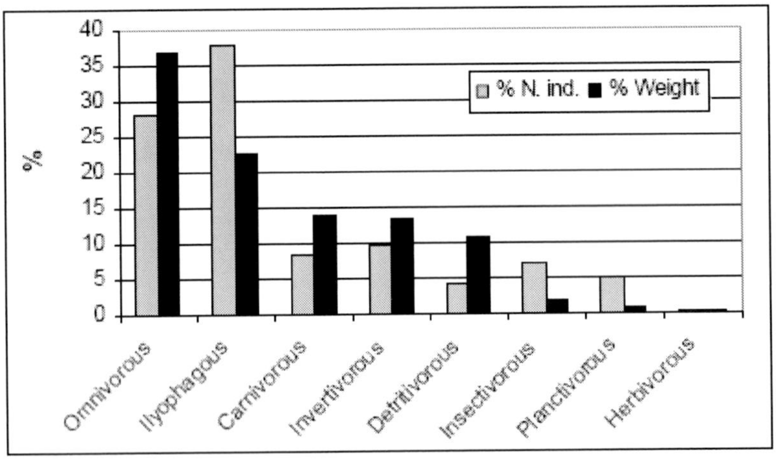

Figure 23. Relative frequencies in number and weight of the trophic categories.

In relationship to the weight, the larger contribution is of the omnivorous (36.6%) and iliophagous (22.5%). In this case, invertivores and carnivorous, each contribute in the measure of 13% and herbivores, insectivorous and planktivorous are the quantitatively less important categories.

Reproductive Structure

This information is available for 80% of the species sampled. The analyses of the reproductive structure shows that the reproductive period spans from September/October to March/April according to Vazzoler and Menezes (1992). These authors, studying the Characiforme family, observed that the increase of water temperature and the beginning of the flooding period are the factors starting the reproduction in the Paraná River basin.

The partial spawn strategy is the most common reproduction type observed among the 15 most abundant species in the Barra Bonita reservoir (figure 24). Only two species, *Leporinus obtusidens* and *Liposarcus anisitsi*, have total spawning. The success of these species in the reservoir can be explained considering that *L. obtusidens* undertakes spawning migrations and *L. anisitsi* exhibits parental care.

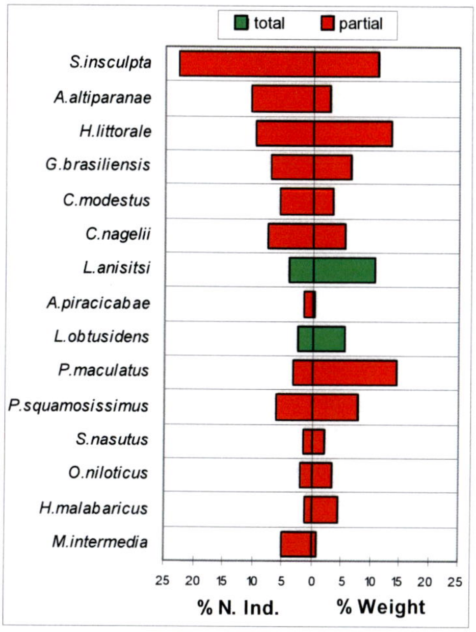

Figure 24. Frequency in number and weight of the most abundant species and its reproductive strategy (total and partial spawning).

Taking the reproductive strategies into account, we observed that 31% of the species belongs to the "non-migratory without parental care" category. The migratory and the parental care categories are also well represented by respectively 26% and 23% of the total species. These observations indicate that in the Barra Bonita reservoir, conditions still exist for the conclusion of the total vital cycle of these species. In the first case, this can be related to the presence of relatively long fluvial tracts and in the second with the reaching of a "maturity stage" of the reservoir environment (Agostinho *et al.*, 1999).

Regarding the resilience (figure 25), the "mean" category is the most common with 46% of the species. In this case the populations duplicate in 1.4-4.4 years. Only two species, *Hoplias malabaricus* and *Liposarcus anisitsi,* belong to the "low" category (duplication time 4.5-14 years) and can, therefore, be considered "vulnerable". On the other hand, 31% of the species show "high" resilience with a duplication time less than 15 months. The resilience information is not available for 17% of the species sampled.

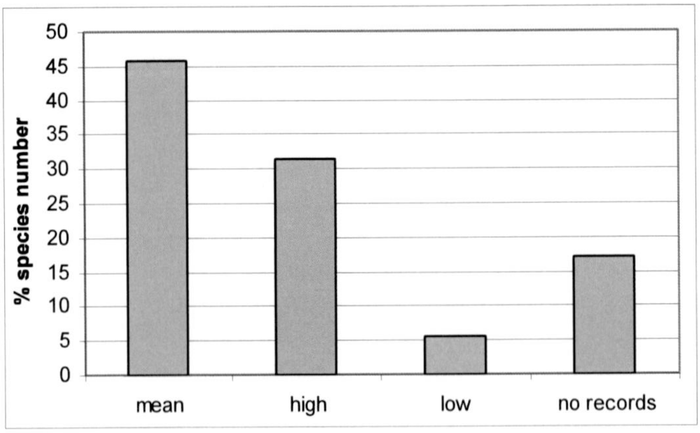

Figure 25. Resilience relative frequencies of the fish assemblage of Barra Bonita reservoir.

The overlap between the reservoir hydraulic cycle (emptying and flooding periods along the year) with the fishes reproductive periods, shows that most species spawn from September to March/April (figure 26). This period can, therefore, be considered critical for the success of reproduction. In relationship to the hydraulic management, the period of September/November corresponds to the emptying final phase when the reservoir water level reaches its minimum. As observed in the preview hydraulic analysis, the emptying phase is abrupt and continuous thus the spawning that occurred in the shallow habitats is subjected to emersion risk with the consequent loss of the reproduction products. In the period December/May the hydraulic flooding phase of the reservoir occurs. In this period, factors of interference on the reproductive success of the species, are also associated to the worst conditions of water quality due to the turbulence and polluted input from the upper portion of catchment basin. In this context, it is known that in tropical and sub-tropical regions, rainy events are frequently related with fish mortality (Straskraba, 1999). In synthesis, the species with a long reproductive period or with special reproductive strategies such as migration or parental care are favored in reservoirs.

Hydraulic cycle of Barra Bonita Reservoir												
Species	Jun.	Jul.	Aug.	Sep.	Out.	Nov.	Dec.	Jan.	Feb.	Mar.	Abr.	May
L. lacustris			x	x	x	x	x	x	x			
L. obtusidens					x	x	x	x	x			
S. intermedius						x	x					
S. fasciatus					x	x	x	x	x	x		
S. nasutus						x	x	x				
A. lacustris				x	x	x	x	x	x	x		
A. altiparanae	x	x				x	x	x	x	x	x	x
A. fasciatus				x	x	x	x	x	x			
A. schubarti							x	x	x	x	x	x
M. intermedia					x	x	x	x	x	x		
S. hilari					x	x	x	x	x			
T. paranensis												
H. eques												
P.mesopotamicus						x	x	x				
S. maculatus												
S. spilopleura					x	x	x	x	x	x	x	
M. maculatus												
C. modestus			x	x	x	x	x	x	x	x	x	x
C. nagelii					x	x	x	x	x			
S. insculpta			x	x	x	x	x	x	x	x	x	
H. malabaricus				x	x	x	x	x	x			
A. piracicabae				x	x	x	x					
P. lineatus						x	x	x	x			
H. littorale						x	x	x	x	x	x	
L. anisitsi						x	x					
H. ancistroides				x	x	x						
P. maculatus					x	x	x	x	x	x		
Pimelodella sp.												
R. quelen												
G. carapo					x	x	x	x	x			
C. haroldoi												
G. brasiliensis				x	x	x	x	x	x			
S. jurupari												
O. niloticus				x	x	x						
P.squamosissimus						x	x	x	x			

= flooding   = empty   = periods when most species spawn

Figure 26. Overlap between the reproductive period and the hydraulic cycle of Barra Bonita Reservoir.

Table 26. Reproductive period and strategy, parental care, migration and resilience of Barra Bonita reservoir fish assemblage (?= no record).

Species	Year months 1	2	3	4	5	6	7	8	9	10	11	12	Reproduction strategy	Parental care	Migration	Resilience	Authors
Leporinus lacustris	x	x						x	x	x	x	x	?	no	no	high	Vazzoler *et al.*, 1997; Fishbase , 2004
Leporinus obtusidens	x	x								x	x	x	total	no	x	media	Vazzoler *et al.*, 1997; Nakatani *et al.*, 2001; Fishbase, 2004
Schizodon intermedius											x	x	?	no	no	mean	Fishbase , 2004
Schizodon fasciatus	x	x	x							x	x	x	total	no	x	mean	Santos, 1980; Nakatani *et al.*, 2001; Fishbase, 2004
Schizodon nasutus	x										x	x	parceled	no	no	mean	Vazzoler and Menezes, 1992; Nakatani *et al.*, 2001; Fishbase , 2004
Acestrorhynchus lacustris	x	x	x						x	x	x	x	?	no	no	?	Vazzoler *et al.*, 1997
Astyanax altiparanae	x	x	x	x	x	x	x				x	x	parceled	x	(small) x	high	Vazzoler and Menezes, 1992; Braga,1999; Gennari and Braga, 1996; Nakatani *et al.*, 2001; Fishbase , 2004

Table 26. (Continued)

	Year months																
Species	1	2	3	4	5	6	7	8	9	10	11	12	Reproduction strategy	Parental care	Migration	Resilience	Authors
Astyanax fasciatus	x	x							x	x	x	x	total	no	no	high	Vazzoler and Menezes, 1992; Fishbase, 2004
Astyanax schubarti	x	x	x	x	x							x	total	no	(smalls) x	high	Vazzoler and Menezes, 1992; Gennari and Braga, 1996; Fishbase, 2004
Moenkhausia intermedia	x	x	x							x	x	x	parceled	no	no	high	Vazzoler and Menezes, 1992; Gennari and Braga, 1996; Fishbase, 2004
Salminus hilarii	x	x								x	x	x	total	no	x	mean	Vazzoler and Menezes, 1992; Nakatani *et al*., 2001; Fishbase , 2004
Triportheus paranensis													?			high	Fishbase, 2004
Hyphessobrycon eques													parceled	no	no	high	Nakatani *et al*., 2001; Fishbase , 2004
Piaractus mesopotamicus	x										x	x	total	no	x	mean	Vazzoler and Menezes, 1992; Nakatani *et al*., 2001; Fishbase , 2004

Table 26. (Continued).

	Year months																
Species	1	2	3	4	5	6	7	8	9	10	11	12	Reproduction strategy	Parental care	Migration	Resilience	Authors
Serrasalmus maculatus													parceled			high	Fishbase , 2004
Serrasalmus spilopleura	x	x	x	x						x	x	x	parceled	x	no	high	Vazzoler and Menezes, 1992; Vazzoler *et al.*, 1997; Agostinho, 2003; Nakatani *et al.*, 2001; Fishbase, 2004
Metynnis maculatus													?			high	Fishbase , 2004
Cyphocharax modestus	x	x	x	x	x			x	x	x	x	x	parceled	no	no	?	Vazzoler and Menezes, 1992; Vazzoler *et al.*, 1997;
Cyphocharax nagelii	x	x								x	x	x	parceled	no	no	?	Vazzoler *et al.*, 1997;
Steindachnerina insculpta	x	x	x	x				x	x	x	x	x	parceled	no	no	?	Vazzoler and Menezes, 1992;
Hoplias malabaricus	x	x							x	x	x	x	parceled	x (nest)	no	low	Vazzoler and Menezes, 1992; Nakatani *et al.*, 2001; Fishbase , 2004
	Year months																
Espécie	1	2	3	4	5	6	7	8	9	10	11	12	Reproduction strategy	Parental care	Migration	Resilience	Authors
Apareidon piracicabae									x	x	x	x	parceled	no	no	mean	Nakatani *et al.*, 2001; Fishbase, 2004

Table 26. (Continued).

Species	Year months												Reproduction strategy	Parental care	Migration	Resilience	Authors
	1	2	3	4	5	6	7	8	9	10	11	12					
Prochilodus lineatus	x	x									x	x	total	no	x	mean	Nakatani *et al.*, 2001; Fishbase, 2004
Hoplosternum littorale	x	x	x	x							x	x	parceled	x (nest) no		mean	Nakatani *et al.*, 2001; Fishbase, 2004
Liposarcus anisitsi											x	x	total	x	no	low	Fishbase, 2004
Hypostomus ancistroides									x	x	x		total	no	no	mean	Fishbase, 2004
Pimelodus maculatus	x	x	x							x	x	x	parceled	no	x	mean	Fishbase, 2004
Pimelodella sp.													?			?	
Rhamdia quelen													parceled	no	x	?	Nakatani *et al.*, 2001;
Gymnotus carapo	x	x								x	x	x	parceled	x	no	mean	Nakatani *et al.*, 2001; Fishbase, 2004
Crenicichla haroldoi													?	no	no	high	Fishbase, 2004
Geophagus brasiliensis	x	x							x	x	x	x	parceled	x	no	mean	Fishbase, 2004
Satanoperca jurupari													parceled	no	no	mean	Fishbase, 2004
Oreochromys niloticus									x	x	x		parceled	x (nest) no		mean	Nakatani *et al.*, 2001; Fishbase, 2004
Plagioscion squamosissimus	x	x									x	x	parceled	no	no	mean	Braga, 1997; Nakatani *et al.*, 2001; Fishbase, 2004

Historical Analyses of Barra Bonita Reservoir Fish Assemblage

Most of the studies on the fish assemblage of Barra Bonita reservoir can be classified in three categories: 1. monitoring of professional fishing; 2. biology of the species; 3. ecology of the fish assemblage. The most representative among the first category belong to the work of Torloni *et al.* (1993) that analyzed the daily fishing production of five groups of professional fishermen, located in the Tietê and Piracicaba rivers zones. The collection of data is from August 1989 to December 1991. The authors relate the presence of 39 species observing that in comparison with the other cascade reservoirs Barra Bonita present the largest productivity justified by the presence of spawning areas located in the tributary rivers (CESP, 1996).

The second category, groups studies on the biology of the principal species commercially exploited by fishing such as: *Moenkhausia intermedia* (Braga and Gennari, 1991; Costa and Braga, 1993), *Astyanax altiparanae* (= *A. bimaculatus*) (Costa and Braga, 1993; Gennari and Braga, 1996; Braga, 1999), *Astyanax schubarti* (Costa and Braga, 1993; Gennari and Braga, 1996) and *Plagioscion squamosissimus* (Braga, 1997, 1998; Castro, 1994). In these cases the issue was reproduction, diet or mortality.

The works of Castro (1997), Barrella (1997), Freitas (1999) and Smith (2004) belong to the third category where the objective was to characterize the fish assemblage in terms of diversity and abundance. The sampling methodology was the same: experimental fishing with gillnets of different mesh size between opposite knots.

The frequency of collection was monthly for Castro (1997): from August 1992 to July 1993; bimonthly for Freitas (1999): from February 1998 to December 1998) and in two period of the year (dry and rainy season) in the case of Barrella (1997) and Smith (2004). Observing the distribution of the sampling stations for each author, it can be seen that most of the works considered only the Tietê and Piracicaba rivers zones (figure 27). The remaining reservoir, as well as the smaller tributaries, was only partially sampled. Considering that the Barra Bonita reservoir is over 40 years old, it is also observed that the works on its ichthyofauna, are recent (dating between 1992 and 2004), and show a notable lack of information in relationship to the previous and initial period (filling phase) of the impoundment.

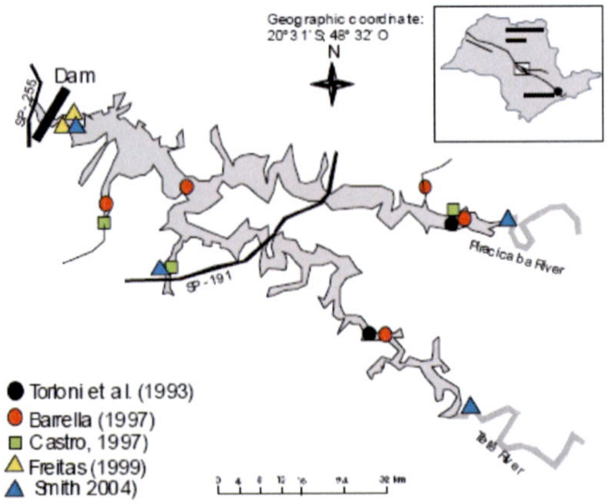

Figure 27. Location of the sampling stations for the historical analyses.

The list of species in the Barra Bonita reservoir, with the inclusion of those collected in the present work is presented in table 27. This reveals the presence of 18 families belonging to six orders, for a total of 64 species. This value is in accordance with Smith (2004) that, analyzing the historical registers at the Zoological Museum of São Paulo, listed the presence of 60 species.

Analyzing the total species recorded in Barra Bonita reservoir, it appears that eleven are common to all the lists. Twenty three species presents the Dajoz constancy index higher than 50% allowing us to consider this group as "constant" in the Barra Bonita reservoir fish assemblage. In this group one species (*Iheringichthys labrosus)* were not recorded in the samples of the present work.

The group of accessory species is formed of 20 fishes, including migratory species with low rate of capture. Twelve species of this group, mainly Anostomidae, did not occur in our samples.

The last group is composed of 21 accidental species, which include rare, difficult to capture, of recent introduction (i.e. *Metynnis maculatus* and *Satanoperca jurupari*) or fish with doubtful taxonomic classification. Five species of this group occurred only in our samplings (*Schizodon fasciatus, Hyphessobrycon eques, Serrasalmus maculatus, Metymnis maculatus* and *Liposarcus anisitsi*). The fact that three of these are alochthonous (*Schizodon fasciatus, Metymnis maculatus* and *Liposarcus anisitsi*) probably indicates that their introduction happened in recent times.

The historical analyses of fish records, not considering Torloni *et al.* (1993) list, show the transformations occurred in the Barra Bonitra reservoir fish assemblage in the last decade (figure 28). In particular, the largest transformations can be noticed in the group of the accessory species where 10 species (83%), mainly belonging to Anostomidae and Characidae families, passed to the accidental category (*Leporinus cf.paranensis, Leporinus octofasciatus, Leporellus striatus, Galeocharax knerii, Salminus maxillosus, Characidium fasciatum, Piaractus mesopotamicus, Rhinelepis strigosa, Pimelodella sp. e Cichla sp)*. Finally seven species, of this last category, completely disappeared from all the lists. In this group we find the most important fishing species belonging to the families of Pimelodidae (*Pseudoplatistoma corruscans* and *Pinirampus pirinampu*), Doradidae (*Rhinodoras dorbignyi*), Ciprinidae (*Cyprinus carpio*) and Loricariidae (*Hypostomus regani* and *Loricaria vetula*), indicate that the impoundment impact principally affected the big migratory species (Pimelodidae and Douradidae families). The transformation of the Tietê River into a cascade of reservoirs, does not allow the natural displacement of these species in the basin, thus their occurrence is strongly dependent on human action (such as artificial repopulation and fishing regulation). The absence of the carp (originating from Asia) can be related to the probable failure of the adaptation process in the reservoir or to the lack of regular introductions. The lack of registrations of these species, in the reservoir experimental fishing, might indicates the probable their local extinction.

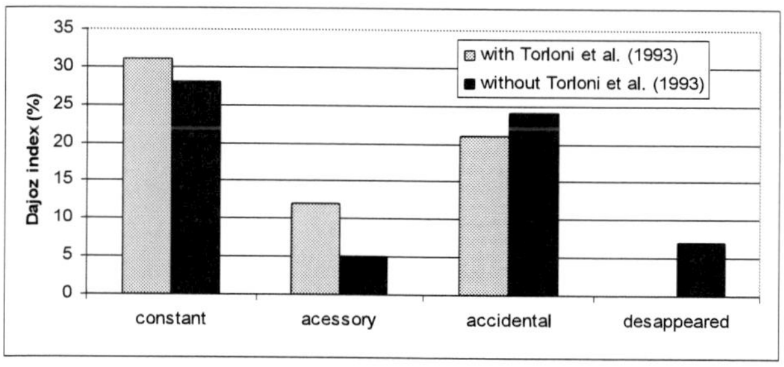

Figure 28. Dajoz index of the historical records of Barra Bonita fish assemblage.

Table 27. Historical records of fish species in Barra Bonita Reservoir (in bracket is reported the name species with doubt identification) and Dajoz index; ([1])= introduced species

Ordem	Family	Espécie	Torloni *et al.* (1993)	Castro (1997)	Barrella (1997)	Freitas (1999)	Smith (2004)	Petesse (2006)	Dajoz Índex
Characiformes	Anostomidae	Schizodon nasutus	x	x	x	x	x	x	100
Characiformes	Characidae	Astyanax altiparanae (=A.bimaculatus)	x	x	x	x	x	x	100
Characiformes	Characidae	Moenkhausia intermedia	x (M. dichroura)	x	x	x	x	x	100
Characiformes	Characidae	Serrasalmus spilopleura	x	x	x	x	x	x	100
Characiformes	Curimatidae	Cyphocharax modestus	x	x	x	x	x	x	100
Characiformes	Curimatidae	Cyphocharax nagelii	x	x	x	x	x	x	100
Characiformes	Curimatidae	Steindachnerina insculpta	x	x	x	x	x	x	100
Characiformes	Prochilodontidae	Prochilodus lineatus	x	x	x	x	x	x	100
Siluriformes	Callichthyidae	Hoplosternum littorale[1]	x	x	x	x	x	x	100
Siluriformes	Pimelodidae	Pimelodus maculatus	x	x	x	x	x	x	100
Perciformes	Scianidae	Plagioscion squamosissimus[1]	x	x	x	x	x	x	100
Characiformes	Anostomidae	Leporinus obtusidens	x		x	x	x	x	83
Perciformes	Cichlidae	Geophagus brasiliensis		x	x	x	x	x	83
Siluriformes	Loricariidae	Hypostomus ancistroides		x (Hypostomus sp.)	x	x	x (Hypostomus sp.)	x	83
Characiformes	Erythrinidae	Hoplias malabaricus	x	x	x	x		x	83
Characiformes	Acestrorhynchidae	Acestrorhynchus lacustris	x	x	x		x	x	83
Characiformes	Parodontidae	Apareidon piracicabae	x	x	x		x	x	83
Characiformes	Characidae	Astyanax schubarti	x	x	x		x (Astyanax sp.)	x	83
Characiformes	Anostomidae	Leporinus lacustris	x	x	x		x	x	83
Characiformes	Characidae	Triportheus signatus [1]	x (T. angulatus)	x (Triportheus sp.)	x (T. angulatus)		x (T.signatus)	x (T. paranensis)	83
Characiformes	Characidae	Astyanax fasciatus		x	x		x	x	67
Characiformes	Characidae	Salminus hilari		x		x	x	x	67
Siluriformes	Pimelodidae	Iheringichthys labrosus	x	x	x	x			67

Table 27. (Continued).

Ordem	Family	Espécie	Torloni *et al.* (1993)	Castro (1997)	Barrella (1997)	Freitas (1999)	Smith (2004)	Petesse (2006)	Dajoz Índex
Characiformes	Anostomidae	Leporinus elongatus		x	x (Leporinus sp.)		x		50
Characiformes	Anostomidae	Schizodon intermedius		x			x	x	50
Gymnotiformes	Gymnotidae	Gymnotus carapo		x	x			x	50
Perciformes	Cichlidae	Satanoperca jurupari[1]			x (Geophagus sp.)		x	x	50
Perciformes	Cichlidae	Oreochromys niloticus[1]				x	x (T. rendalli)	x	50
Characiformes	Anostomidae	Leporinus friderici	x	x	x				50
Siluriformes	Heptapteridae	Rhamdia quelen	x	x				x	50
Perciformes	Cichlidae	Crenicichla sp.	x (C. britskii)	x				x (C. haroldoi)	50
Siluriformes	Loricariidae	Hypostomus strigaticeps			x	x			33
Gymnotiformes	Sternopygidae	Eigenmannia sp.		x		x (S. macrurus)			33
Characiformes	Anostomidae	Leporinus cf.paranensis	x		x				33
Characiformes	Anostomidae	Leporinus octofasciatus	x		x				33
Characiformes	Anostomidae	Leporellus striatus	x		x				33
Characiformes	Characidae	Galeocharax knerii	x	x					33
Characiformes	Characidae	Salminus maxillosus	x		x				33
Characiformes	Characidae	Characidium fasciatum	x		x				33
Characiformes	Characidae	Piaractus mesopotamicus	x					x	33
Siluriformes	Loricariidae	Rhinelepis strigosa	x		x				33
Siluriformes	Heptapteridae	Pimelodella sp.	x (P.gracilis)					x	33
Perciformes	Cichlidae	Cichla sp.[1]	x (C. ocellaris)			x (C. monoculus)			33
Characiformes	Anostomidae	Schizodon fasciatus[1]						x	17
Characiformes	Anostomidae	Schizodon altiparanae		x					17

Table 27. (Continued).

Ordem	Family	Espécie	Torloni et al. (1993)	Castro (1997)	Barrella (1997)	Freitas (1999)	Smith (2004)	Petesse (2006)	Dajoz Índex
Characiformes	Acestrorhynchidae	Oligosarcus pintoi		x					17
Characiformes	Characidae	Roeboides paranensis			x				17
Characiformes	Characidae	Hyphessobrycon eques						x	17
Characiformes	Characidae	Serrasalmus maculatus						x	17
Characiformes	Characidae	Metymnis maculatus[1]						x	17
Characiformes	Erythrinidae	Hoplias lacerdae				x			17
Characiformes	Parodontidae	Apareidon affinis		x					17
Siluriformes	Loricariidae	Liposarcus anisitsi[1]						x	17
Siluriformes	Loricariidae	Hypostomus variipictus		x					17
Siluriformes	Loricariidae	Hypostomus tietensis		x					17
Symbranchiformes	Symbranchidae	Symbranchus marmoratus					x		17
Characiformes	Anostomidae	Leporellus vittatus	x						17
Cypriniformes	Cyprinidae	Cyprinus carpio[1]	x						17
Siluriformes	Loricariidae	Hypostomus regani	x						17
Siluriformes	Loricariidae	Loricaria vetula	x						17
Siluriformes	Doradidae	Rhinodoras dorbignyi	x						17
Siluriformes	Pimelodidae	Pseudoplatistoma corruscans	x						17
Siluriformes	Pimelodidae	Pinirampus pirinampu	x						17
			39	35	34	23	26	35	

CONCLUSION

Hydraulic Management

From the analyses of the morphologic characteristics, rain regimen and hydraulic management of Barra Bonita reservoir, it can be seen that the reservoir is characterized by a great vulnerability, due to the high sinuosity of shoreline that increases the tendency to nutrient accumulation. The retention time (RT=69 day) shows a rapid hydraulic dynamic of the water but the tendency of yearly reduction of RT, is symptomatic of reservoir aging due to loss of usable volume caused by the sediment filling. The mean yearly maximum range (YMR) is superior to the considered reasonable range of 2.5-4 m, endangering the fish reproduction areas in the reservoir. In this context, Clark et al. (1998) observed that little level fluctuations limit the reproductive success of the Centrarquidae in the Brownlee reservoir (Idaho-Oregon).

The characterization of rain regimen of the area shows the presence of two periods: dry (April – September) and rainy (October – March). Related with this, the hydraulic management of reservoir presents a flooding phase from December to May and an emptying phase from June to November.

The comparison of river inflows and dam discharge shows the alteration in the natural hydraulic cycle. In particular, the attenuation of flooding and drying periods can be observed. The flooding period is restricted to January and February while the dry period is attenuated maintaining the discharge from April to November quite constant. These observations enable us to define as potentially critical, for the fish fauna, the periods from December to February and from September to November due to the rainy period and to the emptying phase respectively.

Fish Assemblage

The fish assemblage of the Barra Bonita reservoir belongs to the Characiformes (66 %) and Siluriformes (17%) orders, in accordance with Lowe-McConnel (1999) on the ichthyofauna evolution in the Neotropical region. Most of the species are sedentary and well adapted to the reservoir environment. In the Piracicaba river zone 97% of the species were found, while in the Tietê river zone only 58%. This difference, also observed by Castro (1997), shows the worst condition of the Tietê River strongly degraded by the industrial and domestic sewage coming from the metropolitan region of the Great São Paulo (Barrella and Petrere, 2003).

The diversity index (H'_N and H'_W) shows that the fish assemblage is sufficiently diversified with homogeneous values between the two sampling periods. This is in accordance with Castro (1997) and confirmed by the analyses of variance that reveals significant differences only among habitats, in the case of H'_N, and among zones and habitats in the case of H'_W. The centre habitat, in particular, is characterized by the minor diversity index values (H'_N and H'_W), indicating that the available resources in this habitat can be explored by few species, depending on their adaptation to the limiting conditions here present (low dissolved oxygen, thermal stratification). The H'_W revealed that in the Tietê River zone the centre and shoreline habitats are dominated by small body-sized individuals that contribute little to the total biomass. In this context, Freitas (1999), studying the effect of the artificial reef on the fish diversity, observed the tendency to reduction of diversity (H'_N and H'_W) and species richness with the increase of depth. The author verified that the greatest values of diversity were found at 5 m from the shoreline.

These observations show that in reservoirs, the seasonality effect on the diversity of fish assemblages is reduced and put in evidence the importance of shoreline and mouth of tributary habitats for the diversity maintenance.

The effect of seasonality, on the other hand, is revealed by the analyses of $CPUE_W$. In this case, Agostinho *et al.* (2007) observed that the shoreline zone of the reservoir is occupied by non migratory and small body-sized fishes, while the pelagic zone, which is the main part in the reservoirs, is the less explored by the South American ichthyofauna. This is also the case of Barra Bonita reservoir, where we recognize a clear transversal gradient showing the increase of $CPUE_W$ from the centre to the mouth of tributary and reservoir shoreline, especially during the rainy season. This result can be interpreted thinking that during the summer the level of the reservoir begins to rise, and alimentary sources, not available in the winter, are now more abundant and accessible. Also in this period the reproduction of most of the species occurs, thus the transversal migrations to the mouth of tributary and reservoir shoreline are favored. This is related with the fact that those are important nursery and shelter areas for the juvenile fishes and with the worst environmental conditions of the deep centre areas of the reservoir. In the summer, the reservoir hypolimnion is exposed to the risk of deoxygenation, thermal stratification and to the increase of organic and inorganic matter carried in by the strongest water floods of the summer rains.

The analyses of covariance reveal that the unit of measure of diversity (number or weight) influences the selection of the environmental variables. In the case of H'_N, the factor landscape and the covariates depth, transparency and temperature are selected. For the diversity in weight (H'_W) the analyses selected the factors landscape and macrophytes and the covariates depth, transparency and conductivity. These results show that the landscape and macrophytes are the most important element for the maintenance of diversity in reservoir. In particular, the landscape is related to the complexity of the environment around the impoundment area emphasizing the functional contribution of allochthonous organic matter to the aquatic ecosystem.

Barrella *et al.* (2000) and Smith *et al.* (2003), in the case of macrophytes, reveal how these increase the structural complexity of the environment acting as nursery and shelter for the ichthyofauna. For this reason, the macrophytes banks are attracting large-sized fish too, mainly predators and specialists in the hunting strategy of "wait - attacks". In this context, Thomaz (2005) points out that the management of macrophytes is one of the gears especially used to increase fishing production and to promote the biodiversity conservation in the reservoirs.

The negative correlation of diversity with depth shows that the shoreline and mouth of tributary habitats are the most explored by the ichthyofauna. This is in accordance with Craig (2005), for which the diversity and the fish production is reduced due to the decline of the extension of the ecotone water/earth. In this context, Barrella *et al.* (2000) assert that in reservoirs, the water level fluctuation and the deforestation of the riparian forests contributes to the erosion of the margin, structural simplification of the aquatic environment and reduction of the alimentary contributions of terrestrial origin.

Trophic Structure

The data analyses display the predominance of the omnivorous and iliophagous categories which, together, constitute 66% of the abundance and 59% of the weight of the fish community. The eutrophy condition of Barra Bonita reservoir (Barbosa *et al.*, 1999), added to the structural environmental simplification and to the high stress related with the hydraulic management, promotes the increase of the small-sized species and the loss or reduction of those that are more demanding or with a specialized diet. Karr (1981) established that a community with less than 20% of omnivorous individuals can be considered "good"; on the contrary, environments where the percentage is higher than 45% are considered "strongly degraded" (Araújo, 1998). In the case of Barra Bonita reservoir the omnivorous percentage is 28%, therefore it can be classified in an intermediate position between the two.

Pereira *et al.* (2002) justifies the predominance, in the reservoirs of Medium and Low Tietê, of generalist species with the need to adapt themselves to the lentic conditions of the impoundment due to the presence of qualitatively less abundant alimentary resources in comparison with the fluvial environment.

The absence of predominantly planktivorous species in the Neotropical reservoirs can be explained with the proliferation of poisonous algae (Cianophycea and Dinoflageladae species) and oxygen deficit (Ribeiro-Neto *et al.*, 2004). This is also the case of Barra Bonita reservoir where periodic death of fish associated with blooms of Cianophycea species can be observed (Barrella, 1997).

Considering the results of trophic structure, in the Barra Bonita reservoir the fish assemblage seems unbalanced with the predominance of the more opportunistic trophic categories and the rarity of specialized groups. In this context, the consistent presence of the iliophagous (Curimatidae family) can be interpreted as a signal of environmental aging due to the accumulation of organic sediment on the reservoir bottom.

Reproductive Structure

The fish assemblage in Barra Bonita Reservoir is, at present, constituted mainly of species with mean-high resilience that seems well adapted to the conditions and environmental stress of the reservoir. Granado-Lorencio (1991) shows that in an unstable environment like the reservoirs, fish communities do not depend upon the availability of food resources, but they are the result of their own biology, emphasizing the importance of special reproductive strategies like partial spawning (parceled), parental care and migration, in the performance of the species. In Barra Bonita reservoir the most common type of reproduction is the parceled. This strategy is particularly appropriate in reservoirs where the cycles of flooding and emptying are considerably different from the fluvial natural cycle. In particular, only the species with great resilience capacity and reproductive compensation such as: *S. insculpta, C. nagelii, C. modestus, P. squamosissimus* and *M. intermedia* (Vazzoler and Menezes, 1992; Gennari and Braga, 1996; Vazzoler *et al.*, 1997; Braga, 1997; Nakatani *et al.*, 2001) and those with parental care such as: *A. altiparanae, H. littorale, G. brasiliensis, H. malabaricus, L. anisitsi* and, *O. niloticus* (Vazzoler and Menezes, 1992; Gennari and Braga, 1996; Braga, 1999; Nakatani *et al.*, 2001; Fishbase, 2004) are well succeeded in the reservoir. In this context, tilapia (*O. niloticus*) seems to be favored by the impoundment as observed by Craig (2005) that found a positive correlation between water level fluctuation and tilapia production. The success of tilapia in reservoirs is also justified by the great rusticity and capacity of adaptability of the species, with its wide alimentary flexibility, parental care and gregarious juvenile behavior.

FISH ASSEMBLAGE MODIFICATIONS

The temporal changes observed in the fish assemblage of Barra Bonita Reservoir in the last 15 years, can be related to environmental aging. This is the result of a natural process that is much faster in reservoirs than in natural lacustrine habitats. Aging is still poorly studied, because most reservoirs are still "young" in Brazil and because frequently there is not sufficient data on fish assemblage before or in the filling phase of the reservoir (Agostinho *et al.,* 1999), which is the case of Barra Bonita Reservoir.

Regarding the events resulting from the aging process, Fernando and Holcik (1991) affirm that the ichthyofauna in reservoirs is initially recruited from riverine species pre-adapted to the lacustrine conditions. Subsequently a population depletion is noticed, which mainly affects the large body-sized fish, usually migratory, with long life span and low reproductive potential (k-strategists species). Opportunistic species, on the contrary, characterized by small body-sized, sedentary, with high reproductive potential (r-strategists), short life and for whose food availability in the reservoir is high, become rapidly dominant (Agostinho, 1995; Lowe-McConnel, 1999).

With regard to Neotropical reservoirs, Agostinho *et al.* (1999) highlights several events, whose main points are summarized in table 28. It is clear that the major changes in Barra Bonita Reservoir may have happened 40 years ago, during the filling phase of the reservoir. The absence of surveys in the area before and in the filling phase, do not allow us a complete analysis of the changes that took place. However, considering the predominance of opportunistic (omnivorous and iliophagous) over specialist (herbivorous/planktivorous) categories, the predominance of small body-sized species with high reproductive compensation, fast growth and well adapted to environmental variability, the reservoir can be placed in an intermediate phase between colonization and aging. This also explains the absence of most Anostomidae species in the surveys of the 90´s. Despite the fact that in the last 15 years a general stability in species richness was observed, an alteration of the fish assemblage composition is evident, showing that the dynamics of the community is faster in this environment.

Table 28. Aging process in reservoirs (modify from Agostinho *et al.*, 1999)

	Closing stage	Colonization stage	Aging stage
Definition	It is the period between the reservoir's closing and the ordinary operating conditions. It usually varies between 8 and 80 days.	From the end of closing stage to the age signs.	It is the period characterized by strong environmental degradation and simplification of living communities.
Abiotic environment	Increase in the retention time, increase of nutrients concentration, presence of anoxia and thermal stratification, increase in transparence and turbulence reduction.	The abiotic characteristics are strongly related to the input of water quality and to reservoir's hydraulic management. The range of level fluctuation affects the general environmental productivity. Transparency reduction.	Nutrients accumulation, high rate of sedimentation, decrease in retention time, habitat deterioration particularly of the shoreline habitat.
Biotic environment	Phytoplankton increase	Gradual increase of macrophytes. Phytoplankton increase	Phytoplankton blooms, with presence of poisonous algae; evidence of negative effect of eutrophy (deoxygenation, thermal stratification); reduction of benthic organisms.
Fish	Two phases: Fast river species diffusion in the reservoir at all the habitat types. High capture rate. Moving to the reservoir shoreline and mouth of tributary areas due to the formation of anoxic strata in the pelagic habitat.	High production and diversity in shoreline reservoir habitat; absence of pre-adapted species to colonization of pelagic habitat (these organisms require of special morphological and behaviorist adaptations). The community is dominated by sedentary and medium-bodied species. The diet is constituted mainly of material of autochthonous origin. In general, detritivorous and iliophagous species are rare. On the other hand, herbivorous and zooplanktivorous increase. The shoreline habitat is colonized by generalist species and with wide habitat tolerance. Reduction of migratory species (especially big migratory); Increase of species with parental care and more elaborate reproductive strategies.	Reduction of predators. Increase in small-bodied and opportunistic species. Increase in short life cycle species, fast growth and reproductive compensation. Dominance of adapted species to high turbidity conditions and low oxygen concentration.

References

Aes-Tietê. Home page: available from: *www.aestietê.com.br*, 12 December 2003.

Agostinho AA; Gomes LC; Pelicice F. Ecologia e manejo de recursos pesqueiros em reservatórios do Brasil. *Maringá: EDUEM;* 2007.

Agostinho, AA. Considerações sobre a atuação do setor elétrico na preservação da fauna aquática e dos recursos pesqueiros. In: Comase/Eletrobras editor. Seminário sobre fauna aquática e o setor elétrico brasileiro. 4. Rio de Janeiro: 1995; p. 8-19.

Agostinho, AA; Ferreira, HJ; Gomes, LC; Bini, LM; Agostinho, CS. Composição, abundância e distribuição espaço-temporal da ictiofauna. In: Vazzoler, AE, Agostinho, AA, Hahn, NS editors. A planície de inundação do Alto Rio Paraná: Aspectos físicos, biológicos e socioeconômicos. *Maringá: EDUEM;* 1997; p.179-208.

Agostinho, AA; Ferreira, HJJ. Peixes da Bacia do Alto Paraná. In: Lowe-Mcconnell, RH editor. Estudos ecológicos de comunidades de peixes tropicais. São Paulo: *EdUSP;* 1999.

Agostinho, AA; Miranda, LE; Bini, LM; Gomes, LC; Thomaz, S; Suzuki, HI. Patterns of colonization in Neotropical reservoirs and prognoses on aging. In: Tundisi, JG, Straškraba, M editors. Theoretical Reservoir Ecology and its applications. São Carlos: *IIE;* 1999; p.227-265.

Araújo, FG. Adaptação do Índice de integridade biótica usando a comunidade de peixes para o rio Paraíba do Sul. *Rev. Brasil. Biol. = Braz. J. Biol.,* 1998, v. 58, n. 4, p. 547-558.

Barbanti, L; Calderoli, A; De Bernardi, R; Giussani, G; Guilizzoni, P. Acque lacustri. In: Marchetti, R editor. Ecologia applicata. Milano: *CittáStudi;* 1993; p.220-262.

Barbosa, FAR; Padisak, J; Espindola, ELG; Borics, G; Rocha, O. The cascading reservoir continuum concept (CRCC) and its application to the river Tietê - Basin, São Paulo State, Brazil. In: Tundisi, JG; Straškraba, M editors. Theoretical Reservoir Ecology and its applications. São Carlos: *IIE;* 1999; p.425-438.

Barrella, W. Alterações das comunidades de peixes nas bacias dos rios Tietê e Paranapanema (SP), devido à poluição e ao represamento. 1997. Doctoral thesis, Instituto de Biociências, *UNESP,* Rio Claro, Brazil, p.115.

Barrella, W; Petrere, M Jr.; Smith, W; Montag, LFA. As relações entre as mata ciliares, os rios e os peixes. In: Rodrigues, RR; Leitão, HF editoes. Matas ciliares: conservação e recuperação. São Paulo: *EdUSP;* 2000.

Barrella, W; Petrere, M Jr. Fish community alteration due to pollution and damming in Tietê and Paranapanema rivers (Brazil). *River Res. Applic.,* 2003, n.19, p. 59-76.

Bowen, SH. Detritivory in neotropical fish communities. *Environ. Biol. Fishes,* 1983, v.9, n.2, p. 137-144.

Braga, FMS; Gennari, O. Estudos sobre a fecundidade, desova e mortalidade natural de M.intermedia (Characidae, Tetragonopterinae) na represa de Barra Bonita, Rio Piracicaba, SP. *Naturalia,* 1991, v.16, p. 55-68.

Braga, FMS. Biologia reprodutiva de P. squamosissimus (Teleostei, Scianidae) na represa de Barra Bonita, Rio Piracicaba (SP). *Revista UNIMAR,* 1997, v.19, n.2, p. 447-460.

Braga, B; Rocha, O; Tundisi, JG. Dams and the environment: the Brazilian experience. *Water Resource Development*, 1998, v.14, n. 2; p. 127-140.

Braga, FMS. Alimentação de Plagioscion squamosissimus (Osteichthys, Scianidae) no reservatório de Barra Bonita, Estado de São Paulo. *Iheringia, Ser.Zool.,* 1998, n. 84, p. 11-19.

Braga, FMS. Idade, crescimento e taxas de mortalidade de A. bimaculatus (Characidae, Tetragonopterinae) na Represa de Barra Bonita (SP). *Naturalia,* 1999, n. 24, p. 239-250.

Britski, HA. A fauna de peixes brasileiros de água doce e o represamento de rios. In: Comase/Eletrobras editor. Seminário sobre fauna aquática e o setor elétrico brasileiro. Rio de Janeiro; 1; 1994; p. 23-30.

Britski, HA; Silimon, KZ; Lopes, BS. Peixes do Pantanal. Manual de identificação. Brasília; *Embrapa*; 1999.

Camussi, A; Moller, F; Ottaviano, E; Sari Gorla, M. Metodi statistici per la sperimentazione biologica. Bologna: Zanichelli; 1986.

Carneiro, S. Alimentação da Tuvira, Gymnotus "aff" carapo em duas represas do Estado de São Pulo (Osteichthyes, Gymnotidae). 1998, M.Sc. thesis. Instituto de Biociências UNESP, Rio Claro, Brazil.

Castro, ACL. Ictiofauna do reservatorio de Barra Bonita – SP: aspectos ecologicos da comunidade e dinâmica populacional da corvina Plagioscion squamosissimus (Heckel, 1940) (Pices, Perciformes). 1994; Doctoral thesis. Esc. Eng. São Carlos, Universidade de São Paulo (USP), São Carlos, Brazil.

Castro, ACL. Aspectos ecológicos da comunidade ictiofaunística do reservatório de Barra Bonita, SP. *Rev. Brasil. Biol. = Braz. J. Biol.,* 1997, v. 57, n.4, p. 665-676.

Castro, RM; Arcifa, MS. Comunidades de peixes de reservatórios no sul do Brasil. Rev. *Brasil. Biol. = Braz. J. Biol.,* 1987, v. 47, n.4, p. 493-500.

Celi Fedatto Abelha, M; Agostinho, AA; Goulart, E. Plasticidade trófica em peixes de água doce. *Acta Scientiarum,* 2001, v. 23, n. 2, p. 425-434.

CESP. Santa Maria da Serra Empreendimento hídrico. Relatório ambiental preliminare. São Paulo, CESP , 1996.

CETESB. Relatório de qualidade das águas interiores de estado de São Paulo. São Paulo, *CETESB,* 2001. v.2, 138 p.

Chatterjee, S; Price, B. Regression analysis by example. New York, 2 edition: John Wiley and Sons, 1991.

Clark, ME; Rose, KA; Chandler, JA; Richter, TJ; Orth, DJ; Winkle, W Van. Simulating smallmouth bass reproductive success in reservoirs subject to water level fluctuations. *Environmental Biology of Fishes,* 1998, n. 51, p.161-174.

Cohen, Y; Radomsky, P. Water level regulations and fisheries in Rainy Lake and the Namakan reservoir. *Can. J. Fish. Aquat. Sci.,* 1993, v. 50, p. 1934-1945.

Costa FES; Braga, FMS. Estudo da alimentação natural de A. bimaculatus, A. schubarti e M. intermedia (Characidae, Tetragonopterinae) na represa de Barra Bonita, rio Piracicaba (SP). *Revista UNIMAR,* 1993, v.15, n.2, p. 117-134.

Craig, JF. Large dams and freshwater fish biodiversity. Available from: *www.dams.org.* 12 February 2005.

Dajoz, R. Ecologia geral. São Paulo: Editoria Vozes Limitada; 1973.

Draper, N.R.; Smith, H. Applied regression analysis. 2 edition. New York; John Wiley and Sons; 1981.

Esteves, FA. Fundamentos de limnologia. Rio de Janeiro; Interciência, 1998.

Fausch, KD; Lyons, J; Karr, JR; Angermeier, PL. Fish communities as indicators of environmental degradation. American Fisheries Society Symposium. 1990. n. 8, p.123-144.

Fernando, CH; Holcik, J. Fish in reservoir. *Int. Revue ges. Hydrobiol.*, 1991, n. 76, p. 149-167.

FISHBASE, Home page, *www.fishbase.org.*, 20 December 2004.

Freitas, CEC. O efeito de recifes artificiais sobre as associações de peixes do Rio Tietê, na área de influência do Reservatório de Barra Bonita (Estado de São Paulo - Brasil). 1999. Doctoral thesis - Escola de Engenharia.de São Carlos, da Universidade de São Paulo (USP), São Carlos.

Gennari, O; Braga, FMS. Fecundidade e desova de A. bimaculatus e A. schubarti (Characidae, Tetragonopterinae) na represa de Barra Bonita, Rio Piracicaba (SP). Revista UNIMAR, 1996, n.18 (2), p. 241-254.

Granado-Lorencio, C. Fish communities of Spanish reservoir system: a non-deterministic approach. *Verh. Internat. Verein. Limnol.*, 1991, n. 24, p.2428-2431.

Hahn, NS; Andrian, I; Fugi, R; Almeida, VL. Ecologia trófica. In: Vazzoler, AE; Agostinho, AA; Hahn, NS editors. A planície de inundação do Alto Rio Paraná. Aspecto físicos, biológicos e socioeconômicos. Maringá, EDUEM ;1997.

Jackson, D; Marmulla, G. The influence of dams on river fisheries. 12 June 2005. Available from: www.dams.org.

Karr, JR. Assessment of biotic integrity using fish communities. *Fisheries,* 1981, v. 6, n.6, p. 21-27.

Kennedy, RH. Reservoir design and operation: limnological implications and management opportunities. In: Tundisi JG; Straškraba M editors. Theoretical reservoir ecology and its applications. São Carlos: 2 edition; 1999, 592p.

Krebs, CJ. Ecological methodology. 2. edition., Menlo Park CA: Benjamin Cummings CA, 1998.

Lima, SE. Dieta e condição de Pimelodus maculatus (Osteichthyes, Pimelodidae) nos rios Piracicaba e Mogi-Guaçu, SP. 2000, M.Sc. thesis, Instituto de Biociências UNESP, Rio Claro, Brazil.

Lopes, FR. Análise do conteúdo estomacal e aspectos da anatomia externa e do trato digestivo de peixes da represa do Lobo, Itirapina, SP. 1997, Monografy bacharelado, Instituto de Biociências, UNESP, Rio Claro. Brazil.

Lowe-McConnell, RH. Fish communities in tropical freshwaters. New York: Longman Inc.; 1975.

Lowe-McConnell, RH. Estudos ecológicos de comunidades de peixes tropicais. São Paulo, EdUsp, 1999.

McGarigal, K; CUSHMAN, S; STAFFORD, S. Multivariate statistics for wildlife and ecology research. New York, Springer-Verlag, 2000.

Musick, JA. Criteria to define extinction risk in marine fishes. *Fisheries.* 1999, v. 24, n.12, p. 6-14.

Nakatani, K; Agostinho, AA; Baumgartner, G; Bialetzki, A; Sanches, PV; Makrakis, M; Pavanelli, C. Ovos e larvas de peixes de água doce: desenvolvimento e manual de identificação. Maringá: EDUEM, 2001.

Nomura, H. Dicionário dos peixes do Brasil. Brasília: Editerra, 1984.

Nomura, H; Taveira, ACD. Biologia do sagüiru, Curimatus elegans Steindachner, 1874 do rio Mogi Guaçu, São Paulo (Osteichthyes, Curimatidae). *Rev. Brás. Biol.,* 1979, v. 39, n. 2, p. 331-339.

Okada EK; Agostinho AA; Petrere M Jr; Penczak T. Factor affecting fish diversity and abundance in drying pond and lagoons in the upper Paraná River basin, Brazil. *Ecohydrology and Hydrobiology.* 2003, 3 (1), p. 97-110.

Okada, EK; Agostinho, AA; Petrere, M Jr. Catch and effort data and the management of the commercial fisheries of Itaipu reservoir in the upper Paraná river, Brasil. In: Cowx IG editor. Stock assessment in inland fisheries. Bodmin: Blackwell Science; 1996.

Pereira, CC; Smith, WS; Espindola, EL; Rocha, O. Alterações tróficas nas espécies de peixes em decorrência da construção de reservatórios em cascata no Médio e Baixo rio Tietê. In: Programa de Pós-graduação em ciências da Engenharia Ambiental: Recursos hidroenergéticos: usos, impactos e planejamento integrado. São Carlos: RiMa Editora, 2002, p.29-41.

Petrere, M Jr. Fisheries in large tropical reservoirs in South America. *Lakes and Reservoirs: Research and Management,* 1996, n. 2, p. 111-133.

Petrere, M Jr; Agostinho, AA; Okada, EK; Júlio, HF Jr. Review of the Fisheries in the Brazilian Portion of the Paraná/Pantanal basin. In: Cowx, IG Editor. Management and ecology of lake and reservoir fisheries. *Bodmin: Blackwell Science,* 2002. p.123-143.

Reis, RE; Kullander, SO; Ferraris, CJ Jr. Check list of the freshwater fishes of South and Central América. Porto Alegre: EDIPUCRS, 2003.

Ribeiro-Neto BF; Ishikawa-Ferreira L; Höfling JC. A comunidade de peixes do reservatório de Salto Grande. In: Espindola ELG; Leite MA; Dornfeld DC Editors: Reservatório de Salto Grande (Americana, SP): caracterização, impactos e propostas de manejo. São Carlos: RiMa; 2004; 484p.

Ricker, WE. Computation and interpretation of biological statistics fish populations. Bulletin of the fisheries research board of Canada. Department of the environmental fisheries and marine service. Ottawa, 1975.

Rodriguez, MA; Lewis, WM Jr. Regulation and stability in fish assemblages of neotropical floodplain lakes. *Oecologia,* 1994, n. 99, p. 166-180.

Santos, G. Estudo da reprodução e hábitos reprodutivos de Schizodon fasciatus, Rhytiodus microlepis e Rhytiodus argenteofuscus (Pices, Anostomidae) do lago Janauacá. *Acta Amazônica.* 1980, n. 102, p.391-400.

Sazima, I; Pombal, JP Jr. Mutilação de nadadeiras em acarás, Geophagus brasiliensis, por piranhas, Serrasalmus spilopleura. *Rev. Brasil. Biol= Braz. J. Biol.,* 1988, v. 48, n.3, p. 477-483.

Signorini, CE. Alimentação de Hoplosternum littorale Hancock (Callichthyidae, Osteichthyes) do rio Piracicaba e rio Corumbataí, Estado de São Paulo. 1999. M.Sc. thesis. Instituto de Biociências, UNESP, Rio Claro, Brazil, 1999.

SIGRH. Home page, available from: www.sigrh.sp.gov.br., 10 December, 2003.

Smith SW. A importância dos tributaries, da fragmentação artificial de rios e da introdução de espécies na comunidade de peixes dos reservatórios do médio e baixo Tietê (São Paulo). São Carlos, Doctoral Theses; Esc. Eng. São Carlos, Universidade de São Paulo (USP), Brazil, 2004, 294p.

Smith, WS; Espindola, EL; Pereira, CC; Rocha, O. Impactos dos reservatórios do médio e baixo rio Tietê (SP) na composição das espécies de peixes e na atividade de pesca. In:

Programa de Pós-graduação em ciências da Engenharia Ambiental: Recursos hidroenergéticos: usos, impactos e planejamento integrado. São Carlos: RiMa Editora, 2002, p.57-72.

Smith, WS; Pereira, CC; Espindola, EL; Rocha, O. A importância da zona litoral para a disponibilidade de recursos alimentares à comunidade de peixes em reservatórios. In: Henry R (Org.). Ecótonos nas interfaces dos ecossistemas aquáticos. São Carlos: RiMa Editora; 2003; p.233-248.

Straškraba, M. Retention time as a key variable of reservoir limnology. In: Tundisi, JG; Straškraba, M. editors. Theoretical reservoir ecology and its applications. São Carlos: IIE; 1999. p.385-410.

Tejerina-Garro, F; Fortin, R; Rodriguez, MA. Fish community structure in relation to environmental variation in floodplain lakes of the Araguaia Rivers, Amazon Basin. *Environmental Biology of Fishes*, 1998, n. 51, p. 399-410.

Thomaz, SM. Fatores que afetam a distribuição e o desenvolvimento de macrófitas aquáticas em reservatórios: uma análise em diferentes escalas. In: Nogueira MG; Henry R; Jorcin A.(Org). Ecologia de reservatórios: impactos potenciais, ações de manejo e sistemas em cascada. São Carlos: RIMA Ed.; 2005; 472 p.

Torloni, CE; Corrêa, A; Carvalho, AA Jr; Santos, J J; Gonçalves, J; Gereto, E; Cruz, JA; Moreira, JA; Silva, DC; Deus, EF; Ferreira, A. Produção pesqueira e composição das capturas em reservatórios sob concessão da CESP nos rios Tietê, Paraná e Grande no período de 1986-1991. Serie: Produção Pesqueira, 1993, n. 1.

Torloni, CE; Santos, J J; Carvalho, AA Jr; Corrêa, A. A pescada do Piauí Plagioscion squamosissimus (Heckel, 1840) (Osteichthyes, Perciformes) nos reservatórios da companhia energética de São Paulo - CESP. Serie: Pesquisa e desenvolvimento, 1993a, n. 84.

Tundisi JG; Matsumura-Tundisi T; Rocha O. Theoretical basis for reservoir management. In: Tundisi JG; Straškraba M editors. Theoretical reservoir ecology and its applications. São Carlos: IIE; 1999. p.505-528.

Tundisi, JG; Matsumura-Tundisi, T. Limnology and eutrophication of Barra Bonita reservoir, São Paulo State, Southern Brazil. *Arch. Hydrobiol. Beih. Ergebn. Limnol.*, 1990, n. 33, p. 661-676.

Valentin, JM. Ecologia numérica. Uma introdução à análise multivariada de dados ecológicos. Rio de Janeiro: Interciência; 2000, 200p.

Vazzoler, AE; Menezes, NA. Síntese de conhecimentos sobre o comportamento reprodutivo dos Characiformes da América do Sul (Teleostei, Ostariophysi). *Ver. Brasil. Biol.*, 1992, v. 52, n.4, p. 627-640.

Vazzoler, AE; Suzuki, HI; Marques, EE; Lizama, M. Primeira maturação gonadal, períodos e áreas de reprodução. In: Vazzoler, AE; Agostinho, AA; Hahn, NS editors. A planície de inundação do Alto rio Paraná. Aspectos físicos, biológicos e socioeconômicos. Maringá: EDUEM, 1997. p. 249-266.

In: Lake Pollution Research Progress
Editors: F. R. Miranda and L. M. Bernard
ISBN: 978-1-60692-106-7

Chapter 7

POLLUTION IMPACTS AND KEY ANTHROPOGENICALLY-INDUCED PROCESSES IN LAKES OF RUSSIAN EURO-ARCTIC REGION

Tatyana I. Moiseenko*
Water Problem Institute of RAS, Russia

ABSTRACT

Basing on the results of studies in the industrially developed Arctic region - Russian Kola - there are discussed the features anthropogenically-induced processes in lakes under pollution impacts: distribution and fate of metals, features of acidification under influence of local and trans-boundary acid deposition, eutrophication under domestic sewage pollution. Main attention is paid to pollution from copper-nickel smelters and mining industry. The estimation of metal accumulation in lake sediments in a historical retrospective is given. The influence of a combination of pollutants on water quality and aquatic ecosystems is discussed. Critical levels of an integrated toxicity index for Arctic waters, the excess of which creates a risk of fish pathology, are indicated.

* Tatyana I. Moiseenko: tatyana@aqua.laser.ru

INTRODUCTION

The Euro-Arctic region (Barents Sea drainage basin) is a part of the Planet where the territory is covered by a very great number of lakes. The high provision of the Arctic regions with water till recently has not caused a trouble about the state of the latter. At the same time, intensive development of the rich deposits of mineral recourses and trans-boundary transmissions of pollutants lead to a rapid disturbance in the fragile environmental equilibrium already in many urbanized and industrial Arctic regions, which leads to qualitative depletion of the water resources.

The Russian part of the Euro-Arctic region - Kola North is most densely populated and industrially developed. The spectrum of anthropogenic impacts on the lakes is wide: mining, metallurgy, refineries and chemical industries, nuclear power plants, etc. For more than 70 years the lakes are used as a source of technical and drinking water supply, for recreation, tourism and fishery. Considerable industrial expansion in the early 1900s resulted in building of large industrial enterprises in the region. Industrial development of copper- nickel, rich apatite-nephelinite and iron deposits in the Kola Peninsula began in the 1930s. Large amounts of pollutants entered the lakes between 1940 and 1990; the catchment areas were also polluted by airborne contaminants. The main pollutants were heavy metals (predominantly nickel and copper), sulphates, chlorides and nutrients. The main pollution occurred in the northern and central parts of the region. Since 1990, as a result of the economic crisis in Russia, the anthropogenic pressure on the lakes has decreased. The recent recovery of the economy goes on simultaneously with technological modernization and stricter controls of pollutant emissions into the lakes and the atmosphere. The map of the region and location of basic industries in it is shown in figure 1. This region can serve as a model region for understanding key anthropogenically-inducted processes in lakes on local, regional and global scales.

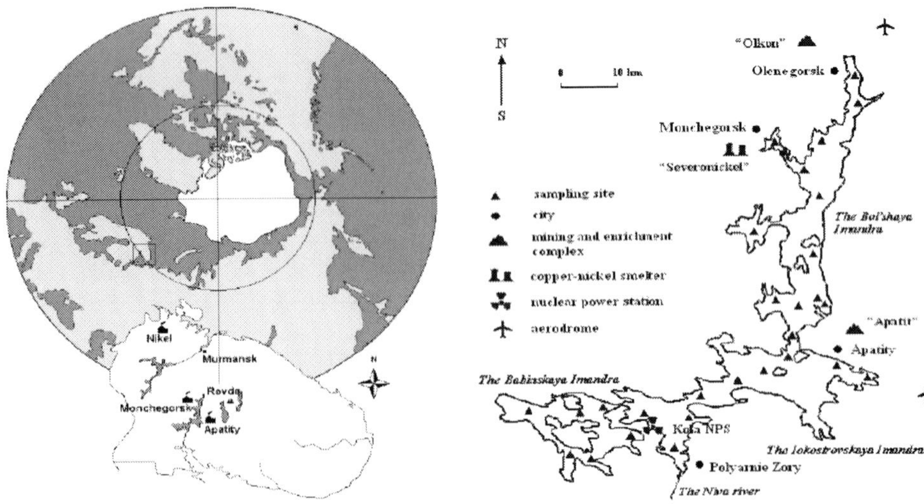

Figure 1. A map of Kola Peninsula within Arctic zone and Lakes Imandra showing the distribution, cities and industrial enterprises on its catchment ("Olkon" complex specialises in iron ore, "Apatit" complex specialises in apatite-nepheline ore).

Our main objectives were as follows:

- to determine reference conditions of water chemistry for the Arctic climatic zone through a retrospective analysis of the published information and own data on unpolluted lakes;
- to analyze pollution and changes in the water chemistry of the lake in the Euro-Arctic region caused by waste waters and airborne pollution from the Kola smelters and minings;
- to characterize key processes in the lake: pollution by heavy metals, acidification and eutrophication.

A considerable amount of works has been done on the lakes. The given paper is based on an analytical review of the related published results and also of more than 30-year investigations of the author in this region (Vereshagin, 1930; Rikhter,1934; Krokhin, Semenovich, 1940; Moiseenko, 1994, 1999, Moiseenko *et al.*, 1996; 1999; 2001a,b; 2006; 2007; Moiseenko, Yakovlev, 1990; Moiseenko, Kudryvtseva, 2002; Dauvalter *et al.*, 2000; 2001; Antopogenic Modification.., 2002). Detailed descriptions of the research methods have been already given in the literature and, therefore, they are not repeated here. In this review attention is focused on the main parameters of water chemistry and sediments indicators. Although much information is available, but there has been no continuous long-term monitoring of the lakes and, therefore, this paper is based on discontinuous information.

We note that the exploitation of mineral resources in all the Arctic regions has increased in recent years, which makes the case study of the Kola lakes important. It is a useful case study in the environment management, particularly for the purpose of avoidance of negative impacts from the Arctic mineral resources exploitation.

Reference Conditions and Vulnerability of Euro-Arctic Lakes

Climatic conditions in the Arctic region determine a number of specific features of the water chemical composition formation there, making the waters vulnerable to anthropogenic impacts (Atlas..., 1971; Year-book dates..., 1961- 2005).

Recharge of lakes and rivers is greatly determined by precipitation - to 75-90% of the annual runoff are provided by spring flood and summer-autumn rainfalls. Accumulation of precipitation in the snow cover takes place during a long winter (6-8 months), rapidly penetrating during a short period of spring flood into drainage basins.

During snow melting the topsoil remains in the frozen state, so its upper layer is actually impermeable during the entire snow melting period. The weak development of vegetation and thin soil cover provide a high drainage of falling precipitation in summer. Predominance of precipitation amount over evaporation and slow mineralization of organic matter results in availability of a large number of surface small-sized logged lakes with a high content of humus and natural acid waters or so-called "Wetland ponds".

Formation of the surface runoff in the conditions of excess wetting causes a low water mineralization and oligotrophic character of the lakes due to that the bedrocks are leached lowly. Quaternary rocks are intensively washed out, and the topsoil cover is thin. Low average annual air temperatures weaken the processes of water erosion, resulting, thus, in low water mineralization. Lack of development of the topsoil cover makes the geochemical composition of the underlying rocks determinant in formation of salt composition.

In conditions of low mineralization the migrating ability of pollutants is high, their cycling in water bodies is longer, ionic equilibrium is unstable, and toxic effects on aquatic inhabitants in lowly-mineralized waters are much higher. The ecosystems are characterized by simplified food webs, low biological diversity, and rapid transfer of material through trophic layers. This can result in rapid migration of pollutants through food webs and, hence, a severe damage to ecosystems.

Thus, in the Kola North predominantly oligitrophic, fresh and ultra-fresh waters are being formed. For large lakes (with an area of over 100 km^2) the reference conditions of water chemistry parameters are very similar (table 1). These lakes are typically ultra-fresh and oligotrophic with low concentrations of suspended material (0,7 – 1.0 mg/l), microelements (<1 μg/l), and nutrients. The concentration of total phosphorus is less than 2 μg/l; phosphates during the vegetation period are practically completely utilized in the production processes.

Table 1. Reference condition of water chemistry of Kola large lakes

Variable	Imandra square =880km^2 Deep max =64m, average =13,4m	Umbozero square =313km^2 Deep max=115m, average =32.5m	Lovozero Square =200km^2 Deep max=35m, average =5,7m
pH	6.4-7.2	6.4-7.2	6.4-7.2
O_2 ,%saturation	60-100	60-100	60-100
Σ ions, mg/l	20-30	15-20	20-30
Ca, mg/l	1.6-4.0	1.4-1.8	1.2-2.2
Mg, mg/l	0.5-1.3	0.4-1.8	0.6-1.6
Na + K, mg/l	2.5-7.5-	2.2.-3.2	3.5-7.6
HCO_3, mg/l	13-18	5-9	10-15
SO_4, mg/l	1.0-3.0	0.5- 1.0	1.0-2.0
Cl, mg/l	1.4-1.8	1.0 -1.5	1.0 -1.5
PO_4, μgP/l	0-8	0-2	0-18
P_{tot}, μg/l	2-10	0-3	5-12
NH_4, μgN/l	4-8	4-8	10-12
NO_3, μgN/l	0-10	0-10	10-35
N_{tot}, μg/l	10-100	10-50	10-100
Si, mg/l	0.3-0.8	0.3-0.6	0.3-0.8
COD, mgO/l	3-6	2-3	4-10
Ni, μg/l	≤1	≤1	≤1
Cu, μg/l	≤1	≤1	1-2
Sr, μg/l	10-30	40-60	30-50
Al, μg/l	10-20	10-20	30-50
Fe, μg/l	≤15	6-20	70-120
Mn, μg/l	≤5	1-3	8-22
Zn, μg/l	≤2	≤2	≤2

This table is compiled from the following dates: Voronikhin(1935); Poretskij *et al.* (1934); - Krokhin and Semenovich (1940); Moiseenko and Yakovlev(1990).

High saturation of waters by oxygen (up to the bottom) is due to mountainous ice-free rivers flowing into the lakes. Water transparency is about 8 m (Rikhter, 1934; Krokhin, Semenovich, 1940; Moiseenko, Yakovlev, 1990).

For small lakes (with an area of 0.4 to 10 km^2) the water chemical composition has a high variability (Henriksen *et al.*, 1998). In mountainous areas the lakes are characterized by low contents of organic carbon, salts and nutrients. However, the water in many lakes has a high colour degree and high contents of dissolved organic carbon (DOC). In swamped areas the high saturation of water with humic acids indicates to lake dystrophy. The distribution of main water chemistry parameters of small lakes is given in table 2. Judging from the above-mentioned data, it can be concluded that naturally the major part of lakes are characterized as oligotrophic and dystrophic.

Table 2. Distribution of the main parameters of water chemistry (weighted cumulative frequency) of Kola lakes

Parameters	2.5%	10%	25%	50%	75%	90%	97.5%
pH	4,49	4,80	5,92	6,45	6,79	7,11	7,34
(Ca + Mg + Na + K), μeq/l	28	51	95	172	277	391	1019
Alkalinity, μeq/l	<0	<0	32	79	170	266	558
SO_4, μeq/l	1	10	20	36	65	106	352
DOC, μeq/l	2,22	3,51	4,86	7,62	11,82	17,36	26,9
N_{tot} μg/l мкг/л	66	94	135	201	299	478	848
P_{tot} μg/l	1	2	3	6	10	20	49

Anthropogenic Impacts

The largest influence on the state of lakes in the Russian part of the Euro-Arctic region is exerted from mining and metallurgical enterprises, hydro- and heat-power electrostation, as well as from the (concentrated in this region) objects of nuclear power station, objects of tourism, etc. The basic types of impact are schematically shown in figure 2. The industrial objects are distributed very unevenly within the region. The basic industries are concentrated in the central part (drainage area of Lake Imandra) and the north-west (drainage area of the Patso-Ioki River). At the same time, there are widely spread the territories that are actually not touched by human activity. Therefore, in order to understand the key processes in the lakes, impact zones of heavy pollution are distinguished: north-western, central, buffer zones, and reference territories.

The main industrial potential is concentrated in *the impact zones*. Deposits rich in useful minerals are found on the Kola North (copper- nickel, apatite-nepheline, iron, manganese ore) served as the basis for developing big mining industries. In 1930-40s a start was given to mining and enrichment of the apatite–nepheline ores; the "Apatit" industrial complex was built. In the 1940-50s – "Severonickel" and "Pechenganikel" enterprises were built for nickel extraction. In the 1950-60s – Olenegorsk mining operations started, producing iron-manganese ores. In the 1974 – Nuclear Power Station were built. Basic types of impact upon the lakes within these zones are as follows:

- waste waters of metallurgic and mining and processing manufactures containing heavy metals, oil products, phenols, fluorine, fine-disperse suspensions, products of ore flotation, etc.;
- pollution of atmospheric air (oxides of sulphur and nitrogen, benz(a)pyrene, nickel, mercury, carbon fluoride, aluminium, strontium, radionuclides, dust, oil products and others);
- tail depositories, dumps of stripped rocks, sludges, discharge of untreated waste waters;
- pollution of ground and surface waters (organic matter, oil products, heavy metals, flotation reagents, suspended substances, sulphates, chlorides);
- pollution of lands (abandon machinery, unsanctioned landfills), possibly – radioactive contamination.

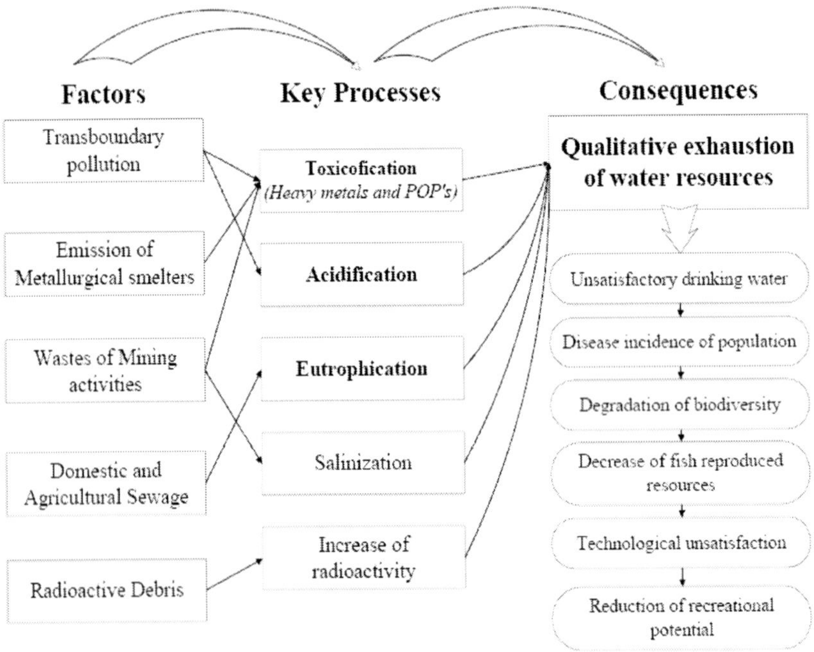

Figure 2. Anthropogenic impacts on lakes of Euro-Arctic region and ecological consequences.

Over many years these anthropogenic loads, which started in 1940s and reached maximum in 1980s, caused serious changes in water chemistry of lakes. The influx of multi-contaminants into the lake led to an increase in salt concentrations, a change in the water ionic composition, a decrease in transparence; pH moved towards alkaline condition as compared with the reference condition. Technogenic sulfates in the ionic composition of waters dominated over hydro-carbonates. The pollution from the metallurgical enterprises has increased the content of heavy metals and especially that of nickel, copper and zinc (table 3). Dynamics of sewage water influx into Lake Imandra from the copper-nickel enterprises and nickel concentrations in the lake during different periods, as indicator pollution in historical retrospective are shown in figure 3.

Table 3. Example of water chemistry change of Kola lakes in impact and buffer zones. Average value and limits of the contents.

Variables	Sewage Kola cupper-nickel smelter	Sewage Kola apatite-nephelin mining (case study Imandra lake)	Airborne impact zone (case study small lakes, n=74)	Buffer impact zone (case study small lakes, n=226)
pH	7.04 (6.3-7.4)	7.5 (6.5-8.0)	6.8 (4.4-10.3)	6.7 (4.2-7.9)
Σ ions, mg/l	78.1 (16.0-168)	88.3 (13.3-176)		
Conductivity, μCm/sm	15.3 (6.1-26.0)	17.7 (14.0-26.7)	4.4 (1.7-8.4	3.6 (2.0-7.4)
Ca, mg/l	4.3 (1.3-8.4)	5.8 (1.6-15.0)	4.0 (1.06-44)	2.5 (0.23-11.6)
Mg, mg/l	2.1 (0.4-5.1)	1.4 (0.8-2.7)	1.2 (0.31-29)	0.99 (0.04-4.68)
Na, mg/l	23.2 (3.7-51.9)	21.9 (17.0-34.6)	2.4 (0.75-9.4)	2.4 (0.5-31.4)
K, mg/l	1.7 (0.5-3.0)	4.7 (2.9-10.9)	0.5 (0-2.2)	0.58 (0.1-6.79)
Alkalinity, mg/l	16.8 (2.4-24.4)	27.2 (8.5-70.5)	167 (0-1013)	193 (0-1345)
SO_4, mg/l	37.0 (7.6-64.5)	29.1 (14.7-37.1)	9.4 (2.7-210)	4.16 (2.2-14.68)
Cl, mg/l	10.4 (1.0-48.0)	8.3 (1.5-12.8)	3.1 (0.9-8.8)	1.95 (0.36-20)
PO_4, μgP/l	26 (10-39)	61 (14-154)	1.8 (0-46)	1.1(0-7)
P_{tot}, μg/l	32 (5-76)	35 (2-176)	11.8 (1-62)	4.5 (2-7)
NH_4, μgN/l	41 (26-75)	37 (35-38)	37(1-590)	12 (1-64)
NO_3, μgN/l	71 (4-257)	469 (55-1271)	30 (4-400)	3 (0-2440)
N_{tot}, μg/l	374 (164-81)	792 (180-1925)	306 (65-1037)	183 (107-259)
Si, mg/l	2.3 (0.6-5.1)	0.9 (0.06-4.7)	1.9 (0.1-5.7)	2.1(1.8-2.5)
COD, mgO/l	3.6 (1.3-4.8)	3.3 (0.9-9.9)	3.9 (2.0-8.6)	5.1 (0-35.2)
Ni, μg/l	82 (6-150)	28 (4-63)	13 (0-480)	1 (0-55)
Cu, μg/l	28 (0-165)	9 (2-45)	5.6 (0-117)	1.55 (0-142)
Sr, μg/l	33 (15-53)	78 (53-149)	25 (12-41)	27 (3-160)
Al, μg/l	26 (9-51)	72 (15-540)	18 (0-444)	20 (0-339)
Fe, μg/l	29 (10-70)	60 (6-645)	615 (21-1800)	262 (7-1600)
Mn, μg/l	17 (4-38)	13 (5-41)	9.5 (1.6-35)	6.5 (0.7-36)
Zn, μg/l	19 (5-113)	17 (1-57)	2.2 (1.1-5.8)	1.9 (0.2-8.5)

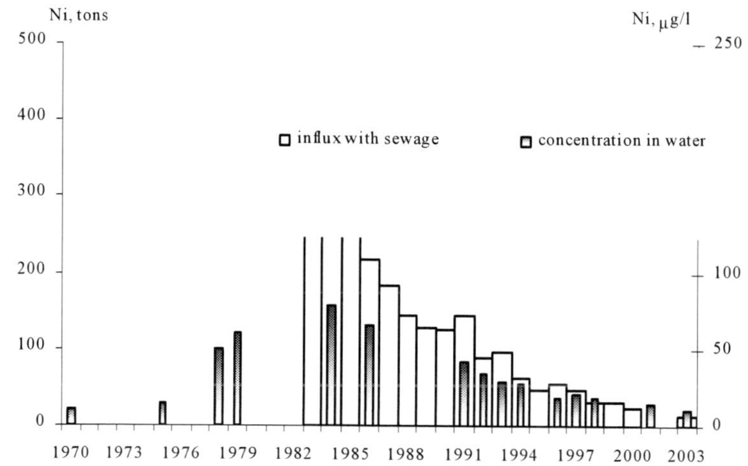

Figure 3. Dynamics of nickel influx (tons) into Lake Imandra and nickel concentration in water (μg/l) of the Imandra lake.

Basic deposition of heavy metals from the emitted industrial dust is observed within a radius of about 30-50 km from the smelters of "Nickel" Industrial Plant (Moiseenko, 1999). These emissions are accompanied by emission of sulphur dioxide into the atmosphere. However, no water acidification in the lakes occurs here; on the contrary, the water is alkalified as a result of dust deposition and dissolution in lakes and also erosion processes at the catchment area. In these zones (around Monchegorsk Town and Settlement Nickel) the "Nickel" Plants activity has led to formation of technogenic barren lands, where actually all the vegetation and the topsoil have been perished (Nikonov et al., 1993), causing a rapid migration of metals to the lakes. The water chemical composition in the lakes of this zone is subjected to the same changes like in the zones of waste water discharge (table 3).

Buffer zones are formed in areas at 30 to 100 km from the industrial sites, depending on wind-roses, and reflect the regional level of pollution. These areas are subjected mainly to airborne pollution by acid-forming substances and heavy metals. There are developed the processes of anthropogenic acidification of the lakes, which are accompanied by heavy metal dispersions and also leaching from the underlying rocks, especially aluminum and other labile metals (table 3).

Reference condition areas are preserved in the eastern part of Kola Peninsula. These water catchment areas reflect those transformations which occur on a global scale in the northern regions due to climate changes and trans-boundary transfers of pollutants to large distances. The air masses, that move from the south toward the north, bring (according to AMAP data), to the Arctic regions acid-forming substances, radionuclides, chlorine-organic compounds, etc (Arctic Pollution Issues...., 1997). The specific feature of an impact of trans-boundary pollutants transfer to Arctic is a more active fallout of micro-admixtures onto catchment areas during the contact of them with colder air.

Distribution and Fate of Metals in Lakes

Three basic processes are distinguished that lead to high contents of metals in the lakes of the Euro-Arctic region: (i) in waste waters of metallurgic manufactures; (ii) distribution with smoke emissions; (iii) acid leaching from surrounding rocks, especially from natural geochemical formations. On the example of detailed investigations in Kola North an understanding is given to the basic regularities of distribution, specific migration and circulation of technogenically-brought metals in the lakes of the Arctic basin. General scheme of metal fluxes and circulation are shown in figure 4 (Moiseenko, 1999).

Distribution of Metals in Waste Waters, Sedimentation and Inactivation

A large spectrum of elements (Ni, Cu, Mn, Sr, Fe, Al, Co, Cr, Cd, Pb и As) with wastes from the metallurgic enterprises and airborne deposition penetrate to the water basins. Near the discharge points the concentrations of ionic forms are the highest, which indicates that they penetrate into the lake with waste waters. In the process of migration the metal species are redistributed due to the complexation with water humus. However, as a rule the content of

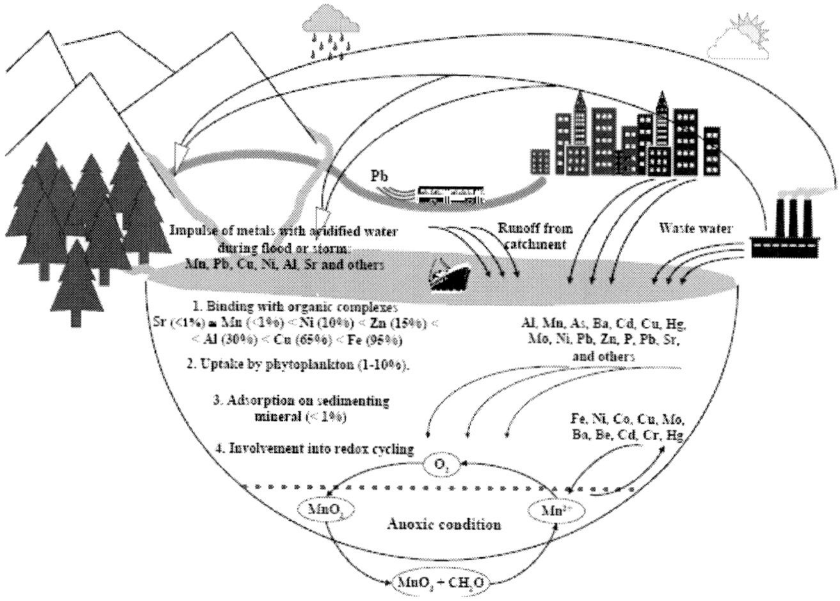

Figure 4. Generalized scheme of metal's input and fate in the Euro-Arctic lakes (Moiseenko, 1999).

labile forms of metals from the smelter discharges considerably exceeds the quantity of metals bound in complexes. This finding indicates the low complexing capacity of the Arctic waters. The main reason is the extremely small contents of suspended and organic matter in oligotrophic water, and respectively – low biomass of algae (Moiseenko *et al.*, 1996). In the lakes of a total area of over 10 km^2 the content of suspended substances is very low (<1 mg/l); the content of dissolved organic carbon (DOC) - 2-6 mg/l; processes of bioproduction are slow at cold latitudes. Absorption of metals by the phytoplankton here even during summer vegetation amounts to not more than 30% of the element content in a suspended fraction which is generally low and varies for different metals.

In small lakes with brown-coloured water, enriched by humic acids, the ability of the water to inactivate metals increases. Wetlands are widespread in north-eastern and south-eastern areas of the Kola Peninsula. The brown, Fe-rich waters are also typical to Karelia, Finland and other northern countries (Henriksen *et al.*, 1998; Manio, 2001). The contents of dissolved organic carbon (DOC) in the Kola surface waters are up to 35 mg/l, with an average value of 7.4 mg/l that is typical to Fennoskandia countries (Henriksen et al., 1998;). There are considerable results on studying the binding of metal ions by aquatic organic matter, including the complexation with different metals. For 64 water samples from lakes with a deferent concentration of DOC the relationship of the sum of nonlabile metals (in μequivalents) from DOC has been obtained (Moiseenko, 1999):

$$\Sigma\text{Metal nonlabile, } \mu eq/l = 0.59\,(\text{DOC, mg/l})^{2.5},\ r=0.96,\ n=64.$$

We assume, that in these water samples all metal-binding sites of organic ligands are occupied by metals, as these water samples contained also free metal ions. The metal-binding capacity is not constant and varies depending on mixture of humic ligands. In large lakes with autochthonic organic matter (2-5 mg/l), the number of metal-binding sites (per 1 mg/l of

DOC) is very low. Brown water with high concentration of allochthonic organic matter possesses a very high capacity for metals binding. If DOC concentration is equal to 10 mg/l, the metal-binding capacity is up to 30 μeq of metal ions. This implies that with DOC increase, the number of metal-binding sites for metals grows exponentially. According to our data, the percentage of aquatic humic complexes of various metals in surface waters of Kola North is as follows (Moiseenko *et al.*, 1996):

Fe (99 %)> Cu (65 %) >Al (30 %) >(Ni 25 %) > Zn (10 %) > Mn (< 1 %) > Sr (< 1 %)

It is easy to conclude that if in water the content of organic ligands is low, the metals compete for binding with them. With low concentrations of organic matter Fe will be bound first, for waters of Kola North up to 99 % humic matter is consumed on complexes with Fe, then Cu and Al, and the rest of the elements are present mainly as ions that is confirmed by direct measurements.

Desorption from Bottom Sediments

In conditions of accompanying eutrofication(or rich organic humic matter) of lakes during winter sub-ice period, when an oxygen deficiency appears in the near-bottom water layers, of a sharp significance become the processes of desorption of metals from the bottom sediments, which are due to a redox-cycle. The transport of iron and manganese into water bodies and sediments has received a great deal of attention because of the central role, which these abundant metals can play in the geochemical cycling of other elements (Davison, 1985).

The deficiency of O_2 in near-bottom layers forms a barrier for burial of metals. The alteration of oxidation-reduction conditions near the bottom gives rise to redox cycles, which are well studied for Mn and Fe. Under anoxic conditions near the bottom in winter period the metals are reduced: ($Mn^{4+} \rightarrow Mn^{2+}$ and $Fe^{3+} \rightarrow Fe^{2+}$). In the latter case, they are diffused from the bottom in a dissolved form to layers enriched with O_2, where they are again reacidified and converted into the insoluble form. A large group of other elements studied (Cd, Hg, Cu, Mo, Ni, Pb, Zn, Cr, Co, Ba, Ga, U) are also involved in this process (Moiseenko *et al.*, 1996; Moiseenko, 1999). The oxides of Mn and Fe at the boundary of oxicline (2-3 m from the bottom) are able to adsorb and concentrate trace elements. Under the anoxic conditions in the bottom-water interface, reduction and desorption of the elements occur. A high positive correlation ($r > 0.8$, $***P < 0.001$) of the above elements with Mn and Fe in the suspended and dissolved fractions indicates that they take part in a common cycle at the redox boundary.

It is easy to envisage a situation that anoxic conditions near the bottom can create an extremely toxic medium for bottom fauna caused by acidification of organic compounds and involvement of many metals into the turnover. In the Arctic regions during the long polar winter, this phenomenon aggravates the unfavorable effect of pollution by toxic metals in humic and eutrophicated lakes. Wastewater discharges may pose a much greater hazard to the fauna during the polar winter than in any other time of year.

Distribution with Air Emissions from Smelters and Leaching by Acid Precipitation

Areas with higher concentrations of Ni and Cu in surface water are local – 30-40 km zone around the smelters (figure 5a). The element deposition within 10 km distance from the smelters is characterized as follows: Ni - 50 and Cu - 280 $mg/m^2/year$. In the upper organogenic layer of the soils the metals are bounding: Cu - up to 95% and Ni – up to 90% of the total deposition. In spite of the higher level of Cu deposition as compared to Ni, the migration capacity of the latter in water is higher (Nikonov et al., 1993). The same regularity in distribution is typical to a sufficiently large group of metals which accompany copper-nickel ores - Co, Cr, V, Mo, Cd and others, though in lower concentrations (see table 4). Urbanization, highway roads in impact zones and global airborne transport of pollutants are responsible for an increase of the most toxic metals in the waters in the following concentrations (in μg/l): Pb - 0.5, Hg - 0.003, Cd - 0.1, As - 0.1, Cr - 0.3 (Moiseenko, 1999; Moiseenko, Kudryvtseva, 2002). Metals, that are accumulated during the vegetation period, can serve as the source for secondary pollution of waters in extreme periods of rainfall floods. The same local impact was shown for smelters in Sudbury, Canada (Jeffries et al., 1984).

Acidification favors the release of the same metals. Contents of the ionic forms of metals increase due to their leaching by acid precipitation from the composing rocks and release from the bottom sediments. Figure 5b demonstrates distribution of aluminum concentrations in Kola North. It is obvious that the high concentrations have a mosaic structure connected with development of geological formation vulnerable to acid precipitation. In the eastern tundra lakes of Kola North where industries are not developed, contents of many elements in acid lakes are high as compared with the water bodies where the pH values are close to neutral. This regularity controls the distribution of such elements as Al, Cd, Zn, Sr, etc. This phenomenon is proved for other metals as well (Dillon *et al.*, 1988; Nelson, Campbell, 1991; Johansson *et al.*, 1995; Manio, 2001; Rodushkin *et al.*, 2006).

The high contents of Sr are characteristic of the lakes located near Khibiny and Lovozero Mountains which are mainly composed of ultrabasic alkaline rocks. Predominance of nepheline in the mineral structure, high SrO (0.009 weight %) and Al_2O_3 (20-25 weight %) contents of these rocks are responsible for their intensive weathering and the Sr transfer into surface waters. Acid precipitation increase the content of ionic Sr forms in the water of lakes up to 400 μg/l (Moiseenko, Kudryavtseva, 2002).

Especially critical situation is formed in the lakes in the end of spring floods when thawed snow waters with a high content of accumulated metals and protons swiftly move into drainage basins. Sharp decrease in pH is accompanied by an increase in the contents of many metals in ionic forms. By our estimates, up to 75% of the annual metal input from the catchment basin penetrate into the lakes during the short period of intensive snow melting (Rodushkin *et al.*, 1996).

Table 4. Seasonal change parameters of the trophic status of lake Imandra in polluted areas by domestic sewage water in compare reference condition (numerator – average value, a denominator – limits of the contents)

Areas	Season	transparency, m	pH	O_2, %	PO_4, µg/l	P_{tot}, µg/l	NH_4, µg/l	NO_3, µg/l	N_{tot}, µg/l	COD, mg/l	Si, mg/l	Fe, µg/l
Polluted areas	winter	-	7.1 6.6-7.5	58 15-102	9 8-11	25 20-33	23 2-38	131 112-168	225 111-410	2.8 2.7-2.9	3.3 2.6-3.8	19 16-21
	summer	2.5 2.0-3.0	7.3 6.9-8.2	83 75-92	9 0-19	47 6-189	51 4-201	220 3-550	391 180-765	3.6 3.3-3.9	0.5 0.01-2.7	78 6-645
	autumn	3.0 2.3-3.4	7.3 7.0-7.5	72 31-113	7 1-14	50 17-176	22 11-32	245 52-1271	531 236-1925	4.3 3.1-9.1	0.9 0.07-4.7	58 26-231
Reference condition	winter	-	7.0 6.4-7.6	80 63-98	3 0-8	10 4-24	14 1-39	58 33-78	118 97-147	4.6 3.8-5.8	1.1 0.3-3.4	23 7-159
	summer	5.0 4.1-6.6	6.9 6.9-7.4	93 85-101	1 0-6	5 1-24	19 4-73	13 1-49	201 64-694	3.7 2.5-4.7	1.1 0.6-1.6	19 3-85
	autumn	5.3 4.0-6.0	7.0 6.9-7.3	89 85-94	1 0-3	7 4-14	10 8-11	31 23-46	175 99-325	4.7 3.3-6.9	1.3 0.8-1.4	8 6-11

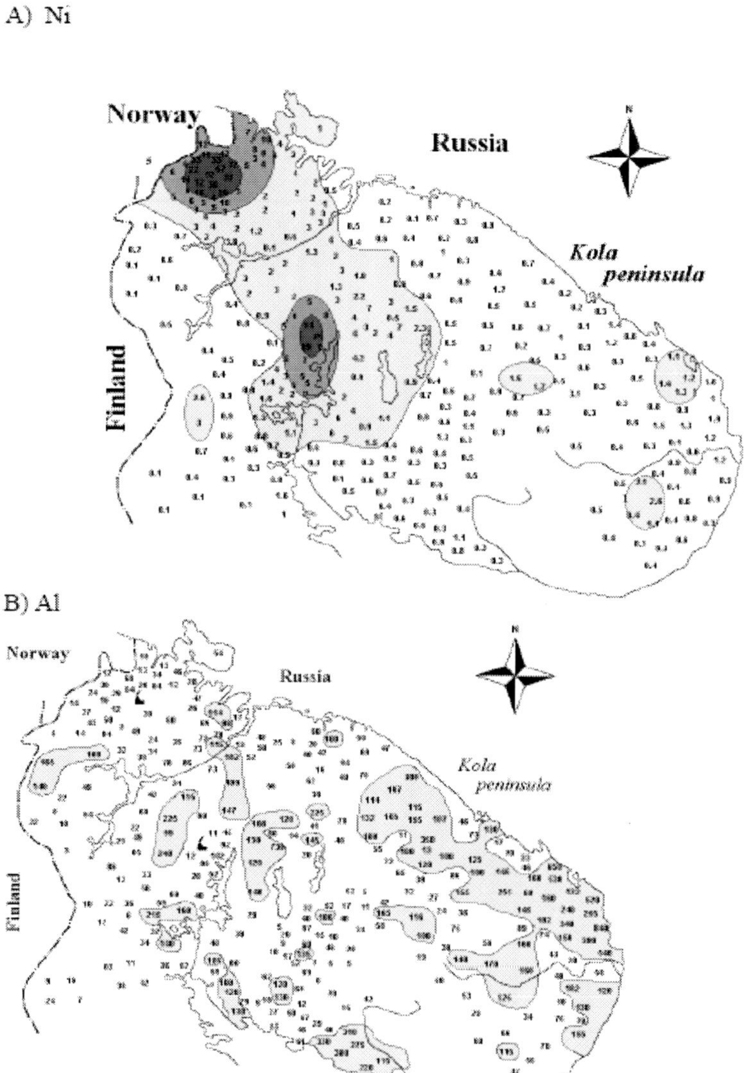

Figure 5. Concentration and territorial distribution: A) Nickel (μg/l) and B) Aluminium (μg/l) in water of Kola small lakes.

Specifics of Ecotoxicological Hazard of Heavy Metal Pollution in Arctic Region

A change in element geochemical cycles in nature under the impact of mining and metallurgical industry results in an increase of their contents in the environment, causes violation of microelementary ratio of alive organisms and occurrence of a number of diseases.

It is known that a number of human diseases are connected with increased metal concentrations. The surplus of trace elements in the human organism results in specific diseases: Hg causes a neurological effect, Cd and Pb have cancerogenic properties, Sr leads to pathologies of bone tissues, Cu - to anemia, etc. (Kovalsky, 1974; Foulkes, 1990; Spry,

Wiener, 1992). New toxicological properties of elements became wide known. As an example there can be Al that is actively leached into water bodies under an impact of acid rains. It was proved that not low pH values but high aluminum content associated with them leads to destruction of water fauna (Rosseland, Staurnes; 1994). Difficulties in determination of dangerous metal levels for vitality are stipulated by the following factors: (i) many elements (Cu, Zn, Co, Sr, Se, Ni etc.) are significant, i.e. inherent to organisms and are present in organisms in microquantities; (ii) poisoning influence of metals is formed both due to direct effects and ability to be accumulated in organisms, causing remote consequences - mutagenic, embriotoxic, gonadotoxic, cancerogenic, etc.; (iii) toxicological properties depend on metal speciation, combinations of elements (phenomena of sinergetism and antagonism) and concomitant factors.

Fish may be a convenient and informative test-object for ecotoxicological assessment of consequences of areas with high metal contents in water. Nephrocalcitosis of fish was revealed for the lakes in impact zones of the Russian Kola. At a Ni concentration in water up to 3-5 μg/l there is a risk for occurrence of kidney pathology. High concentrations of Ni in water lead to accumulation of this element in organs and tissues of fish, with maximum in kidney. The relation between accumulation of Ni in kidney and development of nephrocalcitosis and fibroelastosis was established

High biophilous element contents (Cu, Co, Zn) in the environment not always result in their accumulation in fish organism. In organs with active metabolism - liver - the redistribution of elements can lead to a partial decrease of Cu, Co, Zn contents. It can be connected both with their replacement by the main elements-pollutants (in this case - by Ni), and destruction of ferment systems and pathological degeneration of organs containing these elements. The researches indicate that for elements with high biophility, there is observed the complex picture of their redistribution in fish organisms due to an impact of industrial activity. In some tissues and organs the elements under study are capable to be accumulated (skeleton, gills), but in the organs with high metabolic activity, the concentrations of elements may decrease (Moiseenko, Kudryavtseva, 2002).

Under the acidified conditions of the water environment, more active accumulation of many elements occurs. Being widely known in the scientific literature for Al, this phenomenon has been established for Ni and Cu. Accumulation of such dangerous elements as Pb and Cd is observed in fish in the lakes under acid precipitation. This enables us to use fish as indicators of atmospheric loads in the region. Sr is accumulated mainly in fish bone tissues, which results in development of scoliosis and osteoparosis. In high concentrations Sr is also capable to be accumulated in soft tissues. In Sr provinces its contents in the functionally important organs become comparable with such an essential element as Zn. The reason of Pb bioaccumulation is the global Pb enrichment of the northern hemisphere. Accumulation of Hg and As in fish organism reflects an influence of industrial discharges (Moiseenko, Kudryavtseva, 2002).

Thus, accumulation of metals and violation of the evolutionarily determined elementary ratio of organisms under modern conditions is the consequence of the following reasons: industrialization of the region and direct local emissions, global airborne transport and changes in geochemical cycles of elements under the impact of acid precipitation. Microelementosis results in development of specific diseases, creates a risk for remote genetic consequences. For ecotoxicological assessment of water it is necessary to consider co-distribution of the whole mixture of microelements in various systems of an organism.

Winter Stress Syndrome (WSS), which is a consequence of cold water temperature and short photoperiod, was recognized as important seasonal causes of mortality for fish in North America in polluted lakes (Lemly, 1996). For Arctic region Winter Stress Syndrome will be repeatedly significant. During the polar night, as well as after it, the vulnerability of Arctic biota to toxic impact is extremely higher. It is easy to envisage a situation in which two or more stressors would occur simultaneously, thereby multiplying the risk that it would develop for aquatic life.

Critical Levels of Metals

The results mentioned above show that the behavior of various metals in the Arctic regions is specific for each of them and is determined by an interaction of complex mechanisms taking place in water. Despite a variety of the factors affecting behavior of elements and difficulty of their quantification, we have made an attempt to develop an universal scheme for defining a dose of total metal impact on the biota. In nature, aquatic organisms are exposed to an influence of the combined dose of all metals. It is important to estimate a uniform numerical parameter adequately describing the total metal impact on biota. The suggested method on the estimation of Integrated Toxic Impacts Dose (I_{tox}) is described in work of Moiseenko (1999). Using data about toxicological properties of each metal, we can define the integrated toxic impact dose by summing the excess of real concentration for each of metals to their GC or known threshold of impact as follows:

$$I_{tox} = \Sigma(C_i / GC_i).$$

In this connection only ionic forms of metals are taken into account in these estimations.

According to the survey of 460 lakes in Russian Kola North the visualization of estimated I_{tox} is presented at figure 6. Around metallurgical enterprises the high toxic dose is caused by airborne deposition of Ni, Cu and other metals from the smelters. In the remote tundra areas, where the lakes have low pH, the toxic dose is formed due to a release of metals from metal-rich rocks and soils with acid precipitation. Toxicity in this case is caused, first of all, by ionic forms of Al, Sr and other labile metals. In lakes enriched with humus metals are immobilized and also I_{tox} is low.

Occurrence of fish with pathological deviations is registered at a toxicity index of 1-5When the toxicity index was above 20 - all the fish in the local zone had deviations from the norm (Moiseenko, 1999). Thus, toxic properties of pollution by metals are already obtained by water with a toxicity index equal to one unit.

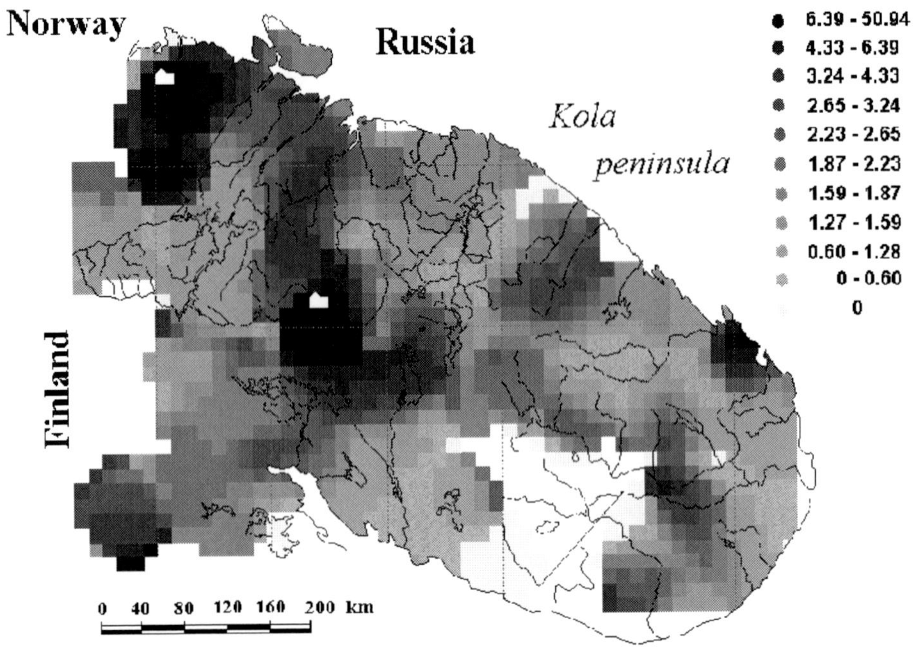

Figure 6. Visualization of toxicity index (I_{tox}) for lakes of the Russian Kola. High parameter I_{tox} around Cu-Ni-smelters is stipulated by dust emission containing heavy metals. Release of metals (Al, Sr, Zn, Mn and others) by acidic depositions is responsible for toxicity in the same remote eastern areas.

Metals in Bottom Sediments as Indicators of Regional and Global Pollution

Bottom sediments accumulate information on past chemical conditions existing in a catchment basin and in a water body itself, therefore reconstructions of the chemical composition (Norton et al., 1992) can enable determination of tendencies and intensity of anthropogenic pollution of a territory. A change in the geochemical composition of the sediments in the Kola lakes occurs due to an influence of: (i) waste waters; (ii) distribution of metals in smoke emissions from local enterprises; trans-boundary migration of substances from the polluted Europe and (iii) and acid leaching from rock and soil (iv).

Accumulation and Forms of Metal Residence in Bottom Sediments in Impact Zones

Lake Imandra can serve as an example of intensive pollution of bottom sediments by waste waters. Due to multi-year pollution by waste waters from the copper-nickel metallurgical enterprises the bottom sediments (BS) have suffered serious transformations in their geochemical composition along the entire water area due to both direct penetration of waste waters from mining-metallurgic enterprises. In the bay of Lake Imandra (Monch-Bay), to where waste waters penetrate directly from "Severnickel" Plant, the Ni concentration in the surface layers of the bottom sediments increased by 320 times as compared with the reference values, Cu – more than by 50 times (figure 7). With distance from the polluting sources the difference in the concentrations in the BS surface and reference layers becomes lower, but even at a distance from the basic sources of pollution the concentration excess in the surface layers reaches several units (Dauwalter *et al.*, 2000).

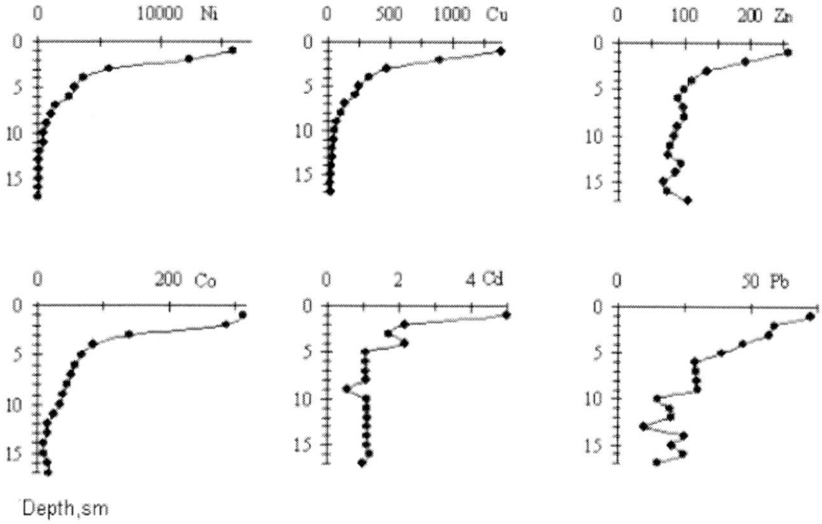

Figure 7. The example of avalanche accumulation of heavy metals in sediments (mg/g d.w.) in impact zone of copper-nickel smelter impact zone lake (case study of Imandra lake).

With increase in Ni and Cu concentrations in the BS surface polluted layers in Lake Imandra, the share of mobile forms increases, increasing, hence, an environmental hazard for hydrobionts and for the lake ecosystems as a whole.

Taking into account the thickness of the BS polluted layer and average concentrations of the elements in this layer in Lake Imandra, V.A.Dauwalter *et al.* (2000) have calculated that over a period of more than 60 years the activities of the mining-metallurgical enterprises resulted in accumulation of metals in the bottom sediments in the following amounts: 4600 t - Ni; 960 t - Cu; 120 t - Co; 250 t - Zn; 11 t – Cd; 200 t - Pb. Distribution of toxic metals in the water and bottom sediments in some stretches of the lake is determined by the location of point discharges of waste waters from the metallurgical enterprises and by scheme of water currents in the lake. Metals, accumulated in the bottom sediments, can present a hazard of water pollution by these metals in future as a source of secondary pollution.

The analogous regularity is characteristic to metal accumulation in lakes located in a radius of 30 km from smelters caused by dust emissions containing heavy metals. Diffusive sources, such as washed-off waters from industrial sites, waste waters from soil dumps are also the sources of lake pollution by heavy metals in impact zones.

Impact Transboundary Pollution on Lake Sediments in Remote Areas

Study of mountainous lakes in conventionally reference regions provides an informative base for estimation of consequences from global and local pollution of the atmosphere. The retrospective analysis of geochemical composition of the bottom sediments of a lake in Chuna-tundras, that is not subjected to any direct anthropogenic impacts, has revealed accumulation of a number of elements during the entire past century (figure 8), in spite of their low contents in the water. It should be noted that our data agree with the results of American scientists who report about the global Pb-enrichment of the northern chemosphere (Norton *et al.*, 1990; 1992).

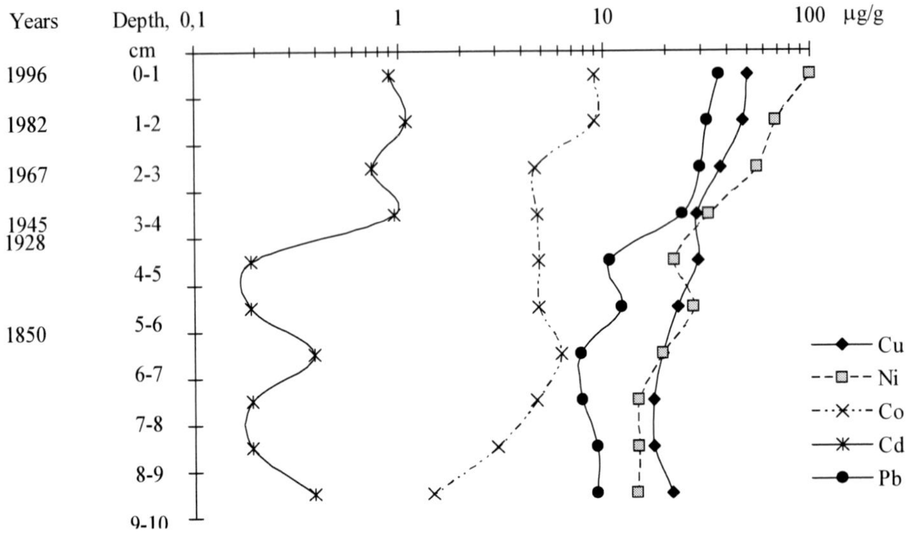

Figure 8. The example of heavy metals accumulation in remote mountain lake in Chuna Tundra: Pb accumulation in the sediments is marked from the end of the last century, accumulation Ni, Cu, Cd Co and other metals are connected with the industrial development of Arctic Region (Moiseenko et al., 2000).

Since the end of XVIII century accumulation of Pb has occurred. In this period Kola North was not a developed territory, so the initial phase of Pb-accumulation could be connected only with migration of polluted air masses from the industrial Europe. Accumulation of Ni, Cu and Co is registered since 1940s and is connected, without doubt, with industrial development of Kola region and beginning of functioning of smelters on processing of copper and nickel ores, which is quite natural. The dynamics of Cd-accumulation is more complicated. By the end of XVIII century the concentration of this element decreases with the next increase by 1930-1940s of XIX century; some decrease is registered by mid 1970s, accumulating then again in the surface layer (Moiseenko *et al.*, 2000). These data vividly demonstrate that the lakes of the Euro-Arctic Region are under the influence not only of regional manufactures, but also of transboundary emissions from Europe.

IONIC EQUILIBRIUM DISTURBANCE AND ACIDIFICATION OF WATERS

Emission of sulphur dioxide and fall-out and dry absorption of acid-forming substances lead to acidification of small lakes in the region. Deposition of anthropogenic sulfate in the impact zone of Kola Peninsula exceeds 2-3 $gS/m^2/year$, but the water acidification happens there not due neutralization of this water by dust, as mentioned above. Over 1/3 of the territory fall-out of anthropogenic sulphur amounts to over 1 $gS/m^2/year$ (Moiseenko, 1994) Experimental data on transport of the torch of emissions from "Severnickel" Plant have shown that the major portion of sulphur emissions is deposited in Kola North and only 20% of them are distributed to far distances (Tuovinen et al., 1993).

As it is known, intensity of acidification is determined by two conditions: 1) a level of anthropogenic load with account of the factor of impact action and 2) natural sensitivity of territory. Local sources of acid-forming substances emitting into the atmosphere in this region are the metallurgical enterprises and heat power plants using coal. Granite-gneiss formations compose the major part of Kola Region, which determines vulnerability of the region to acid rains.

Water acidification has been manifested in three aspects: (i) decrease of buffer capacity of large catchments in historical intervals; (ii) occurrence of episodic acidification during floods in small streams, and (iii) acidification of small lakes (Moiseenko, 1994). Among 460 surveyed lakes in the year 1995 – 10.3% were acidified (pH<6) and 30 % were in a critical state (alkalinity < 50 meq/1) (Henricksen et al., 1998). The paleoecological evidences show that initial acidification of waters in the Arctic regions appeared in the beginning of XX century and was connected with trans-boundary transport of acid-forming substances from industrial Europe. These processes have been accelerated by local industry in mid of century. (Moiseenko *et al.*, 2000; Dauvalter *et al.*, 2002).

Figure 9 visually demonstrates distribution of sulfates and the pH values in the Kola lakes. Near the industrial centers, in spite of considerable sulfate concentrations, the pH is high due to distribution of alkaline aerosols from dust emissions and increased fluxes of cations from water catchment areas, where degradation of the surface ecosystems is observed and erosional processes are developed at a drainage basin. The most critical situation with acidification of lakes, in spite of their being remote from the sources of emitting anthropogenic sulfur, is found in the tundra zone where the geological structure of the territory consists of exposed acid rocks of granite-gneiss formations, topsoil cover is thin or not developed. In such vulnerable drainage basins with a load of over 0.3 $gS/m^2/year$ the lakes are being acidified.

Analyses of multi-year data on the water quality in Kola lakes, carried out in 1990, 1995, 2000 and 2005, have revealed a reliable tendency of a decrease in sulphate contents. This is quite natural, because during the recent years the anthropogenic load on the lakes decreased. In spite of that emissions of anthropogenic sulfur and sulfate contents in the lake waters have decreased during the 20-year period, the pH value and calcium continue decreasing in the lakes, which evidences on deep transformations of the water catchment during the more than 50-year influence of acid-forming emissions upon the catchment (Gashkina, Moiseenko, 2006).

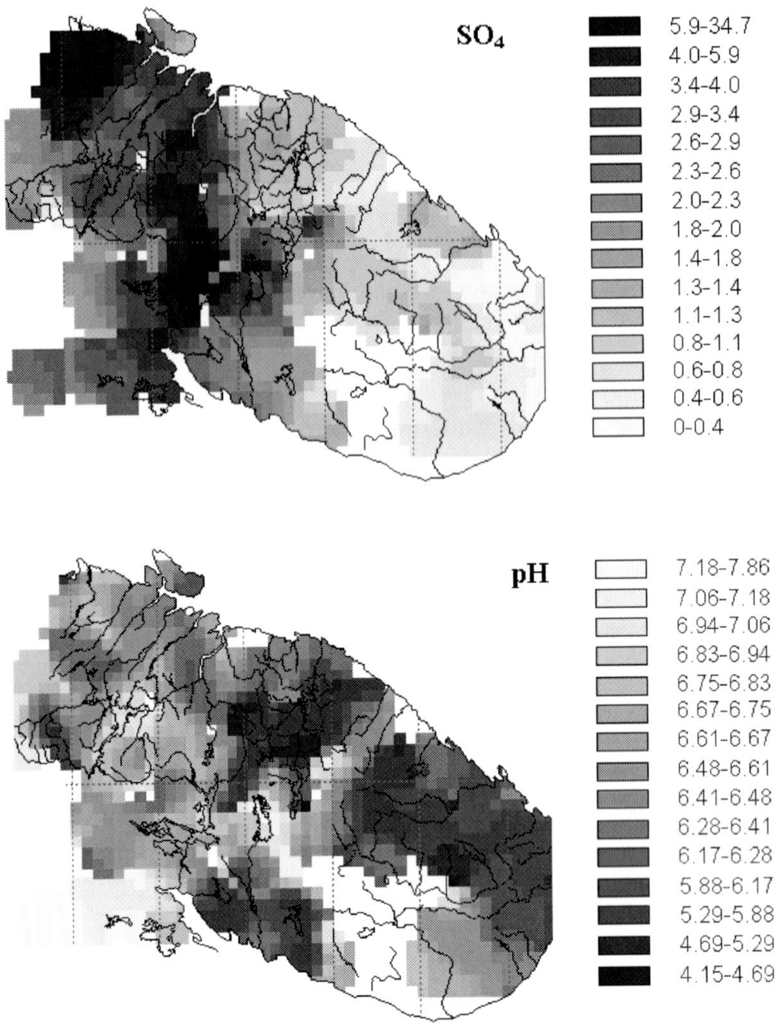

Figure 9. Visualization of pH and nonmarine sulphate concentration (SO_4, μeq/l) after correction on marine salt content. The maximal concentration of technogenic sulfates are formed in the lakes located in impact zones, the concentration witch exceed background extend on 1/3 territories and characterize the lakes located in a buffer zone. Low value pH are connected not so much with high loading of sulfates, how many with wetland catchments in eastern part of territory.

Consequences of Acidification for Fish Organisms and Ecosystems of Lakes

The deposition of acid-forming substances on drainage areas and water acidification have both direct and indirect effects on biological systems, causing changes in individual organisms, communities, and whole ecosystems. At all levels of ecosystems, biodiversity decreases with acidification due to elimination of species that are most sensitive to low water pH. The microflora and destruction processes are inhibited, whereas fungi gain dominance; as a result, coarse organic detritus is accumulated on the bottom of acidified lakes. The trophic structure of bottom communities changes toward the prevalence of gnawing species. Water transparency increases, and macrophytes expand to greater depths, with water mosses resistant to acidity developing most actively. Changes in production processes are ambiguous: they are usually inhibited in strongly acidified waters, but the productivity of communities consisting of acidification-tolerant species often increases due to modification of their trophic structure and alleviation of competition with acidification-sensitive species and predation by fish (Schindler, 1988; Raddum, Skjelkvåle, 1990; Muniz, 1991; Korneva, 1996; Jeffries, 1997; Moiseenko *et al.*, 1999).

The main factor responsible for degradation of fish populations due to acidification is the direct impact of low pH and Al^{3+}, which results in biochemical and physiological disturbances The target systems in fish are gills and sensory organs, and the most vulnerable stages of the life cycle are larvae and fry (Rosseland, Staurness, 1994).

Acidification leads to an increase in metal contents in water. A common regularity in the behavior of a number of metals (Al, Zn, Cd, Pb, etc.), that entails the acidification is an increase in concentrations of metals and transformation of them into a more toxic ionic form, which creates, in combination with a sharp pH decrease, extreme conditions for the aquatic fauna (Rodushkin *et al.*, 1996; Moiseenko, Kudryavtseva, 2002).

A decrease in acid fallout has a favorable effect on biological communities. In recent decades, a tendency toward their recovery has been revealed in water systems with increasing pH values and alkalinity. In some lakes of Canada and the United States, repopulation of biotopes with acidification-sensitive species and increasing biodiversity of aquatic communities have been observed (Gunn *et al.,* 1995).

Critical Loads of Acids and Excess

Figure 10 a,b present schematic maps of acid critical loads and its excess for Kola lakes. Critical loads were determined by the method of A. Henriksen (1998).

Comparison of the obtained results with the geological maps shows that the leading role in water acidification belongs to the geochemical structure of a water drainage area. In remote tundra and mountainous areas from industrial centers and where granite-greiss formations are widely developed at a deposition of 0.4 –0.6 gS/m^2/year, the loads are equal to 0.3 gS/m^2/year. In a radius of 30 to 50 km around the metallurgical plants, where the deposition of technogenic sulphur amounts to 1.2 – 1.8 gS/m^2/year, due to spreading of dust emissions, development of alkaline granites (Khibiny and Lovozersky mountainous massifs) and basalts (to which the plants location is associated) many lakes preserve an acid-neutralizing capacity and the exceeding the critical level is not observed. But here, effects of leaching of cations and toxic metals can be manifested.

Comparison of situations in different countries of the northern Europe shows that the critical loads are exceeded in Norway – for 27% of the lakes, in Finland, Sweden and Denmark - 9%, in Scotland – 1%, and for 17% of the lakes in Kola Region (Henriksen et al., 1998).

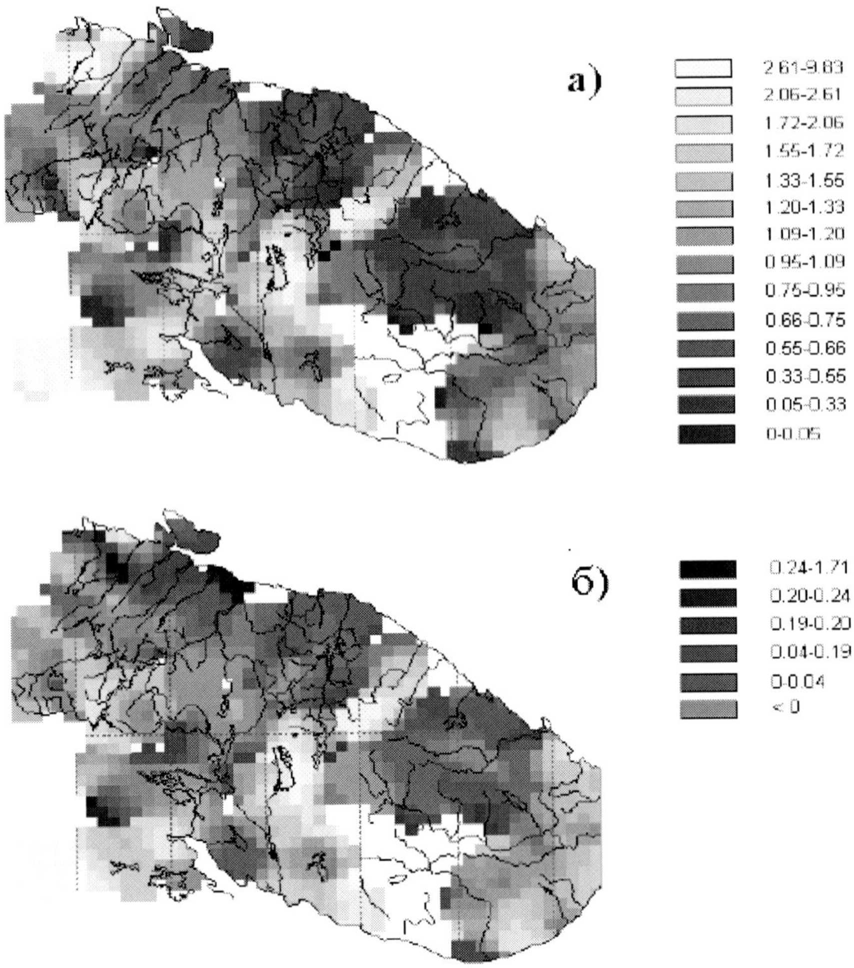

Figure 10. Visualization of Critical loads (a) of and Critical load exceeded (b) for lakes of the Russian Kola, calculated by the steady-state water chemistry method (Henriksen et al.,1992) for level Acid Neutralizing Capacity (ANC limit) 50 μeq/l.

Trophic Status of Lakes and Eutrophication

Water bodies in the Arctic Regions are characterized as oligotrophic and dystrophic. Most lakes with small forest and swampy drainage areas are characterized by the concentration of dissolved organic matter equal to 8.7 mg/l and the color index of >30 grad. Poor water exchange in the swampy tundra maintains high saturation of water with humic acids. This, in combination with the Arctic climate, is responsible for the dystrophic conditions of lakes. Despite the relatively high concentrations of P_{tot} and N_{tot}, the concentrations of their bioavailable mineral forms are very low. The values of the N/P ratio in the lake water indicates that P plays a significant limiting role in production processes (figure 11).

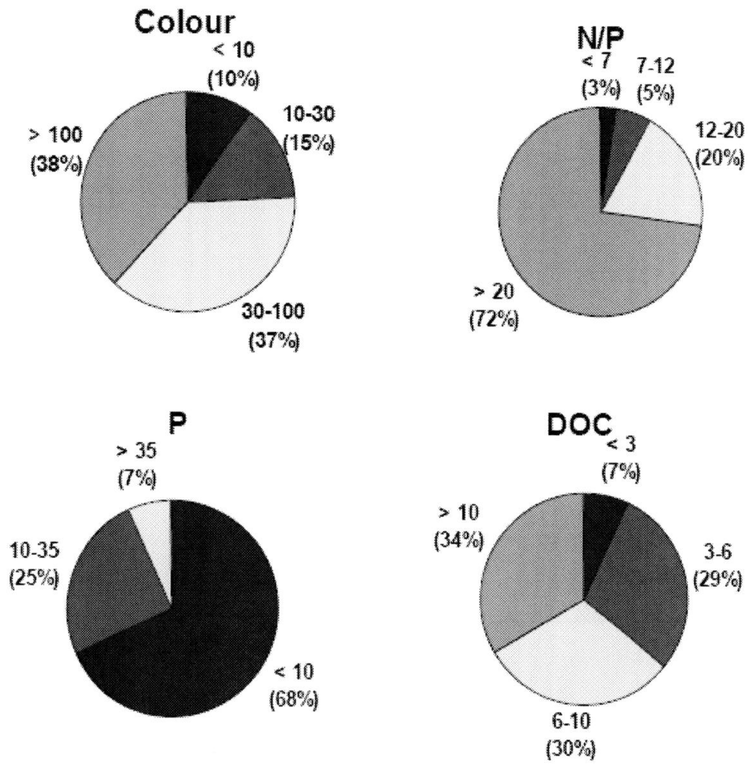

Figure 11. Distribution Kola lakes by criteria trophic status: water colour - Pt°; phosphorus - P, μg/l; dissolve organic carbon - DOC, mg/l; and ratio of nitrogen to phosphorus – N/P.

The development of urbanization of the Arctic region leads to enrichment with nutrients (P, N, organic matter) and thereby provide the prerequisites for anthropogenic eutrophication. But, the problem of eutrophication has not received proper attention, it is even not mentioned in the international program of investigations in the Arctic Regions (Arctic Pollution Issues, 1997). The lack of knowledge about the specific features of eutrophication of natural waters in cold regions may lead to underestimation of the environmental consequences of this phenomenon.

Nutrients (Case Study of Lake Imandra)

Lake Imandra, the largest water body in the study region, demonstrates pronounced anthropogenic eutrophication of water. The development of industrial production on its shores for more than 60 years, the appearance of towns, settlements, stock farms, and agricultural fields have produced a heavy nutrients load on the lake's drainage area.

Currently the annual P input into the lake amounts to 84.4 t and its output with the river runoff, 49.5 t. The annual inputs of P and N due to river runoff are 11.8 and 123.8 t, respectively. These values are much smaller than those associated with the immediate anthropogenic load on the lake due to municipal and industrial (associated with the production of the apatite concentrate) waste waters. And though P in minerals of apatite is firmly bound in crystal lattice, it can be involved into the turnover of substances due to microbiological activity in the lake. Enrichment of water with P and N is an important criterion of the extent of water eutrophication. The amount and ratio of the dissolved and suspended forms of P and N vary with seasons, and their dynamics is to a considerable degree governed by the productive processes and, consequently, by the trophic conditions of the water body (table 4). The highest concentrations of P_{tot} (up to 189 μg/l and N_{tot} (up to 1925μg/l) are registered in the Imandra (Moiseenko et al., 2001 b).

The average concentration of the dissolved organic matter in Lake Imandra is rather low (2.8–4.3 mg/l) and slightly changes with pools and seasons. Despite the input of the dissolved organic matter with PO_4^{3-} in Lake Imandra, this matter might be rapidly utilized in the ecosystem. The principal source of Si compounds in lake bodies is the chemical weathering of silicon-containing minerals. Si concentration in Lake Imandra varies from 0.01 to 4.7 mg/l. The lowest Si concentration is observed in the eutrophied part. The stimulation of the development of diatoms, which use Si for constructing their valves, contributes to lowering Si concentration in water (Moiseenko et al., 2006).

The Formation of Oxygen Deficiency

The dynamics of O_2 in water is in many ways associated with the productive–destructive processes. During the period of the Arctic summer with round-the-clock solar radiation, the development of algae leads to the saturation of the surface water layers with O_2. In winter, the deficiency of O_2 for the process of oxidation of the dissolved organic matter accumulated in the water body is a criterion of the rate of water eutrophication. The natural parameters of Lake Imandra are indicative of the high year-round saturation of waterwith O_2. Even in winter, O_2 concentration in water to a depth of 15–20 m makes about 95% of saturation (Year-book dates.., 1952-2005). This is due to numerous non-freezing mountainous rivers flowing into the lake.

Observations of the distribution of O_2 concentration in the lake water during different seasons over the recent decade (figure 12) show that in winter (over the period of up to three months) and in August (the period of the maximal warming up and density foliation), the zones of high O_2 deficiency are formed in the near-bottom layers in the deep sections of the lake. These are also the zones of the highest P concentration. Figure 12 shows the distribution of O_2 concentration in a deep zone of Lake Imandra. In summer, the zones of O_2 deficiency (up to 10% of the total depth of the water body) are much less pronounced than in winter because of the turbulent mixing. The rest of the water mass is well aerated owing to intensive wind wave effect and convective mixing. With the onset of the autumn thermal cooling of water mass, the area of O_2 deficiency disappears very rapidly, and the weighted means of O_2 concentration exceed the 85%-level of saturation.

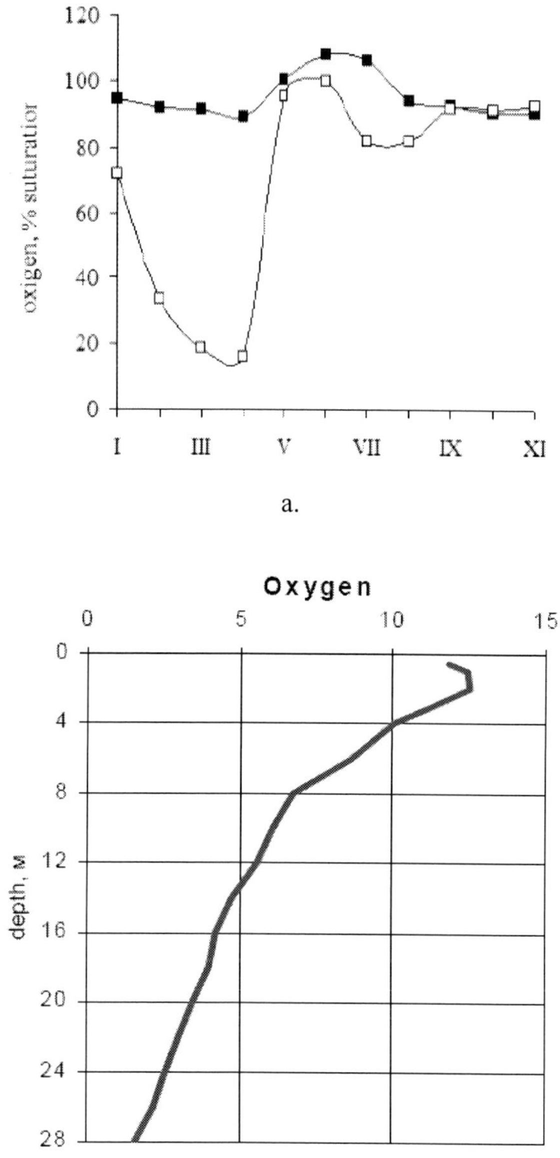

Figure 12. Example of seasonal dynamics (a, % saturation) and deficiency of the oxygen (b, mg/l) near bottom lakes within long polar winter as results of anthropogenesis eutrophication, which occurs in the local sites under impact of domestic sewage water (case study of Imanra lakes).

The relationship between the development of O_2 deficiency and P concentration for Lake Imandra testifies to a considerable O_2 consumption under the conditions of P accumulation (Moiseenko et al., 2001):

$O_2 = 96 - 1.04$ Ptot ($r = 0.61, p = 0.05$).

Analysis of the equation shows that P_{tot} = 90 μg/l is the critical threshold; at a higher P_{tot}, there will be no dissolved O_2 in the water mass. This equation reflects the general features and regularities of the distribution of concentrations of the above components. The above data confirm the high significance of the criterion of the oxygen deficiency for the assessment of the eutrophication of surface water in the Arctic zone and warn about a serious adverse environmental consequence.

Chlorophill "a"

The concentration of chlorophyll ***"a"*** serves as an indicator of productive processes and development of algae. The concentrations of chlorophyll *a* in natural waters in the Kola Peninsula are extremely low (0.5–1 mg/m^3). The concentration of chlorophyll ***"a"*** in the seston of Lake Imandra varies from 2.3 to 8.38 mg/m^3. The areas subjected to the effect of municipal wastewater are the most productive ones. The concentrations of chlorophyll "*a*" in the oligotrophic, mesotrophic, and eutrophic water bodies amount to 0–4, 4–10, and 10–100 mg/m^3, respectively (Vollenweider, 1979). However, other gradations of chlorophyll ***"a"*** concentrations are also proposed (Rast, Lee, 1979): 0–2 for oligotrophic lakes, 2–6 for mesotrophic, and >6 mg/m^3 for eutrophic lakes. This classification is more acceptable for the Arctic waters where the natural productive processes are very slow.

Analysis of the relationship between the Chlorophyll ***a*** and P_{tot} for Lake Imandra shows that they are closely correlated:

Chlorophyll "*a*" = 0.92 Ptot – 0.45 ($r = 0.82, p = 0.05$).

Thus, urbanization in the regions of the Arctic entails the enrichment of surface waters with nutrients and the anthropogenic eutrophication in lakes or their particular areas. This eutrophication is marked off by a number of indices and entails a series of adverse consequences. The comprehensive analysis of the process of eutrophication is needed, because the use of one or a few criteria can result in a distorted picture of the trophic condition of lakes. The role of nutrients in the Arctic water bodies is very high. This is due to their enhanced vulnerability caused by their being covered with ice for long periods.

In winter, the processes of the molecular transfer of substances prevail over the convective mixing. Therefore, despite the low water temperature and rate of O_2 consumption, anoxia develops in the near-bottom layers of lakes by the end of winter. Along with the enrichment of water with nutrients, the intensification of productive processes and diminishing transparency, there is also a tendency for decreasing DOC and Si concentrations in water over the recent 30 years. This tendency can be explained by the high rate of utilization of the above elements at a relatively high levels of organization of aquatic systems and is a specific feature of developing eutrophication of surface water in the high latitudes (Moiseenko et al., 1996a).

Under the conditions of the nutrient load on lakes in the Arctic Regions, a comprehensive integrated approach to assessment of criteria of the trophic condition of water bodies, as well as consideration of the tendency toward the development of their qualitatively new condition (as compared with the natural one) are needed.

For the surface waters in the Arctic, the critical value of Ptot concentration, at which, in accordance with chlorophyll "***a***" concentration, the water body transforms from the oligotrophic into mesotrophic type, is equal to 15 μg/l; the change to the eutrophic status occurs at a P_{tot} concentration of >200 μg/l.

The Environmental Consequences

The eutrophication of lakes lead to development of phytoplankton and changes in the species composition. In Lake Imandra, the portion of green, cryptophyte, and blue-green (Cyanophyceae) algae in the total biomass increases (in particular, species of genera Pandorina, Eudorina, Scenedesmus, Cryptomonas, and Oscillatoria). Among diatoms, *Rhizosolenia eriensis*, *Stephanodiscus astraea var. minutisima*, *Aulakoseira islandica subsp.helvetica* prevail in the biomass. In September, the population of *S. astraea var.minutisima* increases to 11 million cell/l. The total biomass of diatoms increases by the autumn, whereas the biomass of green algae decreases. In all samples, the representatives of genera Fragilaria and Synedra; *Asterionella Formosa* often were met. *Tabellaria fenestrata* and *T. flocculosa*, widespread in Lake Imandra, were virtually absent in plankton near the site of wastewater discharge. Desmidiaceae were not found (Moiseenko *et al.*, 1999).

CONCLUSIONS

The technologies used at the acting mining, smelting and chemical enterprises in Russian Arctic result in pollution of lakes by waste waters, as well as airborne contaminants. As a result of direct sewage dumping into the lakes and airborne pollution of their water catchments, there are appeared some negative processes in water chemistry in the Russian Arctic:

1. Water toxification by heavy metals (Ni > 20 μg/l ,Cu > 10 μg/l) is characteristic of the large lakes, such as Imandra, Pyasino, Kuetsjarvi, where the sewage waters from smelters input and it is similar for the small lakes within the radius of 30 km around the smelter complexes. In the Kola region at a radius up to 30 km the impact zones of pollution have appeared where the concentrations of these elements are equal to toxic levels. Lakes surrounded by Khibiny and Lovozero Mountains are characterized by the high content of Sr. Here large apatite-nepheline mining has resulted in an increase of Sr, Al and other elements. Mineralization and turbidity are increasing in the lakes situated near the areas of mining activity and also subjected to the industrial sewage effects.
2. Water acidification occurs beyond the limits of dust emission effect - at a distance of more than 30 km from the smelters where sulfur load is more than $1gS/m^3yr$ and the area vulnerability is high. Water acidification results in higher mobility of many trace elements, primary of which is Al. The high level of Al concentration is characteristic of acidified lakes (mainly in the eastern parts of the Kola peninsula).
3. Water eutrofication in the Subarctic regions has a local character in the lakes, where the municipal sewage or heated waters from the Nuclear Power Station come. Accumulation of nutrients (P>20 μg/l, N > 200 μg/l) in the arctic lakes does not lead to alga development relevant for these concentrations because of low water temperatures and intensive water

exchange. Only in some shallow well-heated lakes or inlets the phytoplankton development may run into the level of meso- or eutrophic lake.

4. The first three factors may simultaneously develop in all the industrially-developed areas (impacts zones) of Russian Arctic (Norilsk and Kola mining-metallurgical complexes, Severodvinsk production), i.e. toxification, a change in salt content and eutrophication. Beyond the boundaries of industrial centers also acidification and pollution by metals occur, strengthening their negative effects.

5. During the flood period in Arctic region the pulse of metals in combination with low pH values can have a dominant negative effect on the fauna after the long Polar night. The danger is also in rapidly washing out of heavy metals from catchments and release of their ionic forms by acid snow-melt water.

6. During the ice period of the long Arctic winter for eutrophic lakes or lakes rich by humic matter there is the dramatic situation associated with anoxic condition near the bottom layer and recycle of metals at the redox boundary. The eutrophication aggravates the unfavorable effect of toxic metals.

7. In cold regions the pollutants have much more expressed negative effects. At the same time, lakes here get a special value due to high-quality water resources and good fish production. The basic principle of Arctic lake preservation should be given up as follows: priority of the clean water and fish production; refusal from non-limited Arctic water resources due to their high vulnerability to anthropogenic loadings; differential approaches to protection of lakes depending on natural conditions and on lake purpose; preventive maintenance of pollution sources instead of struggle with consequences: withdrawal of toxic substances from industrial discharges; prevention of emergencies connected with burial places of industrial and radioactive wastes, transport of petroleum, localization of non-point contaminated flows from mining activity.

Acknowledgments

The work was supported by grants from the Russian Fund for Fundamental Investigation (Grant 07-05-00302) and the Euro-limpacs International Project (contract no. GOSE-CT-2003-505540). The author is also grateful to Kudryavtseva L., Rodyushkin I. and Platonenkova G. for the large number of analytical measurements.

Reference

Anthropogenic Modifications of the Lake Imandra Ecosystem. 2002. (ed. Moiseenko T.I.). Moscow, *Nauka*. 485p.

Arctic Pollution Issues: A state of the Arctic Environment. 1997. Published by AMAP, Oslo, Norway. 188 p.

Atlas of Murmansk area. 1971. Murmansk. 33 p.

Dauvalter, V., Moiseenko, T., Kagan, L. 2001. Global change in respect to tendency to acidification of subarctic mountain lakes // Global Change and Protected Areas (eds.Visconti G. et al.). Kluwer Academic Publishers, Dordrecht, the Netherlands. P. 187-194.

Dauvalter, V.A., Moiseenko, T.I., Kudrjavtseva, L.P., Sandimirov, S.S. 2000. Accumulation of heavy metals in Imandra lake sediments under industrial pollution // *Water resources.* V. 27 (3). P. 313-321.

Davison, W., 1985. Conceptual models for transport at a redox boundary. // Chemical processes in lakes. (Eds W. Stumm and A. Wiley.). Interscience publication John Wiley and Souns. P. 31-53.

Dillon P.J., Evans H.E., Scholer P.J. 1988.The effects of acidification on metal budgets of lakes and catchments //*Biogeochemistry.* V. 5. P. 201-220.

Foulkes, E. C. 1990. *Biological Effects of Heavy Metals*, Vols 1 and 2. CRC Press, Boca Ration, FL.

Gashkina, N., Moiseenko, T. 2006. Temporal and spatial assessment of water quality around copper-nickel smelters on Kola Peninsula. ICP-Water Report 88/2006: Joint Workshop on Confounding Factors in Recovery from Acid Deposition in Surface Waters. *NIVA,* Oslo, ISBN 82-577-5041-7. P. 33-44.

Gunn, J.M., Keller, W., Negushiti, J., Potvin, R., Beckett, P., Winterhalder, K. 1995. Ecosystem recovery after Emission Reductions: Sudbury. Canada //*Water, Air and Soil Polution.* V. 85. P. 1783-1788.

Henriksen, A., Skjelvale, B.L., Mannio, J., Wilander, A., Moiseenko, T.I. et al. 1998. Northern European Lake Survey, 1995: Finland, Norway, Sweden, Denmark, Russian Kola, Russian Karelia, Scotland and Wales *AMBIO,* V. 27, № 2. P. 80-91.

Jeffries, D.S. Canadian Acid Rain Assessment: Aquatic Effects. 1997. Published by the Authority of the Minister of Environment. Canada. Ontario. Burlington. 270 p.

Jeffries, D.S., Scheider, W.A. and Snyder, W.K., 1984. Geochemical interactions of watersheds with precipitation in areas affected by smelter emissions near Sudbury, Ontario. Offprints from Environmental Impact of Smelters. Kanada. Ontario POA JEO. P. 192-238.

Lemly, A.D., 1996. Waste water discharges may be most hazardous to fish during winter. *Environ. Pollut.* V.93, № 2, P.169-174.

Johansson, K., Bringmark, E., Lindevall, L., Wilanders, A. 1995. Effects of acidification on the concentration metals in running water in Sweden // *Water, Air and Soil Pollut.* V. 85. P. 779-784.

Korneva, L.G. 1996. Impact of Acidification on Structural Organization of Phytoplankton Community in the Forest Lakes of the North-western Russia // *Water Science Tech.* V.3. P.291-296.

Kovalsky, V.V. 1974. Geochemical ecology. Moskow, *Science.* 176 p.

Krokhin, E.M., Semenovich, N.I. 1940. Data on Water Bodies in the Kola Peninsula. Apatity (The collection № 1, manuscript. Funds of Kola science centre of RAN). 151p.

Mannio, J. Responses of headwater lakes to air pollution changes in Finland. Academic dissertation. University of Helsinki. 2001. 226 p.

Moiseenko T. 1994. Acidification and Critical Loads in Surface Waters: Kola, Northern Russia // *AMBIO.* V. 23, no 7. P. 418-424.

Moiseenko, T. I. 1999. A Fate of Metals in Arctic Surface Waters. Method for Defining Critical Levels // *The Science of the Total Environment. V.* 236. P.19-39.

Moiseenko, T., Kudrjavzeva, L., Rodyshkin, I. 2001. The Episodic Acidification of Small Streams in the Spring Flood Period of Industrial Polar Region, Russia //*Chemosphere V.* 42/1, № 362. P. 45-50.

Moiseenko, T..I.,.Rodushkin, I.V, Dauvalter, V.A. 1996. Geochemical migration and covariation of elemets in the Imandra lake, Barents Region. Re-print Lulea of Technology, Sweden. 107p.

Moiseenko, T.I. Dauwalter, V.A., Iljashyk, B.P., Kagan L.J., Iljashuk, E.A. 2000. A paleoecological reconstruction of anthropogenic load// *Doklady Akademii Nauk. V.* 370, № 1. P. 115-118.

Moiseenko, T.I., Sandimirov, C.C., Kudrjavzeva, L.P. 2001. Peculiarities of Eutrofication of Arctic Region Waters //*Water Researches,* V. 3. P. 339-348.

Moiseenko, T.I., Voinov, A.A., Megorsky, V.V. et al. 2006. Ecosystem and human health assessment to define environmental management strategies: The case of long-term human impacts on an Arctic lake // *Science of the Total Environment,* V 369. P. 1-20.

Moiseenko, T. I., Kudrjavzeva, L.P. 2001. Trace Metals Accumulation and Fish pathologies in Areas affected by Mining and Metallurgical enterprises // *Environmental Pollution,* 114(2). P. 285-297.

Moiseenko, T.I.,Yakovlev, V.A. 1990. Anthropogenic transformations of aquatic ecosystems in the Kola North (ed. Rumiancev, V.A.) Leningrad: *Nauka.* 221p.

Moiseenko, T.I., Sharov, A.N. Vandysh, O.I., Yakovlev, V.A., Lukin, A.A. 1999. Changes in biodiversity of surface waters of the North under acidification, eutrophication and toxic pollution. *Water Resources.* V 4. P. 492-501.

Muniz, I.P. 1991. Freshwater acidification: its effects on species and communities of freshwater microbes, plants and animals // *Proceeding of the Royal Society of Edinburgh.* V.97B. P. 227-254

Nelson, W.O., Campbell, P.G.C. 1991. The effects of acidification on the geochemistry of Al, Cd, Pb and Hg in freshwater environments: A literature review// *Environment Pollutution.* V. 71. P. 91-130.

Nikonov, V.V., Lukina, N.V., Derom, D., Petrowa, N.V. and Goryainova, V.P. 1993. Migration and accumulation of Ni and Cu compounds in Al-Fe humic podzoil soils of pine woods (impact zone of "Severonickel" company). *Soil science,* V. 11. P.31-41.

Norton, S.A., Bienert, R.W.J., Binford, M.W., Kahl J.S. 1992. Stratigraphy of total metals in RIPLA sediment cores // *Journal Paleolimnology.* V 7. P. 191-214.

Norton, S. A., Dillon, P. J., Evans, R. D., Mierle, G., Kahl, J. S. 1990. The history of atmospheric deposition of Cd, Hg and Pb in North America: Evidence from lake and peat bog sediments. Sources, Deposition and Capony Interactions. (eds Lindberg S. E. et al.) Vol. III, Acidic Precipitation, Springer-erlag, New York. P. 73-101.

Raddum, G.G., Skjelkvåle, B.L. Critical loads of acidifying compounds to invertebrates in different Ecoregions of Europe // *Water, Air and Soil Pollution.* 2001. V. 130. P. 1131-1136.

Rast, W., Lee, G. 1979. Relationship between summary mean and maximum chlorophyll-a concentration in Lakes // *Environment Science Technoljdy.* V 13. P.869-870.

Rikhter, G.D. 1934. Geographical sketch of Lake Imandra and its catchments. Leningrad. The State technical-theoretical publishing house. Moscow. 144p.

Rodushkin, I.V, Moiseenko, T.I. Kudrjavtseva, L.P. 1996. Changes in trace element speciation in Kola North surface waters during snow melt // *Water, Air, and Soil Pollution.* V. 2. P. 731-736.

Rosseland, B. O., Staurnes M. 1994. Physiological Mechanism for Toxic Effects and Resistance to Acidic Water: An Ecophysiological and Ecotoxicological Approach //*Acidification of Freshwater Ecosystem: Implications for the Future.* P. 227-245.

Schindler, D.W. 1988. Effects of acid rain on freshwater ecosystem // *Science. V.*239. P.149-157.

Skjelkvale, B.L., Andersen, T., Fjeld, E., Mannio, J. et al., 2001. Heavy Metals in Nordic Lakes; Concentrations, Geografical Patterns and Relation to Critical Limits // *AMBIO,* V.30, N1. P.2-10.

Spry, D., Wiener, G. (1991). Metal Bioavailability and toxicity to fish in low-alkalinity lakes: a critical review // *Environmental. Pollution.* V. 71. P. 243-304.

Tuovinen, J.P., Laurila, H., Lattila, A. at al. 1993. Impact of Sulphur Dioxid Sources in the Kola Peninsula on Air Quality in Northermost Europe //*Atmospheric Environment. V.* 27. P. 1379-1395.

Vereshagin, G.U. 1930. Methods of morphometric characteristics of lakes. Works of Olonetskoy scientific expedition. Leningrad. 53 p.

Vollenweider, R.A. 1979. Advances in defining critical loading levels for phosphorous in lake eutrophication // *Met. Ins. Ital. Jdrobion V.* 33. P. 53-83.

Year-book dates of surface water quality formation in Murmansk area: Reports 1961- 2005. HYDROMET: 45 volumes for the period 1951-2005. Murmansk.

In: Lake Pollution Research Progress
Editors: F. R. Miranda and L. M. Bernard
ISBN: 978-1-60692-106-7

Chapter 9

HEALTH EFFECTS OF LAKE POLLUTION

Paul Froom*

Departments of Occupational and Environmental Health, and Epidemiology and Preventive Medicine, School of Public Health and Sackler's Medical School, Tel Aviv University, Tel Aviv, Israel

INTRODUCTION

The USA Environmental Protection Agency (EPA) was established in 1970 to consolidate in one agency a variety of federal research, monitoring, standard-setting and enforcement activities to ensure environmental protection. The EPA's mission is to protect human health and to safeguard the natural environment—air, water, and land—upon which life depends. Since the establishment of the EPA scientific advancements have been made in the determination of health risks from human exposure to pollution. Methodology has been developed that can provide an educated and quantitative guess of the possible effect of pollution on human health.

* email- froomp@gmail.com

The EPA focused on air pollution until the 1980s when their efforts were broadened to lake pollution and other environmental pollution including hazardous waste sites and other water, sediment and soil pollution. The Comprehensive Environmental Response, Compensation, and Liability Act of 1980 (CERCLA), commonly known as the "Superfund" Act, was signed into law by President Jimmy Carter and provided the Congressional with a mandate to remove or clean up abandoned and inactive hazardous waste sites and to provide federal assistance in toxic emergencies. The CERCLA was amended by the Superfund Amendments and Reauthorization Act of 1986, and provided the EPA with the tools to evaluate the health risks and determine interventional levels for environmental pollution on land, in water and sediments. The necessary methodology needed to accomplish their tasks lead to significant advances in hazard analysis, and quantitative risk assessment.

The chain of events leading to these decisions was the media attention to the percieved negative effects of pollution on health. It began with the Love Canal in New York where the press and many scientists concluded that the exposures to pollution caused cancer and other diseases in the local population. Whether or not exposures to pollution in the population living in the area of the Love Canal actually resulted in an increased risk of disease is controversial. Hazard analysis was not used, and it is uncertain whether or not the huge costs involved in the clean up, and compensation of the residents was justified. The Surgeon General of the United States influenced by the "common knowledge" at that time considered that toxic chemicals posed a major threat to health in the United States. Eighty percent of the American people wanted some protective legislation. In response, twenty four Senators joined as sponsors of environmental legislation. Senator Robert T. Stafford worked on the legislation together with the other members of the Senate Committee on Environment and Public Works for nearly three years to put together the laws.

The Senator wrote that "Modern chemical technology has produced miracles that have greatly improved this Nation's standard of living. But the increased generation of hazardous substances associated with these new products has proved to be a serious threat to our Nation's public health and environment". More than 2,000 dump sites containing hazardous chemicals are believed by the Environmental Protection Agency to pose threats to the public health. In a report to the President of the United States, the Toxic Substances Strategy Committee concluded that the cancer death rate in the United States had increased sharply and that "occupational exposure to carcinogens is believed to be a factor in more than 20 percent of all cases of cancer. There is not one adult American who does not carry body burdens, of one or several of these substances, many of which have now been removed from the market because of their dangers". Since then hazard analysis has lead to a drastic decrease in such estimations.

Costly preventive measures included the destruction of hundreds of thousands of hogs, chickens, turkeys and a large quantity of other food stuffs after contamination with PCBs whose manufacture is now banned. PCBs leaked from an out-of-service transformer, and entered the food chain and spread through 19 States and two foreign countries. Other preventive measures included closing parts of the Great Lakes to commercial fishing because of chemical contamination. The question is whether or not these decisions were correct. There are side effects of going public with suspected but unproven public health concerns. Interventions are costly and the money might be better spent elsewhere. Even in developed countries where resources are unlimited, publicity will cause anxiety in the population and may lead to a decrease in self-rated health in those exposed. Some will live in fear believing that it is just a matter of time before they become ill with cancer. Others with cancer will have the added burdon of dealing with their anger whether or not justified. This anger might also be shared by family and friends of the cancer victum. Although theoretically one can never rule out the possibility that exposures caused an individuals cancer, methods were developed over the last 30 years taking into account exposures in order to calculate the actual theoretical probability that a person became ill because of exposure to a pollutant or pollutants. The use of available scientific information to calculate risks is call risk assessment.

The purpose of this chapter is to quantitate risks to human health from exposure to polluted lake water and sediments. We will start with consideration of the major lake pollutants of concern. In order to quantitate risks to human health we will focus on how chemicals are are absorbed and by what routes of exposure, and how chemicals are classified as carcinogens by what route of exposure and for what type of cancers. For added perspective we will discuss the limitations and uncertainties of such classifications. After identifying a chemical as a potential hazard, we will use cancer slopes derived by extrapolation from high exposures in humans and animals to low dose exposures from polluted lakes to calculate risks to the exposed population. Finally we will calculate the personal attributable risk of an exposure to a definite carcinogen, the chance that a patients cancer was caused by his exposure.

LAKE POLLUTION

Besides living organisms, lake pollutants of concern include metals and organic chemicals (table 1). Some of these chemicals are known carcinogens only when inhaled while others have been shown to be carcinogens by ingestion or by skin exposure. Some of these chemicals have been shown to increase the risk of cancer in humans, while others increase the cancer rate only in animal studies. Table I summarizes the chemicals that might increase the risk of cancer in humans exposed to polluted lake waters. Carcinogenicity is considered the capability of an exposure to increase the incidence of malignant neoplasms, reduce their latency, or increase their severity or multiplicity. The International Association for Research on Cancer (IARC) classifies chemicals into five groups; I – a definite carcinogen, IIA – a probably carcinogen, IIB – a possible carcinogen, III- not enough information to classify, and IV- not a carcinogen. If there is consistent evidence in humans that exposure increases risk, the chemical is considered to be a definite carcinogen. If there is good evidence in animals only, then the chemical is considered to be a possible carcinogen. A probable carcinogen requires good evidence in animals and suggestive evidence in humans. There are other factors that enter into the decision that we will discuss later. In any case the classification only implies that under certain conditions the chemical can increase the risk for cancer, but the IARC does not define the conditions. Therefore this is only the first step in determining the potential risk to human health. The IARC working groups provide scientific qualitative judgments on the evidence for or against carcinogenicity by considering available data. These evaluations represent only one part of the body of information on which public health decisions may be based.

What we know about cancer comes from natural human experiments in workers exposed to high concentrations of carcinogens, from exposures of animals to high concentrations of chemicals, from follow-up studies of humans with various risk factors, from the study of cellular biology, from studies of DNA damage found in cancer cells, and from studies of genetic damage in humans and animals of various ages and with various diseases. The IARC considers that there is sufficient evidence of carcinogenicity when "a positive relationship has been observed in which chance, bias and confounding could be ruled out with reasonable confidence. After classifying a chemical, the IARC identifies the target organ(s) or tissue(s) where an increased risk of cancer was observed in humans. This does not preclude the possibility that the agent may cause cancer at other sites". Risk assessment on the other hand considers the target organ and the conditions of exposure in order to quantify the risks.

Table 1. Lake pollution of concern to human health

Chemical	Carcinogen classification	Route causing cancer Respiratory GI Dermal			Type of cancer associated with exposure
Metals					
Arsenic (As)	I	Yes	Yes	No	Skin, Liver, kidney , bladder, Lung
Cadmium (Cd)	I	Yes	No	No	Lung
Chromium (Cr)	I	Yes	No	No	Lung, Nasal septum
Lead (Pb)-inorganic Organic	2A III	Yes	Yes	No	Kidney (animals only) Limited evidence-man
Nickel (Ni)	I	Yes	No	No	Lung , nasal septum
Mercury (Hg)	IV				None
Copper (Cu)	IV				None
Zinc (Zn)	IV				None
Organic chemicals					
Benzene	I	Yes	Yes*	Yes*	Acute non-lymphocytic leukemia
Poly aromatic hydrocarbons	I	Yes	Not shown	Yes	Lung by respiratory route Skin by dermal exposure
PCBs/PBBs	2B				Liver (animals only) Gastric (animals only)
DDT**	2B	Yes	Yes	Yes	Liver (animals only)
Dioxins	2B	Yes	Yes	Yes	Liver (animals only)
Aldrin	2B	Yes	Yes	Yes	Liver (animals only)
Chlordane	2B	Yes	Yes	Yes	Liver(animals only)

*not shown in human studies but thought to apply because of toxicological considerations.
**DDT- p,p'-Dichlorodiphenyltrichloroethane.

The route of exposure is important when determining potential risks. For example cadmium chromium and nickel are definite carcinogens when inhaled, but have not been shown to be carcinogenic by either dermal or gastrointestinal absorption. Inorganic arsenic on the other hand is carcinogenic by either inhalation or gastrointestinal absorption. Dermal absorption of these metals is very limited, making it unlikely that absorption by this route could increase the risk of cancer.

Another important factor is the type of cancer(s) found for the various exposures. Although the IARC does not claim that such associations rule out the possibility that the carcinogen can also cause cancer at another site, it can be seen that the sites are limited in animals and humans exposed to high doses of the chemicals. DDT, the dioxins and Aldrin have been shown only to cause liver cancer in animals, whereas PCBs can increase the risk in animals for liver and gastric cancer. In humans inhaled certain forms of nickel and chromium can increase the risk for lung and nasal septal cancers, whereas exposure to inorganic arsenic increases the risk for skin , liver, kidney, bladder cancers, and lung cancer (evidence primarily for inhalation exposure).

In the Great Lakes and other lakes around the world, a major concern is for chemicals that do not break down easily, persist in the environment, and bio-accumulate in aquatic biota, animal and human tissue. They are called persistent bio-accumulative toxic chemicals (PBTs). These chemicals are lipid soluble, can be absorbed through the skin and are found primarily in lake sediments, a sink for metal and organic chemicals.

Similar to plants growing in soils and forming the base of the terrestrial food chain, benthic organisms are the base of the aquatic food chain. Benthic organisms lack cell walls, and containing high proportions of lipids and fats in cell membranes and other organelles and are a source of lipophilic compounds to the aquatic food chain and, potentially to humans and other fish eaters.

Benthic organisms spend most or all of their lifecycle in the sediment of the lake. Some fish are benthic feeders or spend most of the time near the bottom; others eat organisms that have spent part of their lifecycle as benthos. Finally, birds and mammals prey on these same fish. Each organism has bio-accumulated contaminants during its lifecycle, and the effect magnifies as one moves up the food chain (bio-magnification) There are species used as indicators of this phenomenon including the bald eagle and herring gull. Food is the primary route of human exposure to these PBT chemicals, and consumption of Great Lakes fish is the most important source of exposure originating directly from the lakes. Sources from air, soil/dust, and water constitute a minor route of exposure. Exposure to polluted sediment might also be a significant route of exposure in some cases since fat soluble chemicals are well absorbed through the skin, and tend to accumulate in the body in fat rich tissues like the human brain and in breast milk.

It has become clear that the accumulation into the food chain of persistent bio-accumulative toxic (PBT) chemicals, such as PCBs, dioxins and furans, and mercury (as methyl-mercury), is not solely dependent on their concentration in sediments. Characteristics of the sediment such as organic content, microbial environment, pH, and presence of sulfates and sulfides can all affect the potential for PBT chemicals to be bio-accumulated. Furthermore, although interactions are dynamic, sediment reactions are typically characterized and studied as static systems. This is an important area of future research.

Note in table 1, that the organochlorines have been shown in animals and not in humans to increase the risk for cancer. At high doses they consistently increase the risk for liver cancer. This might suggest that these chemicals are not actually carcinogenic, but it is the damage to the animal's liver at the high concentrations that results in the observed increased incidence of liver cancers in the animals. We will discuss later the significance of animal models for determining the carcinogenic potential of chemicals. Nevertheless these chemicals are of concern because they accumulate in fish and subsequently in human tissues and have a very long half-life. They are also present in breast milk and therefore are a source of exposure to infants. Most of the health effects studies on PBTs in the Great Lakes have focused on fish consumption. On the other hand because the sediment acts as a sink, lipid soluble substances are found only in low concentrations in water. In the Great Lakes basin, measured levels of these persistent toxic chemicals in drinking water are below the Maximum Contaminant Level (MCL) and therefore they are not considered to be a human health concern for drinking water.

CHEMICALS OF CONCERN

DDT

DDT was produced at the end of World War II, to control malaria and typhus and the commercial peak in the USA was in 1963. DDT was banned in 1973. Dr. Paul Muller was awarded the Nobel Prize in Physiology and Medicine in 1948 for discovering its insecticidal properties. DDT is a good example of the dilemmas in public health. One needs to balance the positive effects of using DDT versus the negative ones. On the positive side DDT saved many lives by decreasing the incidence of malaria. On the negative side, although not shown to cause morbidity or mortality in humans, it is very persistent, and accumulates throughout the food chain. Agricultural use of DDT has since been banned in North America following the discovery that DDT and its breakdown products were causing widespread reproductive failures in eagles and other wildlife species. DDT breaks down into toxic metabolites, primarily DDE. Although DDT continues to be used in other parts of the world, levels of DDT in the North American environment have decreased significantly since this pesticide was banned, and species impacted by DDT, such as the bald eagle, are said to be recovering.

DDT is a suspected carcinogen especially for breast cancer, but studies in humans, including those with occupational exposure are mostly negative and inconsistent. Some have shown higher DDT levels in the body of humans ill with breast cancer whereas others have not found such results. DDT is an animal carcinogen with liver cancer consistently shown at high levels of exposure. The minimal risk level was fixed at 0.0005 mg/kg/day based on neurological developmental effects in animals at 0.5 mg/kg/day, the lowest observed adverse effect levels (LOAEL). However, workers exposed to up to 42 mg/day of DDT did not show any adverse neurological effects. Lung absorption is minimal since DDT crystals are large and therefore swallowed with concentrations peaking in three hours and back to baseline at 24 hours. DDT is also not well absorbed in the skin, 3.3% over 24 hours. Two liters of DDE ingestion per day at a concentration of 1 μg/L for an entire lifetime is estimated to increase the lifetime risk of cancer by risk 1 per million and for DDT this dose has been estimated to increased the risk by 1/100,000. DDT is poorly soluble in water and has very low rates of bio-degradation. Therefore the concentration of DDT in water is very low (few ppb most of the time with maximal values up to 100 ppb). DDT is transported via the atmosphere and accumulates in sediments and biotic tissues.

Diedrin

Diedrin was used a pesticide and was also banded because it is an animal carcinogen and a chemical that bio-accumulates. In the recent past most Lake Ontario fish have been found to have concentrations ranging from approximately 0.005 to 0.030ppm, higher than those recommended.

Dioxins and Furans

There is no known use of dioxins and furans (Chlorinated Dibenzo-p-dioxins, or CDDs); they are by-products of processes involving chlorine, organic chemicals and heat, including incineration, pulp and paper bleaching with elemental chlorine, and chemical manufacturing. Dioxins are extremely toxic. They are animal carcinogens and may be an important endocrine disrupter. Dioxins are furans are still identified as critical pollutants in the Great Lakes because levels of these contaminants exceed human health standards in some lake fish and because these chemicals may limit the full recovery of the bald eagle, mink, and otter populations by reducing the overall fitness and reproductive health of these species.

Mirex

Mirex is well known as a pesticide but is also used as a flame retardant and has also been found to exceed human health standards in some lake fish. It is also an animal carcinogen increasing in animals the risk for liver cancer.

Polychlorinated Biphenyls (PCBs) and Polybrominated Biphenyls (PBBs)

PBBs a flame-retardant material, was introduced into the food chain in Michigan in 1973 due to a manufacturing and distribution mistake. Polychlorinated biphenyls (PCBs) were widely used in electrical equipment such as transformers and capacitors. PCB oil-filled electric switches eliminated electric sparking problems that could trigger explosions at petroleum refineries. PCB oils were used in electrical transformers as a non-flammable electrical insulating fluid. PCBs were also used as industrial lubricating oils to replace earlier types of hydraulic oils that could more easily catch fire under conditions of high pressure and temperature. PCBs were manufactured between 1929 and 1978. PCBs are also animal carcinogens and probable endocrine disrupters. The production of PCBs was halted following the discovery that PCBs released into the environment were bio-accumulating in a wide range of organisms. The hazards posed by PCBs were discovered in the 1960s when ranch mink, that had been fed a diet of Great Lakes fish, experienced reproductive failures. The investigations that followed determined that Great Lakes fish were contaminated with PCBs at levels that warranted human fish consumption advisories. Levels of PCBs in fish and wildlife continue to exceed human health standards for food consumption and they also may still pose health and reproduction problems for bald eagles, mink, and otter.

Mercury (Hg) and Methyl-Mercury

Mercury is widely used in batteries, electrical equipment (switches), medical equipment, thermometers, thermostats, and preservatives. Many former uses (as a fungicide, pesticide, and in latex paint) have been discontinued, but mercury is still needed in some products and processes. Historically, mercury was added to paints as an anti-mildew agent. Its largest U.S. use today is at chlor alkali plants that produce chlorine gas and caustic soda. Small concentrations of mercury that exist in natural materials such as coal, wood and metal ore are released when these materials are processed; because such huge quantities of these materials are processed, much mercury is released. Mercury is also released when garbage is burned, and it vaporizes from landfills. Mercury is not known to be a carcinogen, but is a neurotoxin and toxic to the fetuses of humans and animals.

Metals: Alkylated Lead, Nickel, Copper, Zinc, Cadmium, Chromium

Lead, nickel, copper, zinc, and cadmium are heavy metals common in hazardous waste. They can damage organisms at low concentrations and tend to accumulate in the food chain. These heavy metals are found in sediment and are associated with degradation of benthos, and plankton communities.

There are no oral cancer slopes proposed by the EPA for cadmium, chromium or nickel (ATSDR 1997), USEPA 2005c, d) although quantitative estimates of carcinogenic potency from inhalation are available. This is because these substances have not been shown in animal models with very high exposures to cause cancer by the oral or dermal routes of absorption. Chromium and possibly nickel are essential elements in the human diet and cadmium poisoning in Japanese women exposed to high levels of cadmium in their food during World War Two had an increased total mortality rate in those with proteinuria, but no significant increase rate in cancer over a 15-year follow-up period (Arisawa 2001).

Similar to cadmium, chromium VI is considered a definite carcinogen only by the inhalation route of exposure and only for lung cancer, and the evidence is poor that chromium is a carcinogen either by oral or dermal absorption (Paustenbach 2003) (USEPA 2005d). Because inhalation of chromium VI can cause lung cancer in some persons exposed to a sufficient airborne concentration, questions have been raised about possible hazards associated with exposure to chromium VI in tap water via ingestion, inhalation and dermal contact. If there is such a risk it might be mitigated by an observed threshold of 10 mg/L to 22 mg/L due to the reductive capacity of the stomach, skin and blood (Paustenbach 2003). Chromium VI is reduced to Chromium III that has not been shown to be carcinogenic.

Arsenic

Absorption of arsenic however from drinking water is definitely carcinogenic. There is evidence in humans that intestinal and dermal absorption increases the risk for cancers at various sites. In Bangladesh in order to decrease the risk of drinking infectious water, they deepened the wells that led to naturally high concentrations of arsenic in the drinking water, and widespread poisoning with an increased rate of cancer.

Benzene

Benzene is a definite carcinogen with workers exposed to high concentrations of benzene in the air having an increased risk of nonlymphocytic acute leukemia. The evidence for other types of cancers is limited (ATSDR 2007). Although there is no evidence that benzene can cause cancer by oral or dermal routes of absorption, it is assumed that it can because of similar metabolism to that from pulmonary absorption. Recommendations for swimming and diving in chemically contaminated waters are vague (Amson 1991), and for benzene it is stated that exposure should be "absolutely minimal". This creates a problem for recreational and professional divers since benzene is ubiquitous in recreational surface waters due primarily to motorized recreational boating (Heald 2005), where up to 30% of unburned fuel is discharged into lakes. Furthermore industrial afferents also add to benzene concentrations in surface waters and aquatic sediments. Usually benzene surface water concentrations in lakes are less than the MCL for drinking water (U.S. ATSDR 2007), yet maximal concentrations can reach 100 μg/L (surface waters tested in 571 sites during the years 2003-2005 (U.S. ATSDR 2007)). The maximum concentration of benzene in industrial facilities using or producing benzene tested during the same years ranged from less than 1 μg/L to 179 μg/L. In sediment I could not find values in US lakes over 5 mg/Kg. Yet under anerobic conditions benzene can persist and accumulate in sediments. In general however most of benzene in water is volatilized, and because of a low octanol/water partition coefficient it does not bio-accumulate nor accumulate in sediments to the same degree as polycyclic aromatic hydrocarbons.

PAH (Polycyclic Aromatic Hydrocarbons)

In water most of the polycyclic aromatic hydrocarbons will be non-carcinogenic, whereas in the sediments because of their inert nature, the carcinogenic PAH will accumulate (Shor LM 2004). These levels can be particularly high in harbors. For example in various areas of Hamilton Harbor at the western edge of Lake Ontario total polycyclic aromatic hydrocarbons concentrations in sediment were 20-67, 76-184, 215-650 and even over 1400 ppm. Mixed exposure to various PAH has been shown at very high levels to cause lung cancer in workers inhaling such pollutants (USEPA 2005b), and to cause squamous cell skin cancer, primarily of the scrotum, in those with dermal exposure. It has not been demonstrated in humans that PAH exposure results in cancers outside the area of direct contact, but in some animals with oral exposure to over 10,000 times that found in the human diet, cancers were reported outside the contact area (USEPA 2005b).

Acute Effects of Exposure to Lake Water and Sediment and Human Health

The major human health concern in lake water and sediment is exposure to microbiological contamination (bacteria, fungi, viruses, and parasites) and the only common acute effect of exposure to lake water and sediment is infection. In developed countries we are primarily concerned with fecal bacterial concentrations where as in developing countries both bacterial and parasitic diseases are common.

Many sources contribute to microbiological contamination, including combined or sanitary sewer overflows (CSOs and SSOs), residential and commercial areas without sewage systems, and failing private, household and commercial septic systems. Other possible sources are agricultural runoff especially from animal/pet fecal waste, and a high number of swimmers/bathers in the water. Factors affecting contamination levels are shallow water, hot weather, and winds that can stir up bacteria that are in the sediments.

Local beach closings due to elevated levels of E. coli (or fecal coliform bacteria) are indicative of fecal contamination and the possible presence of enteric (intestinal) pathogens that can pose a potential health risk. Gastro-intestinal disease can result in serious morbidity and even mortality. A serious outbreak of acute diarrhea disease in those who swam in a polluted lake was reported with twelve percent of those who swam in the lake becoming ill with Escherichia coli O121 and three of the 11 ill children developed hemolytic uremia syndrome a potentially fatal disease (McCarthy TA, 2001). Acceptable pollution levels are from 100 to 200 organisms per 100 ml.

Sediment Effects on Human Health

The major potential chronic human health hazard is from exposure to polluted sediment and not from polluted lake water because the pollutants of concern are not very soluble in water and concentrate in the sediment that acts like a sink and can pollute fish through the food chain. Levels requiring intervention and thought to be a significant risk to human health are called maximal tolerated risk (MTR) levels. The MTR is based on potential adverse health effects that are dependent on the degree of exposure, either by ingestion or skin exposure to the sediments resulting in risk to human health.

Lake dwelling animals are effected at pollutant concentrations that are less that the MTR levels (table 1). The terms for chemical pollution of sediments are as follows for effects on organisms; rare effect levels (REL), threshold effects limit (TEL), no observed effects limits-NOEL), probably effects limit (PEL), median effects level (ERM), and severe pollution, effecting most benthic organisms.

The human toxicological definition for "serious soil contamination" is taken as the soil quality resulting in exceeding of the Maximum Permissible Risk for intake (MPRhuman). The MPRhuman along with exposure modeling is the basis for what is called the serious risk concentration in humans (SRChuman) . For genotoxic carcinogens the acceptable excess lifetime cancer risk was set at 1 per 10,000 individuals; for all other compounds the MPRhuman does not result in any adverse health effects during lifetime exposure (70 yr.). Intervention Values used to classify historically contaminated soils were extrapolated to sediments. For deriving human-toxicological risk for sediment the human-toxicological Maximal Permissible Risk (MPR) level was used in combination with the SEDISOIL exposure model (exposure to contaminated sediment), that assumes ingesting 350 mg of sediment per day, with 50% of the body smeared in sediment for 8 hours per day for a total of 30 days per year over a lifetime. In this exposure model it was also assumed that 95% of the body's surface area to the water with 50 ml ingested per day of exposure (1.5 liters per year).

Table 1. Sediment concentrations that negatively affect human and other organisms in ppm (mg/kg).

Exposure	REL	TEL	PEL	Severe**	MTR	Target
As	4.1	5.9	17	30	55-3300****	29
Hg	0.09	0.13	0.49	0.7**	10- 6700	0.3
Cd	0.33	0.6	3.5	10	12-1800	0.8
Cr	25	37	90	110	380-17600	100
Cu	22	16	108	110	73->100000	35
Pb	25	30	91	250	530-3210	85
Ni		16	43	75	210-6700	35
Zn		124	270	410**	620->100000	140
DDT	0.0003	0.001	0.005	0.046**	4-7.3	0.01*
PCB	0.025	0.023	0.18	5.3**	1	
Benzene					1- 5.5	
PAH	1	1.6		40	40***	
Oil					1000 from 100-100000	50

REL- rare effect level, TEL – threshold effect level, PEL – probable effect level, Severe- severe effect level, ERM- median effect level, MTR- maximal tolerated risk,

Target – level that should be achieved after remediation, Early- pre-industrial levels found in St Laurence estuary, As- arsenic, Hg- mercury, Cd- cadmium, Cr- chromium, Cu- copper, Pb- lead, Ni-nickel, Zn- zinc, DDT- includes also DDE and DDD , PCB- , PAH- poly aromatic hydrocarbons.

*includes DDT, DDE and DDD

**is ERM rather than severe,

***includes the sum of only 10 carcinogenic PAHs

****The higher values are called the serious risk concentration which is meant to be equivalent to the MTR, the maximal tolerated risk, for a carcinogenic risk of up to 1 per 10,000 and with the following exposure to sediment: eating 350 mg per day, with 50% of the body smeared in sediment for 8 hours per day for a total of 30 days per year over a lifetime. Water drank per time is 50 ml, and exposure of 95% of the body's surface area to the water.

The human toxicological definition for "serious soil contamination" is taken as the soil quality resulting in exceeding of the Maximum Permissible Risk for intake (MPRhuman). The MPRhuman along with exposure modeling is the basis for what is called the serious risk concentration in humans (SRChuman) . For genotoxic carcinogens the acceptable excess lifetime cancer risk was set at 1 per 10,000 individuals; for all other compounds the MPRhuman does not result in any adverse health effects during lifetime exposure (70 yr.). Intervention Values used to classify historically contaminated soils were extrapolated to sediments. For deriving human-toxicological risk for sediment the human-toxicological Maximal Permissible Risk (MPR) level was used in combination with the SEDISOIL exposure model (exposure to contaminated sediment), that assumes ingesting 350 mg of sediment per day, with 50% of the body smeared in sediment for 8 hours per day for a total of 30 days per year over a lifetime. In this exposure model it was also assumed that 95% of the body's surface area to the water with 50 ml ingested per day of exposure (1.5 liters per year).

RISK ASSESSEMENT FOR CANCER

The EPA established standards to protect human health based on a level of acceptable pollution resulting in a virtually zero risk. The methodology entails extrapolating risks from very high human exposure to carcinogens in the workplace or from very high animal exposures (usually rats and mice) in the laboratory. This calculated risk is theoretical and will remain theoretical because at low exposure levels the potential risk is too small to measure in epidemiological studies, either in humans or rodents. Risk assessement is based on the assumption that the risk is dose dependent even at low levels of exposure. An alternative model is a threshold model that assumes that under a certain exposure level the body can deal with the added stress and that there will be no increased risk.

Some experts believe that this potential risk is a real risk and claim that there is no safe level of exposure to a carcinogenic material and that even a one- time exposure to an unspecified dose can result in cancer. Given this assumption it is possible to quantitate the risk to the individual and to the general population. It is also possible to calculate the probability that a patient with cancer became ill because of a certain exposures (presonal attrributable risk).

Chemicals considered to pose a potential risk for cancer in a linear dose-response manner are those that can mutate the body's genetic material (a mutagen). However, there are thousands of mutations occurring naturally every day in all of us, due to endogenous and exogenous natural and synthetic chemicals. The major exogenous exposure to both natural and synthetic chemicals comes from food. In order for us to survive the potentially dangerous lesions, the body nearly always fixes the damaged DNA or causes the cell dies. However, for a multitude of reasons as the body ages genetic defects accumulate, repair capabilities diminish, and cancer becomes more and more common. Age is by far the most important risk factor for cancer.

"Use of health protective risk assessment procedures as described in these cancer guidelines means that estimates are uncertain but more likely to overstate than under-state hazard and/or risk" (US EPA, 2005e). The National Research Council (NRC), the scientific council that provides the support for EPA decisions, saw the need to treat uncertainty in a predictable way that is "scientifically defensible, consistent with the agency's statutory mission, and responsive to the needs of decision-makers. "Associated with these estimates animal tumor findings are judged to be relevant to humans, and cancer risks are assumed to conform with low dose linearity". These judgments are of uncertain validity. However, estimating thresholds from epidemiological studies are also problematic; for example, a response that is not statistically significant can be consistent with a small risk that falls below an experiment's power of detection. It is the Agency's long-standing science policy position that use of the linear low-dose extrapolation approach provides adequate public health conservatism in the absence of chemical-specific data indicating differential early-life sensitivity or when the mode of action is not mutagenic. When the weight of evidence evaluation of all available data are insufficient to establish the mode of action for a tumor site and when scientifically plausible based on the available data, linear extrapolation is used as a default approach, because linear extrapolation generally is considered to be a health-protective approach. The slope of this line, known as the *slope factor*, is an upper-bound estimate of risk per increment of dose that can be used to estimate risk probabilities for different exposure levels.

The upper-bound estimate of risk per increment of dose might not be protective for a highly susceptible individual or group. Some individuals face a higher risk and some face a lower risk. The use of upper bounds generally is considered to be a health-protective approach for covering the risk to susceptible individuals, although the calculation of upper bounds is not based on susceptibility data. Similarly, exposure during some life-stages can contribute more or less to the total lifetime risk than do similar exposures at other times. Such an estimate, however, does not necessarily give a realistic prediction of the risk. The true value of the risk is unknown and may be as low as zero. The range of risks, defined by the upper limit given by the chosen model and the lower limit which may be as low as zero should be explicitly stated. An established procedure does not yet exist for making "most likely" or "best" estimates of risk within the range of uncertainty defined by the upper and lower limit estimates. It will remain uncertain what happens at the "*point of departure*" (POD). The POD is an estimated dose (usually expressed in human-equivalent terms) near the

Table 2. Oral cancer slopes.

Chemicals	Animal models	Oral Cancer slope* μg/L,
Mercury-inorganic	None	Not carcinogenic
methylmercury	none	Not carcinogenic
Organochlorines		
DDT,	Lung, liver	1
Aldrin/dieldrin	Liver	0.02
dioxins (2,3,7,8 TCDD),	Liver, thyroid skin, other, scarcoma**	0.18 (10-5)
	Liver, thyroid, gall bladder	
Polychlorinated biphenyl	Liver	1
	Liver	
Mirex	Liver, thyroid	1
chlordane		1
toxaphen		0.3
Chlorinated solvents- e.g. TCE	Kidney, liver	0.0007- 0.03
Benzene	Leukemia**	10-100
Benzo[a]pyrene	Skin, lung**	0.05
Arsenic	Skin**, lung**, bladder**, liver**	0.2
Chromium		Not carcinogenic
Inorganic Lead	Renal	None given
organic lead	brain	
Nitrosamines	liver	0.007

* Assuming drinking 2 liters of water per day will increase by 1/100,000. cancers for a lifetime.

** also in humans.

***reference dose for chronic oral exposure.

lower end of the observed range of increased risk without significant extrapolation to lower doses. The oral cancer slope can be expressed as the concentration that would lead to one extra lifetime cancer in 100,000 people given the fact that 2 liters of water is ingested every day. Table 2 gives the oral cancer slopes for the chemicals of concern.

Uncertainties in the Models Used in the Classification of Chemicals as Carcinogens

Some aspects of model uncertainty that should be addressed in an assessment include the use of animal models as a surrogate for humans, the influence of cross-species differences in metabolism and physiology, the use of effects observed at high doses as an indicator of the potential for effects at lower doses, the effect of using linear or nonlinear extrapolation to estimate risks, the use of using small samples and subgroups to make inferences about entire human populations or subpopulations with differential susceptibilities, and the use of experimental exposure regimens to make inferences about different human exposure scenarios (NRC, 2002).

Animal models are used to help in classifying the potential of chemicals to increase the risk of cancer in humans. Without these models, we are left with natural experiments in humans exposed to high concentrations of chemicals in the work place. Examples of carcinogens that definitely cause cancer include, benzene causing acute nonlymphocytic leukemia, asbestos causing malignant mesothelioma, vinyl choride causing angiosarcoma of the liver, and arsenic causing lung cancer. These exposures have been nearly exclusively from breathing polluted air and even then, chronic exposures leading to a definite increased risk of cancer are not common in the workplace despite exposure to tens of thousands of chemicals. The only definite chemical induced cancer caused by other routes of exposure is skin cancer caused by dermal exposure to polycyclic aromatic hydrocarbons. In the general population chronic exposure to carcinogens leading to a definite increase in cancer is even rarer. The only prevalent chemical in water and soil/sediment exposure leading to increased cancer rates is exposure to arsenic in drinking water.

Nearly all of the definite carcinogens as classified by the IARC have required multiple studies demonstrating a clear increase in the risk for cancer. However, recently the IARC has classified some chemicals as definite carcinogens without clear evidence in humans. An example is formaldehyde that has been classified as a definite carcinogen with only suggestive evidence in humans, but definite evidence in animals and theoretical mechanistic evidence. Most studies reported in the literature to support the probable carcinogen classification are for low levels of industrial pollutants, with usually weak and inconsistent associations, and without correction for diet, a potentially large confounding factor. The higher quality studies correct for age, smoking, exercise and body mass, important predictors of cancer and overall morbidity and mortality. There are other important factors such as socio-economic class and ethnicity. The large variation in cancer rates in regions in a single country and between countries is probably due primarily to the major risk factors or to chance.

Using animal models to classify carcinogens is prevalent but controversial. The major reason for using such models is the lack of alternative methods. Bruce N Ames who developed the Ames test pointed out the limitations of animal models (Ames, 2003, Ames,1997). The primary problem is that the dose used in animal studies is high enough to kill cells in the target organs. Thus it is not clear whether the increased cancer rate in animals is do to a mutagenic effect of the chemical or the response of the cells to injury. High doses are required in order to increase the risk of cancer enough to detect it in such studies. That standard is a study with 50 mice in each group exposed to different doses of a chemical in both male and female animals, leading to a minimum of 300 mice per study. (no exposure, exposure and higher exposure in males and females = 6 groups of 50 animals each = 300). A two-year follow-up period is required and the animals are killed and the tissues are examined to determine the incidence of various cancers. Such studies are costly. Given the fact that the lower the dose the lower the incidence of cancer, to detect an increase in risk of cancer at lower doses requires many more rodents in each group and is not economically feasible. Thus animal cancer tests are conducted on synthetic chemicals at the maximum tolerated dose (MTD) of the chemical, and regulatory agencies use the results to predict human risk at much lower levels of exposure.

Using the standard high-dose animal cancer test results in the classification of 50% of chemicals, both natural and synthetic as rodent carcinogens (Ames 2003) and include commercial pesticides, drugs in the Physician's Desk Reference (PDR), natural pesticides found in vegetables, roasted coffee, and burnt material produced in usual cooking practices. It has been suggested that this might be due to cell division caused by the high dose itself, rather than and carcinogenic property of the chemical at lower doses.

The high proportion of natural and synthetic rodent carcinogens leads to the conclusion that either the animal cancer tests are of no value or that we are exposed to a large number of natural and synthetic carcinogens. Certainly it has been shown that in each cell in the human body there are hundreds to thousands of damaging events per day, including those resulting in oxidative damage, depurination, depyrimidination, deamination, nitric oxide interactions, and endogenous exocyclic adducts. Most DNA damage is repaired but not all, and mutations accumulate with age. The reason for a major cancer protective effect of a diet rich in fruits and vegetables in uncertain but might be due to an improved balance between the damage and repair mechanisms of DNA and other cellular components. Nevertheless animal studies have not been shown to correlate with daily DNA damage at low doses of exposure to potential carcinogenic chemicals. Thus extrapolation to low dose exposures in humans is theoretical and conservative.

Alternatives to costly animal models includes the Ames test that tests mutations in bacteria after adding a chemical and liver enzymes; testing for the carcinogenicity of both the chemical and its metabolites. This test is also called the Salmonella mutagenicity assay where mutagenic Salmonella grow better in a histamine poor culture medium, than do the natural strain. Such assays however predict carcinogenicity rather poorly (Cohen SM, 2004) as do other more modern tests based on structure activity relationship models with end points such as DNA reactivity, mutagenesis, and receptor interactions. Further developments are sorely needed.

Acceptable Risk

The EPA recognized that scientific opinion will be divided on these issues, and that there are those who will claim that the results of the linear model give actual risks, even though they were intended to provide a high degree of public safety and define a “virtually zero” risk.

If the acceptible risk of cancer at the maximum limiting concentration (MLC) of the EPA is calculated to be 1/100,000 increased cases of cancer over a lifetime then if we extrapolate this risk to a population of 200 million, there would be an additional 2000 cases of cancer. If there was a100 fold increased concentration of a carcinogen in the drinking water over the MLC, we can calculate 200,000 additional cases of cancer! How can we justify causing cancer in 200,000 people! In fact how can we justify even one extra case of cancer in those exposed to carcinogens in the drinking water? Therefore without looking at the actual potential risk, and realizing that these are meant to be "virtually zero" risks, public outcry and protests are understandable and justifiable.

Risk assessment is the process of quantifying the probability of a harmful effect to individuals or populations from certain human activities. However before beginning assessments, regulatory agencies need to determine what risk is acceptable. The FDA required in 1973 that cancer-causing compounds must not be present in meat at concentrations that would cause a potential cancer risk greater than 1 in a million lifetimes, and the idea of not increasing lifetime risk by more than one in a million is common place even today. How consensus settled on this particular figure however is unclear. The figure provides a numerical basis for what to consider a negligible increase in risk, or a virtually zero risk. Others have considered "acceptable" potential individual risks from environmental pollution of below one in ten thousand increased lifetime risk. Even a potential lifetime risk of 1 in 1000 is often considered acceptable in the workplace. In anycase the risk is clearly small relative to the typical 30% risk of getting cancer by age 75 in developed countries. The major risk factors for cancer are age, gender, ethnicity, nationality, cigarette smoking, alcohol intake, lack of regular physical exercise, obesity, diets low in vegtables and fruit, socio-economic class, and a positive family history. Of course the known risk factors vary according to the type of cancer and in some cases can explain up to 50% of the reasons people become ill with cancer without taking into consideration age, nearly always the most significant risk factor. Today most experts estimate of the proportion of cancers caused by environmental exposures to be 2% or less for the general population. Therefore the effects of environmental pollution even if real are likely to be lost in the sea of major known risk factors. Also any positive findings in studies of populations exposed to low levels of pollutants are likely to be due to the fact that major risk factors were not identical in the comparison groups.

Individuals may be tempted to advocate the adoption of a zero-risk policy. After all the 1 in a million policy would still cause the death of hundreds or thousands, of people in a large enough population if the risks were real. There are those who advocate not only the zero-risk policy but also the precautionary principle. Risk assessment may be used to justify hazardous practices, whereas the goal of the "precautionary principle" is to use science to avoid or minimize dangers, not to justify hazards. "When an activity raises the threat of harm to human health or the environment, precautionary measures should be taken even if some cause and effect relationships are not established scientifically." (Wingspread Conference on Implementing the Precautionary Principle, January 1998 Massachusetts Precautionary Principle Project Fall 1999). Industry introduces thousands of new chemicals each year and there are tens of thousands of other chemicals that haven't been tested for carcinogenic potential. If we wait until the increased risks are conclusively defined we are too late for many of those exposed who are already ill due to exposure.

In practice however, a true zero-risk is possible only with the suppression of the risk-causing activity. More stringent requirements, or even the 1 in a million may not be technologically feasible at a given time, or so expensive as to render the risk-causing activity unsustainable. In the interest of public health, the risks vs. benefits of the possible alternatives must be carefully considered.

For example it may be decided that eating vegetables that have residual pesticides even at very low levels are not allowed. Some pesticides are suspected carcinogens. "Organic" vegetables are more expensive, and this will lead to less intake of vegetables, leading to an increased incidence of some cancers. However, vegetables have many natural pesticides and based on animal models, the exposure to these pesticides have been calculated to be potentially more dangerous than the artificial pesticide residuals. We do not know whether or not artificial pesticides are more dangerous than natural pesticides. The potential risks in either case are extremely low whereas the benefits of eating fruits and vegetables have been found consistently in numerous studies, and are measurable and substantial.

To put acceptable risk into perspective, one can ask what would be the EPAs maximal concentration limits (MCL) for alcohol given the fact that alcohol is a definite carcinogen with 2.5 standard drinks per day increasing the lifetime risk for cancer by 30%. That means that instead of 30,000 cases of cancer per 100000 people (table 3), there will be 39,000 cases in 100,000 people drinking 2.5 standard drinks per day or an increase of 9000 cases per 100,000 people. However, we will accept only 1 extra case per 100,000 people over a lifetime. Therefore we need to decrease the dose by a factor of 9000 so as to have only 1 extra case per 100,000 people (2.5 drinks X 365 days divided by 9000 = 0.1 per year, or one alcoholic drink (beer, or whisky or wine) every 10 years. So if we had an alcoholic beverage once a week we will be 520 times over the recommended MLC. Obviously this will lead to an unmeasureable potential increased risk for cancer, and in fact the risk is probable for all practical purposes zero. The patient with cancer who drank one alcoholic beverage per week could be calculated to have a 1.7% chance that his cancer was due to drinking[1]. This demonstrates how hypothetical and safe the EPA standards are for exposure to polluted waters, and that we would need an extreme exposure to cause an increased risk in the exposed

---

[1] for those interested in the actual calculation the overall increased potential general risk is 30,000 plus 512 or 30,512 divided by 30,000 or 1.017. For the individual with cancer, the chances that he got his cancer from such exposure is 1.017- 1 divided by 1.017 or 1.7%

population that is measurable. It also seems implausible that an exposure to pollutants even hundreds of times over the MLC really increases the risk for cancer. Actual recommendations for alcoholic intake are between 1-3 drinks per day in the USA and Europe for low risk, and Sweden defines "not dangerous" as 4 drinks per week. At 4 drinks per week we are talking about an additional risk of at least 520 X 4 or an extra 2080 cases of cancer per 100,000 population. If we adopt the USA maximum recommendation we arrive at an extra 520 X 3 X 7 = 10,920 cancer cases per 100,000 subjects. In a population of 10 million, a small country, we would be accepting a theoretical increase of 10,920 cases per 100,000 people or 1,092,000 extra cancer cases over a lifetime in the country!

Table 3. SEERS- Cummulative risk of cancer- all cancers except nonmelanoma skin cancer (white males).

Age	Risk
45	2
50	3
55	6
60	9
65	15
70	22
75	30
80	36
85	41
90	43
95	45

Another example might be drinking coffee, where it can be shown that the potential risk to become ill with bladder cancer is 1 per 100,000 if one drinks one cup of coffee every five days (results not shown). For even the increased risk of lung cancer alone from cigarette smoking, we would be allowed one cigarette about every three years, and one cigarette a week would represent an exposure level 150 times above the acceptable limits (results not shown). This puts such calculations into perspective, and shows how strict the standards are in order to protect public health.

How the Risk Is Determined

In 1983 the National Research Council published Risk Assessment in the Federal Government: Managing the Process, commonly known as the "Redbook." This established the 4-step process that has become the dominant paradigm for risk assessment, also used today by the major textbooks of Cancer Epidemiology (Schottenfeld D, 2006). Risk assessment integrates the disciplines of toxicology and exposure assessment to attempt to understand and measure what types of harm humans or ecosystems might experience from exposure to a chemical or pollutant. It uses available scientific evidence as well as assumptions, mathematical modeling and policy judgments, to attempt to estimate risk. In human health terms, risk is a measure of the chance that a person or population will experience injury, disease or death (a hazard) under certain circumstances or exposures. It is a combination of: the probability that an undesired event (exposure to a toxic chemical) will occur, and; the consequences that occur as a result of that event (injury, disease or death).

The first step, Hazard Identification, aims to determine the qualitative nature of the potential adverse consequences of the contaminant (chemical, radiation, noise, etc.) and the strength of the evidence it can have that effect. This is done, for chemical hazards, by drawing from the results of the sciences of toxicology and epidemiology. The International Agency for Cancer Research (IARC) classifies chemicals at this stage. Their decisions are that a certain substance is carcinogenic under certain circumstances, but they don't define the circumstances. "A cancer 'hazard' is an agent that is capable of causing cancer under some circumstances, while a cancer risk is an estimate of the carcinogenic effects expected from exposure to a cancer hazard. The Monographs are an exercise in evaluating cancer hazards, despite the historical presence of the word 'risks' in the title. The distinction between hazard and risk is important, and the Monographs identify cancer hazards even when the risks are very low at current exposure levels, because new used or unforeseen exposures could engender risks that are significantly higher". (IARC, 2006).

The IARC also stipulates the end-organ at risk in animal models and in humans , but this doesn't rule out the possibility that other end-organs are also at increased risk. Their expert decisions take into consideration the available studies in exposed animals and humans by using the following major criteria - strength of association, consistency of the association, biological plausibility, and dose response.

Strength of association is the degree that which the exposure increases the risk of the disease. Usually at least a two-fold risk is considered support for causality. The problem with smaller degrees of increased risk is that small differences in major factors that increase the risk for disease mentioned above in those exposed and not exposed can cause results that are not real. Such a problem is called directional bias due to confounders. There is also the problem of publication bias, in that journals are only likely to publish studies that demonstrate a risk.

Environmental exposures have rarely been shown to increase cancer rates, despite millions of dollars that have been spent to try to explain variations in cancer rates in the US. Once cancer rates became available in various geographical regions in the United States, it was found that there are great variations in such rates between various areas in the USA (table 4). One might be tempted to claim that this is due to environmental exposures to carcinogens. In fact such data has lead to many expensive attempts to explain such differences. For example in Long Island, an attempt was made to determine why one area had an increased incidence of breast cancer. No biologically plausible reason was found. The major suspected chemicals were organochlorides in food. In the world there have been only a few instances where an environmental cause of cancer has been found. An example is in Turkey where a community was exposed to dust that included particles similar to asbestos, and an increased incidence of mesothelioma was found, a cancer that is nearly always due to asbestos or similar type of fiber exposure. It is also very difficult to use changes in the cancer rates over time to support or reject the importance of the influence of environmental exposures on cancer rates. The problem with secular changes in cancer rates is that smoking rates have changed and cardiovascular disease mortality has decreased. The major influence on lung cancer rates is the prevalence of smoking that has changed over the years. Another confounding factor is that over the last 30 years of the 21st century cardiovascular disease has decreased signficiantly, so we would expect that even after age adjustment that the incidence of cancer would increase since we have to die from something (bias of competing causes of death). Most cancers have remained stable over the last 30 years (table 5). Those with increasing incidences are melanoma and non Hodgkin's lymphoma, breast cancer, prostate cancers and kidney cancers. The reason for these increases are uncertain, but might be due in part to increased screening, detection and more exact diagnosis. For breast cancer for example, the recent Cockran review has shown that there is a 33% increased detection rate in those screened versus those not screened with mamography, but that screening did not decrease the risk of dying from cancer or from breast cancer (Olsen O, 2001). Kidny cancers are often incidental findings on abdominal CT or ultrasound studies that are prevalent tests today. In any case secular changes in cancer rates do not support an increased chemical carcinogenic effect in the general population.

Table 4. Cancer rates per 100,000 age adjusted according to 2000 from the years 1995-2004 (SEERS in selected areas in the United States)

Area	Whites	Blacks	Hispanics
Total	556	686	420
San Francisco	553	637	433
Conneticut	588	656	540
Detroit	636	775	449
Hawaii	601	438	365
New Mexico	497	470	267
Utah	486	520	451
Difference – %*	31%	77%	102%

*(highest – lowest) divided by lowest.

In epidemiological follow-up studies it is often very difficult to control for the major predictive factors mentioned above. Sometimes the information is not available, and othertimes the factors are difficult to define. These risk factors are called confounders since they can confound the results. Another major type of epidemiological study is called a case-controlled study. Here the researcher takes one group who has the disease and compairs them to a control group that doesn't have the disease. The reason this type of study is done is that most often the diseases are not common enough to study by a follow-up study of exposed individuals. A follow-up study requires enough people with and without exposure and enough people who develop the disease to study the possibility of an increased risk. The case-control study however is very problematic. The first problem is to make sure that the control group is the same in every way to those with the disease. The second problem is what is called "recall bias". The patient with a disease is much more likely to recall exposures than a healthy person without disease. This is because it is human nature to try to explain why we became ill. On the other hand patients might forget exposures leading to negative findings despite a real association. If the epidemiological data are consistently showing an increased risk, then toxicology studies the exposures and toxic effects on the body to determine if the increased risk is biologically plausible. The interpretation of epidemiologic data in both regulation and litigation is often challenged and debated.

Table 5. Secular changes in cancer rates from 1975 through 2004 (SEERS)- cancers most strongly associated with exposures in the workplace (age-adjusted rates/100,000).

Cancer and group at risk	1975	1985	1995	2004
Total*				
WM	470	530	570	540
WF	370	410	425	420
BM	525	640	740	660
BF	360	390	400	400
Lung				
WM	89	97	85	72
WF	24	40	55	52
BM	114	150	129	100
BF	25	46	50	55
Bladder				
WM	36	39	39	39
WF	9	10	10	10
BM	16	20	18	20
BF	6	8	7	8
Kidney				
WM	11	14	15	18
WF	5	6	8	9
BM	8	12	21	22
BF	4	6	9	10
Leukemia				
WM	17	18	19	16
WF	10	11	11	10
BM	16	17	13	14
BF	8	9	7	8
Cancers and cohorts	1975	1985	1995	2004
Prostate				
WM	92	114	164	155
BM	140	170	280	240
Breast				
WF	107	128	137	128
BF	94	111	124	119
Stomach				
WM	16	13	11	10
WF	7.5	6	5	5
BM	25	22.5	19	17.5
BF	11	11	8	10
Lymphoma				
WM	13	19	26	26
WF	10	14	16	18
BM	7.5	12	23	23
BF	5	8	11	13
Melanoma				
WM	9	17	25	29
WF	8	13	18	20
BM	1	1	2	1
BF	1	1	1	1

*WM = white male, WF = white female, BM = black male, BF= black female.

Hazard identification also needs to consider the route of exposure and the form of the recognized carcinogen (table 6). For example cadmium, chromium and nickel are carcinogenic if inhaled, but are not been shown to be carcinogenic if ingested or absorbed through the skin. On the other hand benzene and arsenic are considered carcinogenic by any route of absorption. We ingest primarily organic arsenic that is not considered carcinogenic, but in areas with high concentrations of inorganic arsenic in the water there is a definite increase in cancer rates, primarily squamous cancer of the skin.

For risk assessments in patients exposed to carcinogens the type of cancer and the time from first exposure until the diagnosis of the disease also needs to be considered. The time from the beginning of exposure until the diagnosis of the disease is critical and can increase or decrease the probability of causality. For example after exposure to radiation, an increase of the rate of leukemia occurs within 10 years, but usually after 5 years, whereas for solid tumors the latency period is usually longer than 20 to 30 years. If a patient became ill 2 years after first being exposed, it is very unlikely that the association is causal. On the other hand studies of exposed workers should not include those who were followed up for only a few years, since they are less likely in a short time period to become ill with cancer. Including such workers in a follow-up study will dilute the results. Thus usually at least 10 years is required for solid tumors to include the worker as being at risk. If all workers are included, even those with short follow-up periods, this is called dilution bias, and will result in falsely lowering the risks from exposure.

The cancer type is also important when considering causality. There are cancers that are associated or suspected to be associated with exposures and there are other cancers that have not been shown to be carcinogenic in humans despite exposure of workers to high doses. For example, inhalation of certain forms of chromium has been shown to increase the risk for lung cancer, but not most other types of cancer (e.g. bladder, kidney, prostate, stomach, breast). Although we cannot rule out an association, the fact that the other cancers were not detected with exposure to high doses of the carcinogen, decreases the probability that a patient with such a cancer and exposure to chromium became ill due to the exposure.

> For the personal attributable risk analysis, the latent period for most cancers implies that exposures immediately preceding the detection of a tumor would be less likely to have contributed to its development and, therefore, may count less in the analysis. Study subjects who were first exposed near the end of the study may not have had adequate time since exposure for cancer to develop; therefore, analysis of their data may be similar to analysis of data for those who were not exposed. However, for carcinogens acting on multiple stages of the carcinogenic process, especially the later stages, all periods of exposure including recent exposures, may be important.

Finally the biological plausibility of an association needs to be considered. Biological plausibility is based on experimental data trying to explain the mechanism the results in an increased cancer risk. The problem with these considerations is that carcinogenic mechanisms are complex, and only partially understood.

The second step for chemical risk assessment, Dose-Response Analysis, is determining the relationship between dose and the probability or the incidence of effect (dose-response assessment). The complexity of this step in many contexts derives mainly from the need to extrapolate results from experimental animals (e.g. mice, rats) to humans, and/or from high to lower doses in both animals and humans exposed to high doses in the workplace.

Table 6. Hazard identification in a patient with cancer exposed to polluted lake water.

Epidemiological and toxicological studies	Consistent and biologically plausible association
Route of exposure	Proper route of exposure
Type of cancer	Associated with the specific chemical exposure
Latency period	Long enough time from first exposure until the diagnosis of the disease

The second step for chemical risk assessment, Dose-Response Analysis, is determining the relationship between dose and the probability or the incidence of effect (dose-response assessment). The complexity of this step in many contexts derives mainly from the need to extrapolate results from experimental animals (e.g. mice, rats) to humans, and/or from high to lower doses in both animals and humans exposed to high doses in the workplace. Nevertheless such assessments are generally thought to be conservative (see above). The dose-response relationship varies with pollutant, individual sensitivity, and type of health effect. The EPA has previously assumed, for cancer only, that there was no "zero risk," and that any exposure created a risk of cancer. This is called a "non-threshold" model. This assumption may change based on the new proposed guidelines for cancer risk assessment that provide for a "threshold" response for certain types of cancer-causing substances that do not directly disrupt DNA.

The "threshold" model assumes that there is some level where no damage occurs or the contamination results in damage that the body can repair or assimilate without causing harm. For non-cancer health effects, a threshold is generally assumed, although science shows that this assumption may be incorrect. For example, for hazards such as harm to the reproductive system or fetus, or to the nervous system, there may be no threshold. Or, for endocrine disruptors, timing of the dose (a small window during pregnancy or development) and not the dose itself (which may be tiny) may be more important. A recent study of people exposed to arsenic in drinking water, a definite carcinogen, found no increase in risk as the arsenic exposure increased (in fact they found a decreased risk, that might be due to confounders, even though there are those who claim that a small dose of a carcinogen, might in some cases be protective by revving up the bodies defense mechanisms)(Basstrup, 2008).

The "threshold" response type of model may become the norm for cancer assessments involving chemicals that do not cause direct DNA damage, even though they may cause damage indirectly. There is no general agreement on which dose-response model is the correct one for different types of hazards. In any case most of these arguments will remain controversial since the risk is small enough to be unmeasurable.

The third step, Exposure Quantification, aims to determine the amount of a contaminant (dose) that individuals and populations will receive. This is done by examining the results of the discipline of exposure assessment. As different location, lifestyles and other factors likely influence the amount of contaminant that is received, a range or distribution of possible values is generated in this step. Particular care is taken to determine the exposure of the susceptible population(s). For exposure to polluted lake water and sediment, an estimation of days of recreation, the body surface are exposed to polluted sediment and water, hours of exposure, amount of water ingested, and sometimes depending on the carcinogen the amount of fish ingested needs to be considered. For the sediment lake model used to determine interventional values (see above), it assumes eating 350 mg of sediment per day, with 50% of the body smeared in sediment and 95% of the body's surface area exposed to water for 8 hours per day for a total of 30 days per year over a lifetime. It was assumed also that 50 ml of water was ingested per day of exposure.

Finally, the results of the three steps above are then combined to produce an estimate of risk, (risk characterization or quantification). Because of the different susceptibilities and exposures, this risk will vary within a population. The results are probabilistic risk assessments. Conclusions from these types of risk assessments include presenting a distribution of risks that considers a range of estimated uncertainties, not just a single number. They also more explicitly acknowledge some of the uncertainties of assumptions used. While such methods provide a better way to demonstrate how risk estimates change with changes in the assumptions made, these models are still limited by the types of information used.

According to the EPA, "the science of risk assessments has developed considerably from its early roots. However, there seems to be widespread agreement that the data and methodologies needed for precise health risk assessment do not yet exist."

Quantitative Risk Assessment

We can calculate absorption of chemicals using the standard formulas to calculate absorption from water dermal absorption, sediment dermal absorption, and gastrointestinal water and sediment absorption. We can then use the cancer slope to calculate the increased risk. We will use benzene exposure as an example of a quantitative risk assessment.

There are three ways benzene can be absorbed during recreational activity in polluted lakes; water dermal absorption, sediment dermal absorption and gastrointestinal water and sediment absorption. For this model we will consider that the gastrointestinal absorption of benzene from the sediment is identical to that in drinking water. This assumption is conservative since some of the benzene will be bound to the sediment and unavailable for absorption. We will use the sediment lake model to determine the risk, assuming eating 350 mg of sediment per day, with 50% of the body smeared in sediment and 95% of the body's surface area exposed to water for 8 hours per day for a total of 30 days per year over a lifetime. It was assumed also that 50 ml of water was ingested per day of exposure.

The drinking water concentration of benzene that will give one extra cancer per 100,000 people per lifetime is 10- 100 μg/L for a total dose of 20-200 μg ingested per day (2 liters). Nearly all of this is absorbed (see table 7). This is derived from the cancer slope which is estimated to range from 4.4 X 10(-7) to 1.6 X 10 (-6) per μg/L. Cancer slopes for the various chemicals can be found in the Integrated Risk Information System (IRIS) in the EPA web site.

For dermal water absorption the formula is as follows; Concentration of the chemical (μg/L) times the absorption coefficient -Kp (cm/hr) times body surface area (19600 cm2 in an adult) times the number of hours of exposure. For benzene the Kp is 0.0207 (table 7). The highest water concentration was 100 μg/L and in sediment maximum concentrations were slightly less than the intervention value of 5.5 mg/Kg. Since we assume 8 hours exposure for 30 days per year, our average daily exposure is 8/12 or 2/3 of an hour.

Daily dermal absorption at 100 μg/L is 0.100 μg/cm3 times 19600 cm2 times 0.0207 cm/hr times 2/3 hour. Thus the total dermal absorption is equal to 27 μg per day.

For sediment dermal absorption , we have 5.5 mg/Kg for a concentration, with 50% of the body smeared with sediment for an average of 2/3 hours per day. The formula for dermal sediment absorption is; the concentration of benzene in the sediment in μg/g that is the same as mg/kg or ppm times the relevant skin surface area (half of the body surface area of 19,600 or 9800 cm^2) times the absorption coefficient (0.01/24 hrs- see table 7) times the adherence factor (.021g/ cm^2) times the exposure time in hours. The adherence factor can very but here it is assumed to be the same as for a child playing in the mud. This gives us the calculation of; 5.5 μg/g times 9800 cm2 times 0.01/24 hours times 0.021 g/cm2 times 2/3 hours = 0.3 μg.

Now we are left with the gastrointestinal absorption that is nearly 100% with an absorption coefficient of 0.97. In our model we drink 50 cc of water ever 30 days and eat 350 mg of sediment. Our daily intake of water is 50 X 30 days divided by 365 days for an average of 4.1 ml of water. At a concentration of 100 μg/L we absorb 4.1 ml X 100 μg/1000 ml or 0.41 μg.

The absorption of the 350 mg of sediment each of 30 days gives us 28.8 mg per day. The concentration of the benzene in the sediment is 5.5 mg/kg, so 5.5 mg/kg X 0.0000228 kg = 0.000001254 mg or 0.0013 μg. We can see therefore that nearly all the absorption comes from skin absorption of the benzene in the water. It is unclear therefore why the interventional value for benzene in the sediment was set at such a low value (table 8).

Table 8. The absorption of maximal values.

Absorption	Total μg
Dermal – water	27
Dermal- sediment	0.3
Ingestion- water	0.41
Ingestion sediment	0.0013

For argument lets assume that higher risk value or a steeper cancer slope for benzene and assume that 10 μg/L for drinking water results in an increased risk of 1 lifetime case of leukemia per 100,000 people. This value is twice the MLC for benzene. Nevertheless the increased risk is 27 divided by that absorbed from two liters of drinking water (27/20 μg times 0.97) equals around 1.4 lifetime increased cases of leukemia over a lifetime. As for all types of exposure we should put the increased dose of benzene into the prospective of daily exposures. We can see in table 7 that most of our exposure to benzene comes from breathing the air with exposure to around 300 μg per day (range of 140- 580 μg per day)(Wallace 1996) and absorption of 150 μg per day. At work it is permissible to absorb up to 8000 μg/day, and those smoking 1 pack per day are exposed to 1000 μg per day and absorb 500 μg/day. Therefore the increased absorption of 27 μg is minor. There is no evidence that even 8000 μg per day can increase the incidence of acute leukemia.

We have also given the basic data for the PAH (table 7) and metals (table 8).

This methodology can help us set standards for interventional values. For example we can ask do the interventional values for sediment PAH conform with 1/100,000 increased risk proposal. The PAH absorption from sediment is 28.4 μg per day given the SEDSOIL exposure model.

It must be emphasized that there are few human data regarding oral or dermal exposure to benzene. The simple absorption ratio approach taken to route-to-route extrapolation here cannot account for differences in disposition of benzene after it crosses the pulmonary, skin or gastrointestinal barrier. First-pass metabolism of ingested benzene may result in a more efficient production of leukemogenic metabolites. On the other hand, rapid clearance of benzene and metabolites after ingestion may be a mitigating factor (U.S. EPA 1999). Nevertheless the U.S. EPA has recommended extrapolating from the benzene inhalation unit risk extimate to the oral route of exposure (E.S. EPA 1999) in determining water standards. This is based on the finding that inhaled or ingested benzene results in similar toxic effects and metabolites of benzene in animals.

Table 7. Various standards , general exposures for the general population, and workers, and absorption coefficients.

	Benzene	PAH
Water standard (μg/L)	5	0.2
Work standard (μg/m3)	1600	200
Lung retained and absorbed (%).	50	>31
GI absorption (%)	97	31
Kp	0.0207	1.24
ABS-soil	0.01	0.13
Food-absorbed (mg)	Minor	1.6
Air absorption- world average	150 μg*	little
Work absorbed (μg)**	8000	6000
Water standard (USEPA)μg/L % absorbed daily absorbed drinking water (μg)	5 97% 9.7	0.2 31% 0.06

*concentration global average 15 μg/m3 X 50% absorbed X 20 m3 per day

**TLV-TWA or PEL X 10 m3/day X proportion absorbed in the lungs.

Increased risk and Increased Personal Attributable Risk

The lifetime risk in the white male population in the United States for all types of leukemia,(U.S. NIH and NCI, 2007)) is 9 per 1000 or 900 per 100,000 people. We calculated that the exposure above could theoretically increase the number of cases by 1.4 cases per 100,000 people. The increased risk can be calculated by taking the expected number + additional cases divided by the expected number of cases (900 + 1.4 divided by 900 = 1.0015). This is called the relative risk (RR), and in this case it is 1.0015.

Lets say that a patient with leukemia thought that this disease was caused by the exposure to the polluted lake waters. Given the relative risk, we can calculate the personal attributable risk, that equals the RR – 1 divided by the RR (in this case 1.0015 – 1 divided by 1.0015 = 0.15%. That means we can tell the patient that the chances that his exposure caused his disease is maximally 1.5 per 1000, and might be zero. There is over a 99% chance that his exposure did not cause the disease.

Table 8. Water standards, work standards, absorption coefficients, water concentrations, and sediment concentrations.

	Nickel	Cadmium	Chromium	Arsenic	Lead
Water standard (μg/L)	100	5	100	50	15
Work standard (μg/m3)	1000	2	10	100	100
Lung retained and absorbed (%).	>27	10-30	>50	>50	20-40
GI absorption (%)	27	5	2	40-80	15
Kp	0.000329	0.00035	0.001	0.002	0.0001
ABS-soil	0.001	0.001	0.001	0.03	0.001
Food-absorbed (μg)	8	0.5-2.5	2.5	0.8-16	1.5
Minimal absorption – outdoor, non-smoking environment	Little	Little	Little	Little	0.4
Work absorbed (μg)*	27918000	41000	517000	5170000	2060000
Water standard (USEPA)μg/L % absorbed	100 27%	5 5%	100 2.0%	50 80%	15 10%
lifetime absorbed drinking water (μg)	27	0.25	2	40	1.5

*outdoor concentration global average 6 ug/m3 X 50% absorbed X 20 m3 per day
**TLV-TWA or PEL X 10 m3/day X proportion absorbed in the lungs.

Multiple Risks

Life however is never that simple, and we have often multiple risk factors in the exposed patient, where his exposure is only part of the story. If we assume that each risk factor is individually sufficient but not necessary for causation of the outcome , since there is an absence of a detailed biologic model we can calculate the part of each risk factor (McElduff P, 2002). Lets say we have a 50 year old patient with lung cancer whose age increases the risk by a factor of 5 compared to a younger person, and his minimal smoking history increased his risk by a factor of 2. Additionally at work he was exposed to cadmium with a dose that would be expected to increase his risk of lung cancer by a factor of 1.5. What is the contribution of the exposure in the work place to the development of lung cancer. Without corrections of the multiple risk model, we could say that there was a 33% chance that his disease came from exposure at work ((RR-1)/RR = 1.5-1 = 0.5 divided by 1.5 = 0.33).

If we use an additive model we can say that the total RR = 1.5+2+5 or 8.5 and if there is a multiplicative model that assumes a multiplicative synergistic interaction between the risk factors then the total RR equals 1.5 X 2 X 5 or 15. In the first case the model explains 8.5-1/8.5 = 88.2% of the disease, whereas if we use the synergistic model we arrive at 15-1/15 or 93.3%.

For the individual risks the formula is as follows; RR-1 divided by all the other relative risks –1 for each one; or in our case 1.5 –1 = 0.5 divided by (1.5+2 +5 minus 3= 0.5 divided by 5.5 or 0.09. The model doesn't totally explain the illness so we need to multiple the final result by the proportion of the disease explained by the total model (either 88.2% for the additive model or 93.3% for the synergistic model). Thus for the additive model the chances that the exposure at work caused the disease is 9% X 88.2% or around 8%. Note that this is significantly less than the model that doesn't take into account the other risk factors (8% versus 33%).

In summary we need to consider many factors when assessing the effect of exposure to polluted lake water and sediment on the risk for cancer or other chronic diseases. This methodology can be used to aid in establishing interventional values for lake water and sediments during defined activities such as diving (Froom, 2008). It seems that the intervention value for benzene in lake water sediment has been set too low. There has been some progress in our understanding of cancer and risk assessment (see below what we know and do not know and what assumptions are made). Much more needs to be accomplished.

What We Know About Cancer

There is daily endogenous damage to the DNA of cells, thousands of oxidative and other lesions.

These lesions are nearly always repaired by the body.

There are various known repair mechanisms and also the immune system protects the body against malignant cells.

As the body ages unrepaired DNA lesions accumulate and the ability to repair these lesions is diminished.

Damage to the cells includes damage to the DNA, additions to the DNA, and damage to the DNA structural proteins (histones).

Exposure to certain chemicals and physical damage that occurs from factors such as radiation can increase the risk for certain cancers.

Cigarette smoke includes over 40 chemicals and radioactive material that causes an increased risk for some but not all cancers.

It takes time for the cancers to develop, usually over 20 years and at least 10 years for solid tumors, and usually 10 years and at least 3-5 years for acute leukemia and high grade lymphomas (latency period).

At high levels of exposure, there is a dose response for most of the carcinogenic materials, the higher the dose the higher the risk for cancer.

On one hand it takes time for cancers to develop, on the other hand the body can sometimes repair itself over time, and the risk for cancer over time can decrease if exposure is stopped. For example the risk of lung cancer decreases over time in people who have stopped smoking.

At high doses of exposure to both natural and man-made chemicals, animals have an increased risk of developing certain cancers. For example, in vegetables, around 50% of the natural pesticides found in the vegetables will lead to an increased risk of cancer in rats and mice. For suspected chemicals this might be as high as 75% of tested chemicals.

The risk of certain cancers is higher in those who eat less fruits and vegetables.

The risk of certain cancers is higher in those who are fatter.

The risk of certain cancers is higher in those who drink heavily.

The risk of certain cancers is higher in those who smoke cigarettes.

The risk of certain cancers is higher in those who don't exercise.

The risk of certain cancers is higher in those who are poor.

Except for cancers due to cigarette smoke, there is no epidemic of cancer in developed countries.

The most significant risk for nearly all cancers is an older age. Around 30% of men in the Western world will become ill with cancer by age 75. This does not include benign lesions or non-melanoma skin cancers. This also does not include cancers that are not biopsied.

What We Don't Know About Cancer

Why a certain individual becomes ill with cancer.

If chemicals that at high doses increase the risk of cancer also increase the risk of cancer at low doses.

How to determine if exogenous chemicals increase the stress from endogenous chemicals in any significant way.

How to explain in biological terms what results in an increased risk of cancer.

Is one lesion in the DNA enough to increase the risk of cancer or do you need multiple lesions over time.

The minimal time that it takes for individual chemicals and physical agents to cause cancer.

How many lesions are necessary to overcome the natural defense mechanisms of the body.

Why aging increases the risk of cancer.

Conservative Assumptions Made for Risk Assessment

If the chemical has been associated with cancers in man, then the dose-response can be calculated.

It is assumed that there is a real risk even at very low exposures.

Even without an association in man, if animals exposed to a chemical at high doses are at increased risk for cancer, then this can be extrapolated to humans.

Exposures to low doses at work and in the general environment lead to a low but definite increased risk for cancer.

REFERENCES

Albering HJ, Rila JP, Moonen EJC, Hoogewerff JA, Kleinjans JCS (1999) Human health risk assessment in relation to environmental pollution of two artificial freshwater lakes in the Netherlands. *Environ. Health Perpect.* 107:27-35

Ames et al. Cancer prevention and the environmental chemical distraction. In: *Politicizing science: the Alchemy of policymaking*. Michael Gough Ed. Hoover Institute Press, Stanford California; 2003; pp 117-142.

Ames et al. the causes and prevention of cancer: gaining perspective. *Environ. Health Perspect.* 1997; 105(suppl 4): 865-873.

Amson JE (1991) Protection of divers in waters that are contaminated with chemicals or pathogens. *Undersea Biomedical Reearch.* 18:213-219.

Arisawa K, Nakano A, Saito H, et al. 2001. Mortality and cancer incidence among a population previously exposed to environmental cadmium. *Int. Arch. Occup. Environ. Health.* 74:255-62.

ATSDR (Agency for Toxic Substances and Disease Registry) 2007 Toxicological Profile for Benzene (Update). Public Health Service, U.S. Department of Health and Human Services, Atlanta, GA. 1997

ATSDR (Agency for Toxic Substances and Disease Registry) 1997. Toxicological Profile for Nickel (Update). Public Health Service, U.S. Department of Health and Human Services, Atlanta, GA. 1997

Basstrup R et al. Arsenic in drinking-water and risk for cancer in Denmark. *Environ. Health Perspect.* 2008;116:231-7.

Besser JM, Brumbaugh WG, Ivey CD, Ingersoll CG, Moran PW. Biological and chemical characterization of metal bioavailability in sediments from Lake Roosevelt, Columbia River, Washington, USA. *Arch. Environ. Contam. Toxicol.* 2008; 54:557-70.

Carnegie Commission. Risk and the Environment: Improving Regulatory Decision Making. June 1993.

Cohen SM et al. Evaluating the human relevance of chemically induced animal tumors. *Toxicol Sci.* 2004 Apr;78(2):181-6. Epub 2004 Jan 21.

Crane JL, MacDonald DD. Applications of numerical sediment quality targets for assessing sediment quality conditions in a US Great Lakes Area of Concern. – journal???

Duarte-Davidson R, Courage C, Rushton L et al. 2001. Benzene in the environment: an assessment of the potential risks to the health of the population. *Occup. Environ. Med.* 58:2-13.

Froom P. Proposed method for setting standards for recreational divers diving in benzene polluted waters. *Bull. Environ. Contam. Toxicol.* 2008;80:251-4. .

Froom P. Is Cancer In Professional Divers Exposed To Polluted Waters An Occupational Disease? *Hum. Ecol. Risk Assess*. (in press).

Fitzgerald DJ, Robinson NI, Pester BA. 2004. Application of Benzo(a)pyrene and coal tar tumor dose-response data to a modified benchmark dose method of guideline development. *Environ. Health Perspect.* 112:1341-1346.

Giovannucci E et al. Cancers of the colon and rectum. In: Cancer Epidemiology and Prevention (third) edition) Schottenfeld D., Fraumeni JF Jr. (eds), Oxford University Press, New York, 2006. 809-829.

Government of Israel. 2001. Report from the Committee investigation of military activities in the Kishon Harbor and surrounding waters on health of soldiers.

Gronbaek M et al. Type of alcohol consumed and mortality from all causes, coronary heart disease and cancer. *Ann. Intern. Med.* 2000; 133:411-419.

Halmes NC, Roberts SM, Tolson JK et al. 2000. Reevaluating cancer risk estimates for short-term exposure scenarios. *Toxicological Sciences.* 58:32-42.

Heald PC, Schladow SG, Reuter JE, Allen BC (2005) Modeling MTBE and BTEX in lakes and reservoirs used for recreational boating. *Environ. Sci. Technol.* 39:1111-1118.

Hussain M Lifetime health risk assessment from exposure of recreational users to polycyclic aromatic hydrocarbons. *Arch. Environ. Contam. Toxicol.* 1998;35:527-31.

Hyötyläinen T, Oikari A. Assessment of toxicity hazards of dredged lake sediment contaminated by creosote. : *Sci. Total Environ.* 1999 Dec 15:97-105.

IARC (International Agency for Research on Cancer) Monographs on the Evaluation of Carcinogenic risks to humans. Preamble WHO (World Health Organization, Lyon, France 2006

Lijzen JPA Technical evaluation of the Intervention Values for Soil/sediment and Groundwater. Human and ecotoxicological risk assessment and derivation of risk limits for soil, aquatic sediment and groundwater; National Institute of Public Health and Environment. Netherlands, RIVM report. 2007-02-27T12:57:39Z

Lijzen JPA, Baars AJ, Otte PF, Rikken MGJ, Swartjes FA, Verbruggen EMJ, van Wezel AP (2001). Technical evaluation of the intervention values for soil/sediment and groundwater. National Institute of Public Health and Environment. Netherlands, RIVM report 711701 023

Lindeberg C, Bindler R, Renberg I, Emteryd O, Karlsson E, Anderson NJ. Natural fluctuations of mercury and lead in Greenland lake sediments. *Environ. Sci. Technol.* 2006;40:90-5 litigation: studies of auto mechanics and petroleum workers.

McCarthy TA, Barrett NL, Hadler JL, Salsbury B, Howard RT, Dingman DW, Brinkman CD, Bibb WF, Cartter ML. Hemolytic-Uremic Syndrome and Escherichia coli O121 at a Lake in Connecticut, 1999. *Pediatrics.* 2001;108:E59

McCormick CA, Burks SL. 1987. Bioavailablity and toxicity of extracts from acid-mine-waste-contaminated sediments. *Proc. Okla. Acad. Sci.* 67:31-37.

McElduff P et al. Estimating the contribution of individual risk factors to disease in a person with more than one risk factor. *J. Clin. Epidemiol.* 2002 55:588-592.

National Research Council. Frontiers in Assessing Human Exposures to Environmental Toxicants. *Guide.* EPA 450/3-90-024, March 1991.

Olsen O et al. Cochrane review on screening for breast cancer with mammography. *Lancet.* 2001 Oct 20;358:1340-2.

Park C, Snee Ronald. Quantitative Risk Assessment: State-of -the-Art for Carcinogenesis. *Fundam. Appl. Toxicol.* July/August 1983.

Paustenbach DJ, Finley BL, Mowat FS, et al. 2003. Human health risk and exposure assessment of chromium (VI) in tap water. *J. Toxicol. Environ. Health.* 66:1295-339.

Richter ED, Friedman LS, Tamir Y et al. 2003. Cancer risks in naval divers with multiple exposures to carcinogens. *Environ. Health Perspect.* 111:609-617.

Rippey B, Rose N, Yang H, Harrad S, Robson M, Travers S. An assessment of toxicity in profundal lake sediment due to deposition of heavy metals and persistent organic pollutants from the atmosphere. *Environ. Int.* 2008;34:345-56.

Russell M, Gruber M. Risk Assessment in Environmental Policy-Making. *Science.* Vol:266, April 1987.

Schettler T et al. Generations at Risk: Reproductive Health and the Environment. MIT Press, Boston, MA.1999.

Schottenfeld D (ed). et al. Cancer Epidemiology and Prevention (third) edition) Oxford University Press, New York, 2006.

Schijven J and de Roda Husman AM. A survey of diving behavior and accidental water ingestion among Dutch occupational and sport divers to assess the risk of infection with waterborne pathogenic microorganisms. *Environ. Health Perspect.* 2006;114:712-717.

Shor LM, Dosson DS, Rockne KJ, Young LY, Taghon GL. Combined effects of contaminant desorption and toxicity on risk from PAH contaminated sediments. *Risk Analysis.* 2004;1109-1120.

St Leger Dowse M, Bryson P, Gunby A, Fife W (2002) Comparative data from 2250 male and female sports divers: diving patterns and decompression sickness. *Aviat. Space Environ. Med.* 73:743-9.

Stafford RT. Why Superfund was needed? *EPA Journal.* 1981. http://www.epa.gov/history/topics/cercla/04.htm (accessed 3/3/08)

Surveillance Epidemiology and End Result (SEER). US National Institutes of Health, National Cancer Institute SEER 17 registries 2001-2003. Available at http://seer.cancer.gov/faststat

Tickner J, Raffensperger C. The Precautionary Principle in Action: A Handbook. 1999.

Tickner J. Hazardous Exports: U.S. Transfer of Risk Assessment to Central and Eastern Europe. New Solutions, Summer, 1996

Tickner, J. Various lectures on risk assessment. And EPA Office of Air and Radiation. Risk Assessment for Toxic Air Pollutants: A Citizen's Guide.

Two Artificial Freshwater Lakes in The Netherlands. *Environ. Health Prospect.* 1999; 107:27-35.

U.S. ATSDR (Agency for Toxic Substances and Disease Registry)(2007). Toxicological profile for Benzene. US Department of Health and Human Services., Public Health Service. http://www.atsdr.cdc.gov/toxprofiles [assessed 1 November 2007]

U.S. EPA (U.S. Environmental Protection Agency) (2005a). Integrated risk information system. Benzene (CASRN 71-43-2). Available: www.epa.gov/IRIS/ [assessed 1 June 2007]

U.S. EPA (U.S. Environmental Protection Agency) (2005b). Integrated risk information system. Chromium (VI)(CASRN 18540-29-9). Available: www.epa.gov/IRIS/ [assessed 1 June 2007].

U.S. EPA (U.S. Environmental Protection Agency)(2005c). Integrated risk information system. Toxicity and chemical-specific factors data base. Available: http://risk.lsd.ornl.gov/homepage/rap_cnt.shtml [June 2007].

U.S. EPA. (U.S. Environmental Protection Agency)(1999). Extrapolation of the benzene inhalation unit risk estimate to the oral route of exposure. . NCEA-W-05117 Washington DC: National Center for Environmental Health, Office of Research and Development.

U.S. EPA. (U.S. Environmental Protection Agency). (2004). Risk assessment guidance for superfund. Volume I: Human Health Evaluation Manual (Part E, Supplemental Guidance for dermal risk assessment) Final EPA/540/R/99/005 OSWER 9285.7-02EP PB99-963312, Office of Superfund remediation and technology innovation. Washington DC: U.S. Environmental Protection Agency.

U.S. NIH and NCI (National Institutes of Health, National Cancer Institute)(2007). Surveillance Epidemiology and End Results (SEER). SEER 17 registries 2002-2004. Available: http://seer.cancer.gov/faststat/ [June 2007].

USEPA (U.S. Environmental Protection agency) 2004. Risk assessment guidance for superfund. Volume I: Human Health Evaluation Manual (Part E, Supplemental Guidance for dermal risk assessment) Final EPA/540/R/99/005 OSWER 9285.7-02EP PB99-963312, Office of Superfund remediation and technology innovation. U.S. Environmental protection agency, Washington, DC.

USEPA (U.S. Environmental Protection Agency) 2005a Benzene (CASRN 71-43-2) Integrated risk information system. . Available at. http://www.epa.gov/IRIS/

USEPA (U.S. Environmental Protection Agency) 2005b Benzo(a)pyrene (BaP)(CASRN 50-32-8) Integrated risk information system. Available at. http://www.epa.gov/IRIS/

USEPA (U.S. Environmental Protection Agency) 2005c Cadmium; CASRN 7440-43-9. Integrated risk information system, Environmental Protection agency. Available at. http://www.epa.gov/IRIS/

USEPA (U.S. Environmental Protection Agency) 2005d Chromium (VI)(CASRN 18540-29-9) Integrated risk information system, Environmental Protection agency. Available at. http://www.epa.gov/IRIS/

US EPA (environmental protection agency). Guidelines for Carcinogen Risk Assessment. Risk Assessment Forum U.S. Environmental Protection Agency Washington, DC EPA/630/P-03/001F March 2005e

USEPA (U.S. Environmental Protection Agency) 2005f. Toxicity and chemical-specific factors data base. Integrated risk information system. Available at http://risk.lsd.ornl.gov/homepage/rap_cnt.shtml

USEPA . Risk Assessment and Comparative Risk Analysis, History and Background. Internet posting, August 1999

Wallace LA. 1996. Environmental exposure to Benzene: an update. *Environ. Health Perspect.* 104(suppl 6): 1129-1136.

Warner RR, Stone KJ, Boissy YL. 2003. Hydration distrupts human stratum corneum ultrastructure. *J. Invest. Dermatol.* 120:275-284.

WHO (World Health Organization) 2006. Dermal Absorption. (Environmental Health Criteria 235) . WHO Press, World Health Organization. Geneva , Switzerland.

ABBREVIATIONS

ABS soil- absorption coefficient, in proportion per 24 hours
ACGIH American Conference of Governmental Industrial hygienists
ADI acceptable daily intake
AL action level
As arsenic
ATSDR
Cd cadmium
CDDchlorinated dibenzo-p-dioxins, include dioxins and furans
CERCLA comprehensive environmental response, compensation and liability act
Cr- chromium
CSO combined sewer overflows
Cu copper
DDTp.p-dichlorodiphenyltrichloroethane
DWEL drinking water equivalent levels
EPA Environmental Protection Agency
ERM median effects level
Hg mercury

IARC International Association for Research on Cancer
Kg kilograms
Kp water permeability constant in cm/hour
LC50 median lethal concentration
LD50 median lethal dose
LOAEL lowest observed adverse effect level .
LOEL lowest observed effect level
m. Meter.
m3 Cubic meter
MCL maximum contaminant level
mL Milliliter
mg milligrams
mg/L milligrams per liter
mm Millimeter
MP maximal permissible levels
MPRmaximal permissible risk
MRL minimal risk level
MTR maximal tolerated risk concentrations
Ni nickel
NIOSH National Institute of Occupational Safety and Health
NOAEL No Observed Adverse Effect Level.
NOCNot otherwise classified.
NOEL No observed effect level.
NRCNational Research Council
NTP National Toxicology Program
Ppm parts per million
OSHA
PAH polycyclic aromatic hydrocarbons
PBB polybrominated biphenyls
Pb lead
PCB polychlorinated biphenyls
PEL- 1 probable effects limit, 2- permissible exposure limits
POD point of departure
ppb. Parts per billion.
ppm. Parts per million
ppb. Parts per billion. Also pg/L or micrograms per liter.
ppm. Parts per million. Also mg/L or milligrams per liter.
PBT Persistent bio-accumulative toxic chemicals
RBC risk-based concentrations
REL rare effect levels
REL Recommended Exposure Limit
RfC reference dose concentration
RfD= reference dose
RMEG reference dose media evaluation guide
RR relative risk
SCBA. Self-Contained Breathing Apparatus.

SEDISOIL exposure model to contaminated sediments
SRC serious risk concentration
SRCeco ecotoxicological risk limits
SRChuman human risk limits
SSO sanitary sewer overflows
STEL short term exposure limit
TCDD Dioxin (Tetrachlorodibenzo-p-dioxin).
TCE Trichloroethylene.
TEL threshold effects limit
TLV-C threshold limiting value-ceiling
TLV-TWA threshold limiting value- time weighted average
Mg microgram
Mm Micrometer
M micron
VOCvolatile organic compounts
Zn zinc

Terms and Abbreviations

Absorb: To soak up, the incorporation of a liquid into a solid substance, as by capillary, osmotic, solvent, or chemical action.

Absorption factor: The fraction of a chemical making contact with an organism that is absorbed by the organism.

Acceptable daily Intake (ADI): Estimate of the largest amount of chemical to which a person can be exposed on a daily basis that is not anticipated to result in adverse effects (usually expressed in mg/kg/day). It is the same as the RfD.

American Conference of Governmental Industrial Hygienists (ACGIH): An organization of professionals in governmental agencies or educational institutions engaged in occupational safety and health programs. ACGIH develops and publishes recommended occupational exposure limits for chemical substances and physical agents.

Action Level. (AL): The exposure level (concentration in air) at which OSHA regulations to protect employees, is generally half of the permissible level (PEL, or TLV-TWA). At this level actions need to be taken to lower the exposure level.

Additive effect/model: The combined effect of two or more chemicals is equal to the sum of their individual effects.

Adsorb: To attract and retain gas or liquid molecules on the surface of another material yet not necessarily internalized adsorption. Adsorption is the process by which chemicals are held on the surface of a mineral or soil particle.

Ames Test: A bioassay for mutagenesis that uses bacteria (Salmonella) with and without liver enzymes to test potentially carcinogenic compounds.

Attributable risk- personal: The part of causality in an ill patient explained by a risk factor. (The personal attributable risk of smoking in a patient with lung cancer is 90%).

Benthic organisms- Benthos are the organisms which live on, in, or near the seabed, also known as the benthic zone. Benthic organisms, such as sea stars, oysters, clams, sea cucumbers, and sea anemones, play an important role as a food source for fish and humans.Benthos is used in freshwater biology to refer to organisms at the bottom of freshwater bodies of water, such as lakes, rivers, and streams.

Bias: The deviation of results or inferences from the truth, and the collection, analysis, interpretation, publication or review of data that lead to conclusions that are systematically different from the truth.

Bias- Dilution: Including those who are not at risk in the cohort, which dilutes the effects of the study.

Bias- Extrapolation: To take patients not included in the study and assume that the results are also applicable to them.

Bias- Publication: the literature is more likely to publish a study with positive results

Bio-accumulate: The accumulation of a substance in a living organism (such as a pesticide).

Bio-concentration: The process by which a chemical is passed through the food chain from soil to plants and animals, ultimately passed to humans.

Biodegradation: The decomposition of a substance into more elementary compounds by the action of microorganisms such as bacteria.

Biological plausibility: An observed association that fits previous existing biological /or medial knowledge.

Bio-transformation. Conversion of a substance into other compounds by organisms; includes biodegradation.

Body mass index; weight in kilograms divided by square of height in meters.

Cancer: An abnormal multiplication of cells that tends to infiltrate other tissues and spread. Each cancer is believed to originate from a single "transformed" cell that grows (splits) at a fast, abnormally regulated pace, no matter where it occurs in the body.

Cancer Clusters: A cluster is the occurrence of a greater than expected number of cases of a particular disease within a group of people, a geographic area, or a period of time. number of cases of one type of cancer, rather than several different types; a rare type of cancer, rather than common types; or a number of a certain type of cancer cases in age groups not usually affected by that type of cancer.

Carcinogenesis: the process of developing cancer with three proposed stages-initiation: (1) irreversible transformation of a cells' growth-regulatory process with the potential for unregulated growth genetic damage promotion: (2) promoting agent induces an initiated cell to divide abnormally progression; (3) initiated promoted cells undergo unregulated growth and invasiveness

Chloracne: A severe form of skin acne caused by exposure to certain chlorinated chemical compounds. (example-dioxin)

Chlorinated Hydrocarbons: A class of hazardous chemicals, that may be highly toxic, persist in the environment and accumulate in the food chain. Includes many insecticides and industrial solvents.

Chlorinated Solvent: Organic solvent with chlorine atoms

Chromium: a heavy metal that in it's. hexavalent form is a human carcinogens by the respiratory route of exposure. It is also a corrosive material to human cells.

Carbon Monoxide (CO): A colorless, odorless, flammable, and very toxic gas produced by incomplete combustion of carbon compounds and as a by product of many chemical processes. It is a chemical asphyxiate reducing the blood's ability to carry oxygen. Hemoglobin absorbs CO 200 times more readily than it does oxygen.

Carbon Dioxide (CO2): A dense, colorless, gas produced by combustion and decomposition of organic substances and as a by-product of many chemical processes. CO2 does not burn and is relatively nontoxic. High concentrations, especially in confined places, can crate hazardous oxygen-deficient environments that can cause asphyxiation. CO2 is 1.5 times as dense as air, making it useful as a fire-extinguishing agent to block oxygen and smother a fire.

Clustering: Aggregation of events more than the expected

contaminant Any physical, chemical, biological, or radio- logical substance or matter that has an adverse effect on air, water, or soil.

Dichlorodiphenyltrichloroethane (DDT): Most widely used contact insecticide until it was banned in 1972 because of its persistence and potential for bioaccumulation in the environment.

Dioxin: Common name for 2,3,7,8-tetrachlorodi-benzo-p-dioxin (TCDD), a contaminant in defoliants (Agent Orange) in Vietnam. It is a possible human carcinogen.

Dose. A measured amount

Dose-response relationship: A relationship in which a change in amount, intensity or duration of exposure is associated with a change in the risk of a specified outcome.

Drinking Water Equivalent Level (DWEL): is a lifetime exposure level specific for drinking water at which adverse, non-carcinogenic health effects would not be expected to occur.

EPA, (U.S.) Environmental Protection Agency: A Federal agency with environmental protection regulatory and enforcement authority. Website: www.epa.gov).

Epidemic: The occurrence in a community or region of cases of an illness, or any other outcome that is clearly in excess of normal expectancy.

Epidemiology: The study of the relationships between diseases and the various factors that could determine their frequency and distribution in populations.

Etiology: All factors that contribute to the cause of a disease or an abnormal condition.

Eutrophication: Excessive phosphorus concentrations cause increased algae "blooms" and algae decay causes oxygen depletion.

Exposure Limits. The concentration in workplace air of a chemical deemed the maximum acceptable. Meaning that most workers can be exposed at given levels or lower without harmful effects. Exposure limits in common use are; 1) TLV-TWA (threshold limit value - time-weighted average); 2) STEL (short-term exposure limit); and 3) C (ceiling value).

Exposure coefficient: Term which combines information on the frequency, mode, and magnitude of contact with contaminated medium to yield a quantitative value of the amount of contaminated medium contacted per day.

Half-life: The time required for an existing concentration to fall to half its original value.

Hazard evaluation: A component of risk assessment that involves gathering and evaluating data on the types of health injury or disease (e.g., cancer) that may be produced by a chemical and on the conditions of exposure under which injury or disease is produced.

Heavy Metals: Any of several metallic elements with high atomic weights, e.g., mercury, chromium, cadmium, arsenic and lead.

Herbicide: A compound, usually a man- made organic chemical, used to kill or control plant growth.

Heterotrophic microorganisms: Bacteria and other microorganisms that use organic matter synthesized by other organisms for energy and growth.

Heterotrophic plate count (HPC): The number of colonies of heterotrophic bacteria grown on selected solid media at a given temperature and incubation period, usually expressed in number of bacteria per milliliter of sample.

Human equivalent dose: A dose that when administered to humans, produces an effect equal to that produced by a dose in animals.

Human exposure evaluation: A component of risk assessment that involves describing the nature and size of the population exposed to a substance and the magnitude and duration of their exposure. The evaluation could concern past exposures, current exposures, or anticipated exposures.

Human health risk: The likelihood (or probability) that a given exposure or series of exposures may have or will damage the health of individuals experiencing the expo- sures.

Hydrocarbons (HCs): Chemical compounds - often combustible fuels - that contain hydrogen and carbon.

IARC. International Agency for Research on Cancer (IARC): Part of the World Health Organization that considers whether or not a chemical has a potential to cause cancer under certain circumstances.

Interaction: An interaction between two factors is present when there is a difference in the effect on the study factor on risk, according to the level of the other factor. This maybe synergistic meaning that the two risk factors act together in the same direction, or antagonistic, acting in opposite directions. The combined risk may be multiplicative (e.g. risk for lung cancer; smoking- 5 times, asbestos = 10 times, and the combined = 5 x 10 = 50).

Induction period: is the time from exposure to the diagnosis or first manifestations of a disease (other terms; incubation period, and latent period). This assumes the that exposure caused the disease.

Kilo: 1) Kilogram. 2) Kilometer. 3) A prefix meaning "thousand" used in the metric system and other scientific systems of measurement.

Kilogram: one thousand grams.

Latency Period: (see induction period): Latency periods can range from minutes to decades, depending on hazardous material and disease produced.

Lipid Solubility: Measure of the maximum concentration of a chemical that will dissolve in fatty substances. Lipid-soluble substances will disperse through the environment via living tissue.

Lead (Pb): A heavy metal that may be hazardous to health if breathed or swallowed.

Lindane: A pesticide that might cause adverse health effects in domestic water supplies and also is toxic to freshwater and marine aquatic life.

Linearity: (a) How closely an instrument measures actual values of a variable through its effective range; a measure used to determine the accuracy of an instrument. (b) that for every incremental increase in exposure that there is an equal increase in the risk for disease.

Micron: A unit of length. One millionth of a meter or one thousandth of a millimeter. One micron equals 0.00004 of an inch.

Maximum Contaminant Level (MCL): Represent contaminant concentrations in drinking water that EPA deems protective of public health (considering the availability and economics of water treatment technology) over a lifetime (70 years) at an exposure rate of 2 liters of water per day. These are enforceable standards in the United States as stipulated by the Safe Drinking Water Act.

Maximum contaminant level goal (MCLG): The maximum level of a contaminant in drinking water at which no known or anticipated adverse effect on the health of persons would occur, and which allows an adequate margin of safety. Maximum contaminant level goals are non-enforceable health goals.

Mercury: A highly toxic, heavy metal that can accumulate in the environment and in body tissues. Chronic exposure may result in permanent nervous system damage.

Meter (m): The basic metric measure of length; equivalent to 39.371 in.

Milligram per kilogram (Mg/Kg). (a) Dosage used in toxicology testing to indicate a dose administered per kg of body weight. (b) The concentration of a substance in sediment or soil that is equivalent to parts per million.

Milligram per cubic meter of air (mg/m3): equals = parts per million x (molecular weight (MW)) divided by 24.45 at 25 degrees C.

Microgram (μg): One-millionth (10-6) of a gram.

Micrometer (μm): One-millionth (10-6) of a meter; often referred to as a micron.

Micron (μ): See micrometer.

Milliliter (ml): One thousandth of a liter. A metric unit of capacity, for all practical purposes equal to 1 cubic centimeter. One cubic inch is about 16 ml.

Millimeter (mm): 1/1000 (10-3) of a meter.

Milligrams per liter (mg/L): A measure of concentration of a dissolved substance. A concentration of one mg/L means that one milligram of a substance is dissolved in each liter of water. It represents 0.001 gram of a constituent in 1 liter of water. For all practical purposes, this unit is equal to parts per million (ppm).

Mole (mol): The quantity of a chemical substance that has a mass in grams numerically equal to the molecular weight. For example, sodium chloride (NaCl) has a formula mass of 58.5 (Na, 23, and Cl, 35.5) equaling one mole.

Molar: A molar solution consists of the gram molecular weight of a compound dissolved in enough water to make one liter of solution. For NaCl this is 58.5 g (see above).

Molecular weight. The molecular weight of a compound in grams is the sum of the atomic weights of the elements in the compound. The molecular weight of salt (NaCl) is 23 for Na plus 35.5 for Cl or 58.5 g total.

Mutagen: An agent that can cause a permanent genetic change in a cell other than that which occurs during normal genetic recombination.

National Institute of Occupational Safety and Health (NIOSH): The agency of the Public Health Service that tests and certifies respiratory and air-sampling devices. It recommends exposure limits to OSHA for substances, investigates incidents, and researches occupational safety

National Toxicology Program (NTP): A federal agency overseen by the Department of Health and Human Services with resources from the National Institutes of Health the Food and Drug Administration, and the Centers for Disease Control. Its goals are to develop tests useful for public health regulations of toxic chemicals, to develop toxicological profiles of materials.

Non-point source: Diffuse pollution sources that do not have a single point of origin or are not introduced into a receiving stream from a specific outlet. The pollutants are generally carried off the land by storm water runoff. Non-point sources for water pollution include agriculture and urban.

Organic Materials. Compounds composed of carbon, hydrogen with or without other elements and have a chain or ring structure. They include living matter and also manufactures not naturally occurring compounds. Thus the classification of vegetables as organic or not requiring the use of pesticides (often synthesized organic materials) is a misnomer.

Organic chlorine Pesticides: Synthetic organic pesticide containing chlorine. May be highly toxic and workers exposed are primarily at risk for effects on the central nervous system.

Organic metallic Compounds: An organic compound consisting of a metal directly attached to carbon. Some are highly toxic or flammable. Many of them are powerful catalysts used as coordination compounds.

Organophosphates: Synthetic organic compound containing phosphorus used as insecticides. Many are highly toxic; insecticides affect the central nervous system by causing cholinesterase inhibition.

Organic: 1) A term used to refer to chemical compounds made from carbon molecules. These compounds may be natural materials (such as animal or plant sources) or man- made materials (such as synthetic organics)

Occupational Safety and Health Administration (OSHA): Part of the U.S. Dept. of Labor, and the regulatory and enforcement agency for safety and health in most U.S. industrial sectors. Website: www.osha.gov).

Particulate: A very small solid suspended in water which can vary widely in size, shape, density, and electrical charge.

Parts per billion: the number of "parts" by weight of a substance per billion parts of water. Used to measure extremely small concentrations.

Parts per million (PPM): Parts per million parts, a measurement of concentration on a weight or volume basis. This term is equivalent to milligrams per liter (mg/L) for water or sediment.

Permissible Exposure limit (PEL): Established by OSHA. This may be expressed as a time-weighted average (TWA) limit, a short-term exposure limit (STEL), or as a ceiling exposure limit. A ceiling limit must never be exceeded instantaneously even if the TWA exposure limit is not violated. The OSHA PELs have the force of law whereas the ACGIH TLVs and NIOSH RELs are recommended exposure limits that OSHA may or may not enact into law.

Persistent bio-accumulative toxic chemicals (PBTs): chemicals that do not break down easily, persist in the environment, and bio-accumulate in aquatic biota, animal and human tissue.

Pesticide. Any substance or chemical designed or formulated to kill or control weeds or animal pests.

Polycyclic Aromatic Hydrocarbons (PAH): A family of chemical compounds containing only carbon and hydrogen, consisting of three or more carbon ring structures with some carbon atoms in common. They are common in smoke, such as that of vehicle exhaust or tobacco, and are also important industrial contaminants in coal gas or coke manufacture and other processes involving heating of coal tar and pitch. They are definite carcinogens for lung cancer in respiratory exposure, and skin cancer after dermal exposure.

Polychlorinated biphenyls (PCBs): a group of synthetic, toxic industrial chemical compounds used as a heat-transfer medium in electrical transformers, which are chemically inert and not biodegradable. They are found but primarily in sediment where they attach and can remain virtually indefinitely. Therefore, although banned they continue to appear in the flesh of fish and other animals.

Recommended Exposure Limit (REL) (see PEL or TLV-TWA).

Relative risk: The ratio of the risk of disease or death among the exposed to the risk compared to the unexposed; also called the risk ratio.

Risk: The potential for realization of unwanted adverse consequences or events.

Risk characterization: Final component of risk assessment that involves integration of the data and analysis involved in hazard evaluation, dose-response evaluation, and human exposure evaluation to determine the likelihood that humans will experience any of the various forms of toxicity associated with a substance.

Risk estimate: A description of the probability that organisms exposed to a specified dose of chemical will develop an adverse response (e.g., cancer).

Risk factor: Characteristic (e.g., race, sex, age, obesity) or variable (e.g., smoking, occupational exposure level) associated with increased probability of a toxic effect. Risk factor might be more properly called a risk marker, not implying that it is a causal factor.

Risk management: Decisions about whether an assessed risk is sufficiently high to present a public health concern and about the appropriate means for control of a risk judged to be significant.

Risk specific dose. The dose associated with a specified risk level.

Risks

-Absolute risk- The observed probability of an event in a population under study. i.e. -if there are 5 events over 10 years in 100 men , then the absolute risk is 5 per 1000 person years.
-Acceptable risk: benefits outweigh the potential hazards, minimal detrimental effects
-personal attributable risk – see below.
Relative risk – 1
Relative risk

Secular trend: changes over a long period of time; example decrease in coronary heart disease mortality in the past 30 years.

Power: relative frequency where a true difference of specified size between populations would be detected by the proposed experiment or test.

Studies , types:
Case control study: (case comparison, case referent, and retrospective);

Past history of exposure to a suspected risk factor is compared between cases and controls who resemble cases with respect to age and sex but do not have the disease.

Cohort study: syn: concurrent, follow-up incidence, longitudinal, prospective. Study of factors hypothesized to influence the probability of occurrence of a given outcome.

Follow-up study: cohort study; individuals or populations selected at one point in time and followed-up for a defined period.

Synergism: (see interaction).

Self-Contained Breathing Apparatus (SCBA): A respirator that contains its own air supply that the user carries, usually in a tank on his or her back (very similar to scuba gear).

Sedimentation: A water treatment process in which solid particles settle out of the water being treated in a large clarifier or sedimentation basin .

Sediment: usually applied to material in suspension in water or recently deposited from suspension. In the plural the word is applied to all kinds of deposits from the waters of streams, lakes, or seas.

Short-term exposure limit (STEL): ACGIH terminology. maximum concentration for a continuous exposure period of 15 minutes (with a maximum of four such periods per day, with at least 60 minutes between exposure periods, and provided that the daily TLV-TWA is not exceeded).

Threshold Limit Value (TLV): A term ACGIH uses to express the maximum airborne concentration of a material to which most workers can be exposed during a normal daily and weekly work schedule without adverse effects. ACGIH expresses TLVs in three ways:

1. TLV-TWA: allowable time-weighted average concentration for a normal 8-hour workday or 40-hour week;

TLV-STEL: short-term exposure limit (see under short-term exposure limit).

TLV-C: a ceiling concentration not to be exceeded at any time. The ceiling exposure limit or concentration not to exceed at any time, even for very brief times.

Toxicology: the study of the nature, effects, and detection of poisons in living organisms. Included are substances that are otherwise harmless but prove toxic under particular conditions. The basic assumption of toxicology is that there is a relationship among the dose (amount), the concentration at the affected site, and the resulting effects.

Volatile organic compounds (VOC): organic compounds that evaporate very rapidly.

In: Lake Pollution Research Progress
Editors: F. R. Miranda and L. M. Bernard

ISBN: 978-1-60692-106-7

Chapter 9

MEAN RESIDENCE TIMES OF STREAM AND SPRING WATER IN A SMALL FORESTED WATERSHED WITH A THICK WEATHERED LAYER

Naoki Kabeya[1*], Akira Shimizu[2], Yoshio Tsuboyama[1], Tatsuhiko Nobuhiro[1] and Jianjun Zhang[3]

[1] Department of Soil and Water Conservation, Forestry and Forest Products Research Institute, 1 Matsunosato, Tsukuba, Ibaraki, 305-8687 Japan

[2] Kyushu Research Center, Forestry and Forest Products Research Institute, 4-11-16 Kurokami, Kumamoto, 860-0862 Japan

[3] College of Soil and Water Conservation, Beijing Forestry University, Beijing, 100083, China

* Corresponding author: Naoki Kabeya; Forest Hydrology Laboratory; Department of Soil and Water Conservation, Forestry and Forest Products Research Institute; 1 Matsunosato, Tsukuba, Ibaraki, 305-8687 Japan; E-mail: kabeya@affrc.go.jp; Tel: +81-29-829-8233; Fax: +81-29-874-3720

ABSTRACT

The validity of using deuterium excess (d) as a tracer of water mean residence times (MRTs) was tested for a watershed having thick soil and weathered gneiss layers (Tsukuba Experimental Watershed). The MRTs of stream and spring water were estimated to be approximately 1 to 3 years. Transit-time distributions estimated for stream and spring water indicated almost no contribution from rain that fell within the previous 90 days, which can be explained by the time needed for rainwater to permeate the thick soil and weathered gneiss layers and reach the groundwater table. A dispersion model and exponential piston flow model estimated the MRTs of stream water for the whole watershed to be 1.3 years and 2.3 years, respectively, corresponding to respective mobile water storage volumes of 839 and 1454 mm (i.e., $V_m = \tau Q$, where τ and Q are the MRT and annual discharge, respectively). These values are within the limits for the water storage capacities of brown forest soil and heavily weathered gneiss layers (2148 mm).

Keywords: baseflow, deuterium-excess, gneiss-catchment, mean residence time, small headwater catchment.

INTRODUCTION

The mean residence time (MRT) or mean transit time of stream water is a fundamental parameter for characterising watersheds and understanding contaminant transport within a watershed (McGuire and McDonnell, 2006). Over the past 20 years, many studies have estimated residence times in watersheds using stable isotopes of water as a tracer (Vitvar et al., 2005; McGuire and McDonnell, 2006). A number of new river-monitoring projects, such as the Hydrological Observation Network by the Consortium of Universities for the Advancement of Hydrologic Science (Band et al., 2002; Hooper, 2004) and the Global Network of Isotopes in Rivers by the International Atomic Energy Agency (Vitvar et al., 2007), are also focusing on this field of research. Various studies have demonstrated that permeable bedrock plays a major role in the hydrological and biochemical processes of granitic headwater catchments, such as the Fudoji (e.g., Asano et al., 2002; Uchida et al., 2003) and Matsuzawa catchments of the Kiryu Experimental Watershed in Japan (Katsuyama et al., 2005). Using a functional intercomparison approach, Uchida et al. (2006) showed that bedrock permeability is a dominant factor for determining the spatial distribution of the MRTs of subsurface waters on a hillslope. They compared functions between different watersheds using information obtained by both isotope tracer and hydrometric observations. Chappell et al. (2007) noted that it is important to consider the hydrological properties of both soil, regolith and rock when modelling runoff processes in humid tropical regions. However, few studies have looked at small forested watersheds in East Asia, other than the granitic watersheds mentioned above.

In East Asia, the origin of water vapour changes seasonally, causing large fluctuations in the deuterium excess ($d = \delta D - 8\delta^{18}O$) found in rainfall (Kondoh and Shimada, 1997). The d value mainly represents a kinetic effect produced by 'primary evaporation' when water evaporates from the sea surface (Dansgaard, 1964), although it is changed by 'secondary evaporation' from falling rain in some arid regions (Clark and Fritz, 1997). Kabeya et al.,

(2007) estimated the MRT of subsurface water based on seasonal variation in *d* values in a small granitic headwater catchment in Japan. They suggested that the *d* value is probably an effective tracer for estimating the MRT of subsurface water not only in Japan, but also in other East Asian countries influenced by the Asian monsoon system. Kawaraya et al. (1999) estimated the MRT of groundwater from the fluctuations of *d* values in precipitation and groundwater in the Yachi landslide of Akita Prefecture in northern Japan. Recently, Lee et al. (2007) also showed that the *d* value is an effective tracer for estimating the MRT of soil water on Jeju Island, Korea. Thus, although a number of studies have shown the validity of the *d* value as a tracer of MRTs in small forest watersheds, the scope of research is still limited in terms of the types and locations of watersheds studied.

To further check the validity of using the *d* value to estimate MRTs, we tested this tracer in a watershed with geologic and subsurface structures different from those of previous studies. In addition, abundant hydrometric and geological data were available for our experimental watershed, allowing for additional tests that had not been possible in previously studied watersheds. This chapter presents the research results that illustrate the subsurface water pathways in gneiss watersheds of the Tsukuba Experimental Watershed (TEW). The TEW has three features that make it a desirable location for examining the *d* value as a tracer.

- Long-term hydrometric observations have been conducted in the watershed, and over 10 years of data are available.
- The watershed has thick soil and weathered rock layers.
- A number of investigations have examined the subsurface structure and produced detailed information.

This chapter examines the discharge characteristics of a small, mountainous, forested watershed based on stable isotope data, hydrological observations, and subsurface structure. Specifically, the chapter has the following goals.

- To describe the hydrological response of the whole watershed and sub-watersheds based on rainfall and discharge data.
- To clarify the temporal characteristic of *d* values in precipitation and stream water and to estimate the MRT and distribution of baseflow by the modified sine-wave method.
- To compare subsurface structure and the MRT and transit-time distribution of water.

SITE AND METHODS

Study Area

Research was conducted in the Tsukuba Experimental Watershed (TEW), located in southern Ibaraki Prefecture, Japan (36°20' N, 140°18' E; figure 1). In 1978, this watershed was established as an experimental study site for investigations of hydrological processes in a forested mountainous area (Water Resources Laboratory and Flood Control Laboratory, 1993). The watershed drains 3.79 ha. The main slope direction is north, and the mean slope is

25°. In the 10-year period from 1972 to 1981, average annual air temperature was 14.1 °C at Kakioka, the nearest weather station (36°14' N, 140°12' E; altitude 27.7 m). The annual rainfall and discharge from 1979 to 1990 (excluding 1988 when data were lacking) were 1429.1 mm and 641.6 mm, respectively. Although the TEW experiences several snowfalls per year, snow depths above 20 cm are rare, and most precipitation is rainfall.

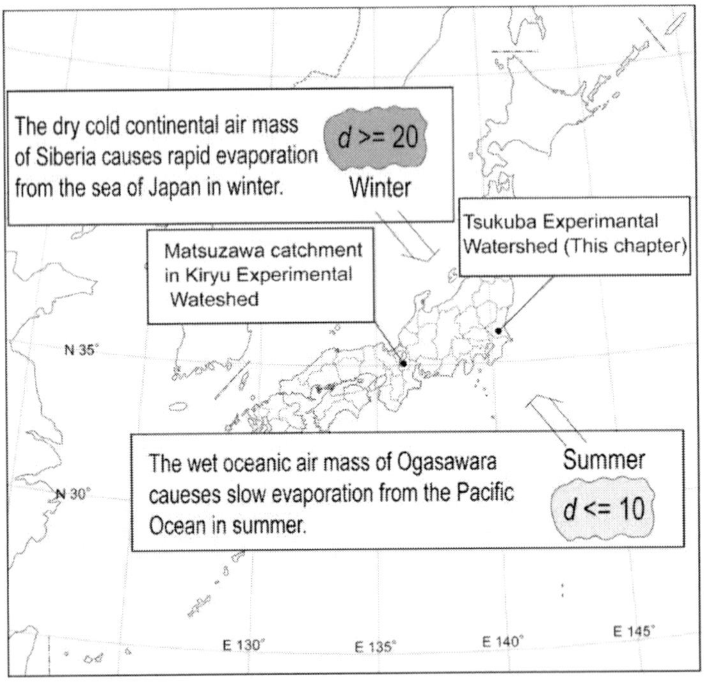

Figure 1. Location of the Tsukuba Experimental Watershed (TEW) and schematic illustration of seasonal air-mass change over Japan.

The stream channel in the TEW originates from three springs (figure 2) and is one of the headwaters of the Koise River that flows into Lake Kasumigaura as part of the Tone River system. To observe discharge from the three spring sub-catchments, 60° V-notch flow-gauging weirs are operated at the three springs (referred to as A, B, and C). The drainage areas of subcatchments A, B, and C are 0.60, 0.93, and 0.36 ha, respectively. The discharge amount from the entire TEW is observed by a 45° V-notch flow-gauging weir at point O. Based on the result of a drilling investigation, an impermeable wall structure was constructed through the subsurface at point O and attached to the base rock (6.0 m depth); the weir was then built to measure the amount of watershed outflow (Water Resources Laboratory and Flood Control Laboratory, 1993). In contrast, the impermeable walls of weirs A, B, and C are only 1.2 m deep and not attached to rock. In addition to these weirs, a temporary flow-gauging weir was placed at point S in the watershed to observe flow and investigate sediment production in the forest watershed; observations were conducted at this weir for about 2 years from May 2003 to May 2005. The S sub-catchment drains an area of 2.97 ha. From that research, Zhang et al. (2005) reported that the discharge amount of suspended sediment was mainly determined by the maximum 10-min precipitation. Each rainfall event also greatly contributed to suspended sediment discharge. The suspended sediment discharge from one storm accounted for about 5% of the total annual sediment discharge and that from a few storms contributed about 30% of the yearly suspended sediment discharge.

At spring points A and B, the flat valley bottoms are very narrow (only several square meters), while the valley bottom at spring C is about 30 m^2. The S and O watersheds contain parts of the river channel (figure 2). Geologically, the watershed is mainly composed of biotite gneiss overlaid by weathered volcanic ash; the soil in this area is brown forest soil, or

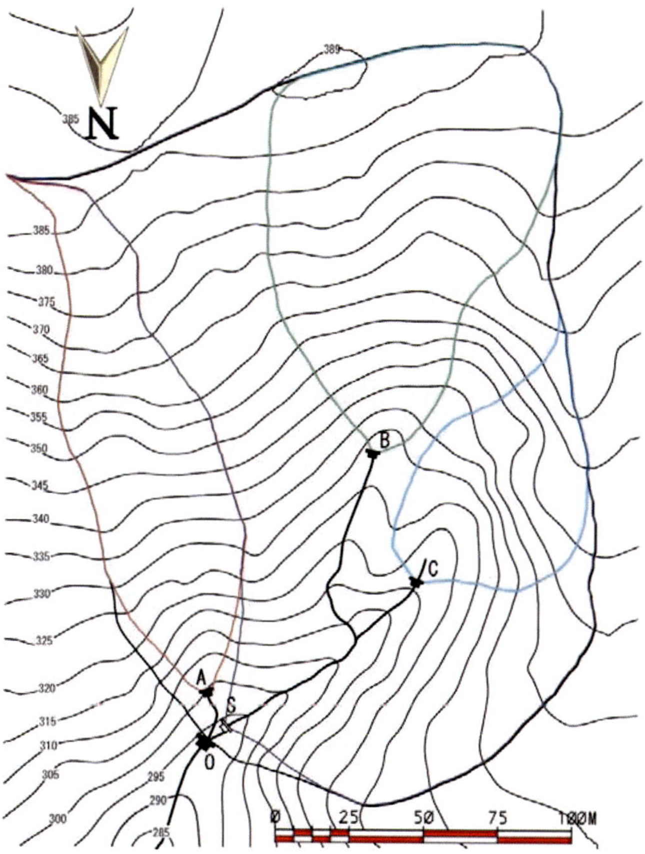

Figure 2. Topography and locations of measurement instruments in the TEW.

Cambisol by the FAO classification (Ohnuki et al., 1999). Plantations of *Cryptomeria japonica* (common name "sugi" or Japanese cedar) and *Chamaecyparis obtusa* ("hinoki" or Japanese cypress) are the main vegetation. *Pleiblastus chino*, a type of bamboo grass, and *Aucuba japonica*, an evergreen shrub, grow on the forest floor.

Subsurface Structure and Soil Depth

Drilling surveys were performed at 11 points in this watershed from 1977 to 1988 to better understand the subsurface structure. Figure 3 summarises the structure revealed by those efforts, as well as results from an on-site permeability test conducted as part of the drilling survey. The results show that the bottoms of the A and B layers of brown forest soil occur at depths of 0.50 m and from 1–3 m, respectively. Under these soft soil layers, heavily-weathered, weakly-weathered, and unweathered gneiss layers are distributed. The bottom of the heavily-weathered gneiss layer was found at 2–6-m depths. Although the bottom of the weakly-weathered gneiss layer was mainly at depths of 10–20 m, the deepest part on the middle slope was about 30 m.

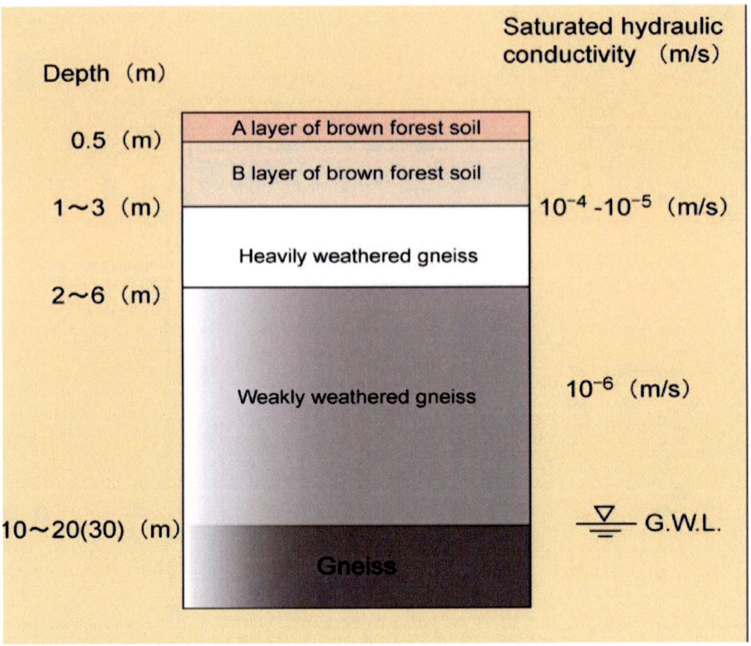

Figure 3. Results of drilling surveys in the TEW.

In addition, surface wave surveys were performed in the TEW in 2003. This survey method to investigate the subsurface structure was established by VIC, Ltd. (Tokyo, Japan). Since this survey technique is non-destructive, implementation is relatively less expensive compared to drilling surveys. Figure 4 schematically illustrates a surface wave survey. A vibration-generating machine includes two detectors (P_1, and P_2) that are arranged on a straight line to the measurement point. The surface wave generated by vibrating the ground in an up-and-down direction with the vibration-generating machine flows to detector P_2 through detector P_1. The surface wave velocity (V_r) is then calculated based on the time taken to pass between P_1 and P_2. The V_r and measurement depth (D) change with the frequency of the wave generated by the machine. Therefore, a D–V_r curve is obtained by performing measurements under various frequencies. Since V_r represents the mean surface wave velocity through the soil surface, a partial velocity value (v_r) can be calculated for each layer by the method described by Kawasaki (1988).

This technique was performed at 35 points in the TEW. At some points, the relationships between the soil layer and surface wave speed were compared to the partial surface velocity (v_r) profile and the drilling core (Kabeya et al., 2005). Spatial interpolation of the subsurface structure in the TEW was performed using this relationship and geographic information system (GIS) software (Shimizu et al., 2007). The mean soil thickness was 7.27 m, within which the brown forest soil (A and B layers) was 3.29 m and the heavily weathered gneiss layer was 3.98 m (Shimizu et al., 2007). The effective porosities of the brown forest soil and heavily weathered gneiss layers were about 0.29 and 0.30, respectively, as determined by soil-moisture retention curves (Ohunuki et al., 1999). Water storage capacity was calculated by multiplying mean soil thickness by effective porosities.

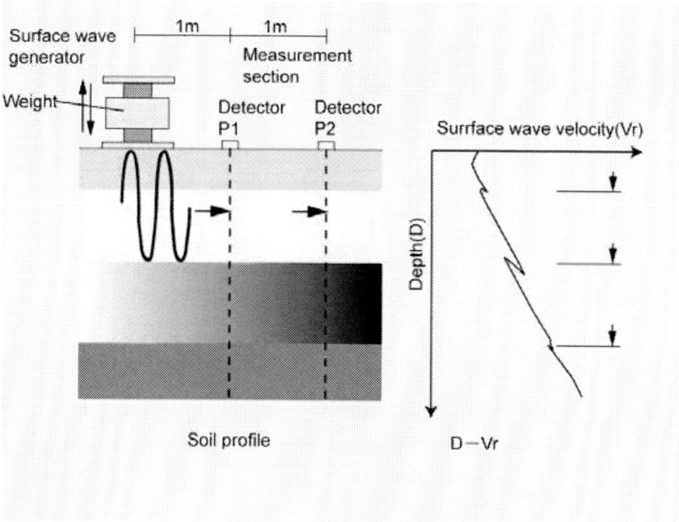

Figure 4. Schematic illustration of a surface wave survey (VIC, Ltd.).

Water Sampling

Rainfall was collected at point O in 30-L polyethylene sampling bottles with a 21-cm diameter plastic funnel attached (figure 2). To prevent changes in the isotope ratio of the rainwater in the bottle due to isotopic fractionation from evaporation, an oil film was poured into the rainfall collector. The sample was protected by this film as soon as it entered the rainfall collector. Spring waters were sampled at points A, B, and C, and stream water was sampled at point O (figure 2).

Sampling has been performed every 2 to 4 weeks since 27 December 2001. In this chapter, the annual data from 27 December 2001 to 27 December 2002 were used. Since most of the sampling was performed on non-rain days, we assumed that the spring water and stream water derived from baseflow. Each water sample was kept refrigerated at 4 °C in an airtight 20-ml glass vial until analysis.

Stable Isotope Analysis

A mass spectrometer (MAT252; Thermo Scientific, Waltham, MA, USA) was used for the hydrogen and oxygen stable isotope analysis of water samples. The H_2–H_2O equilibrium method with a Pt catalyst and the CO_2–H_2O equilibrium method were used to measure the hydrogen and oxygen stable isotope ratios, respectively. The isotope ratio was expressed as the δ value with respect to that of Vienna Standard Mean Ocean Water (V-SMOW), given as

$$\delta D_{sa} = \left(\frac{(D/H)_{sa}}{(D/H)_{re}} - 1 \right) * 1000‰ \text{ V-SMOW}$$

$$\text{or } \delta^{18}O_{sa} = \left(\frac{(^{18}O/^{16}O)_{sa}}{(^{18}O/^{16}O)_{re}} - 1 \right) * 1000‰ \text{ V-SMOW}, \quad (1)$$

where *sa* and *re* refer to the sample and standard reference, respectively. V-SMOW is a standard reference material for measuring stable isotope ratios in water. The *d* value (=δD - 8 $\delta^{18}O$) of each water sample was then calculated using the results. The standard uncertainties of the δD, $\delta^{18}O$, and *d* measurements were ±0.4‰, ±0.02‰, and ±0.43‰, respectively.

Residence Time Models

The input *d* value (d_{in}) in rainfall was transformed by convolution integration into the output *d* value (d_{out}):

$$d_{out}(t) = \int_0^{\infty} g(t')\, d_{in}(t-t')\,dt', \quad (2)$$

where *g(t)* is the system response function specifying the transit-time distribution of water through the system (Zuber, 1986). Here, *t* represents calendar time, and the integration was carried out over transit time *t'*. In the piston flow model (PFM) approximation, the flow lines are assumed to have the same transit time, and the hydrodynamic dispersion and diffusion are negligible (Maloszewski and Zuber, 2002). The system function can be written as

$$g(t) = \Delta(t-\tau), \quad (3)$$

where Δ is the Dirac delta function, and τ is the mean residence time of water. The PFM is applicable only to systems with constant tracer input (Maloszewski and Zuber, 2002). The three most commonly used models are described below.

In the exponential model (EM) approximation, the flow lines are assumed to have an exponential distribution of transit times, *i.e.*, the shortest line of the theoretical transit time equals zero, and the longest line has a transit time equal to infinity (Maloszewski and Zuber, 2002). It is assumed that there is no tracer exchange between the flow lines (Maloszewski and Zuber, 2002). The response function can be written as

$$g(t) = \frac{1}{\tau}\exp(-\frac{t}{\tau}), \quad (4)$$

The EM has only one fitting parameter: τ.

In the exponential piston flow model (EPM), the aquifer is assumed to consist of two parts of a line, one with the exponential distribution of transit times and the other with the distribution approximated by the piston flow (Maloszewski and Zuber, 2002). The response function of the EPM can be written as

$$g(t) = \frac{\beta}{\tau} \exp\left(-\frac{\beta t}{\tau} + \beta - 1\right) \qquad t \geq \tau(1 - \frac{1}{\beta}) \quad , \tag{5}$$

$$g(t) = 0 \qquad t < \tau(1 - \frac{1}{\beta}) \quad , \tag{6}$$

where β is the ratio of the total volume to the volume with the exponential distribution of transit times, *i.e.*, β = 1 indicates the exponential flow model. The EPM has two fitting parameters: τ and β.

The dispersion model (DM) uses the following unidimensional solution to the dispersion equation for a semi-infinite medium as its response function (Zuber, 1986):

$$g(t) = \left(\frac{4\pi D_p t}{\tau}\right)^{-1/2} t^{-1} \exp\left[-\left(1 - \frac{t}{\tau}\right)^2 \left(\frac{\tau}{4 D_p t}\right)\right], \tag{7}$$

where D_p is the dispersion parameter ($D_p = D/vx$), which primarily depends on the distribution of the transit time within the system and is used as the fitting parameter (Maloszewski and Zuber, 2002). However, the physical meaning can be interpreted in terms of the transport process. D is the dispersion coefficient (m^2/s), v is the mean transit velocity of water in the system (m/s), and x is the length of the lines of flow (m). The DM also has two fitting parameters: τ and D_p.

Many previous studies have applied a single-parameter EM to approximate mean residence times (Maloszewski et al., 1983; DeWalle et al., 1997; Soulsby et al., 2000). However, the response function of the EM shows the model to be inapplicable to systems without infinitesimally short flow lines (Maloszewski and Zuber, 2002). On the other hand, the EPM and DM, having two parameters (τ and β or D_p), are flexible and permit wide variation of residence-time distributions. In addition, the parameters (β or D_p) have clear physical meanings (Vitvar and Balderer, 1997). Therefore, the mean residence times presented here are from evaluations using the EPM and DM. The EPM with β = 1 as the best-fit parameter produced the same results as the EM.

Modified Sine-Wave Method

The functions of the following formula were used to express variation in the d value as input:

$$d_{in}(t) = d_0 + A \sin \omega t \,, \tag{8}$$

where d_0 is the mean d value in ‰ V-SMOW at the start time ($t = 0$), A is the amplitude of the d value in ‰ V-SMOW, and ω is the angular frequency of variation in radians per day. The seasonal variation in the output d value becomes

$$d_{out}(t) = d_0 + B\sin(\omega t + \varphi). \tag{9}$$

Here, B is the amplitude of the output d value in ‰ V-SMOW, and φ is the phase lag. Assuming that the system response function is represented by the EPM, the values of B and φ can be obtained by Equations (2), (5), (6), and (8):

$$B = \frac{A}{\sqrt{1 + \frac{\omega^2\tau^2}{\beta^2}}}, \tag{10}$$

$$\varphi = \omega\tau\left(1 - \frac{1}{\beta}\right) - \arccos\left(\frac{1}{\sqrt{1 + \frac{\omega^2\tau^2}{\beta^2}}}\right), \tag{11}$$

using τ and β as variables. Asano et al. (2002) described the mathematical derivation of these equations.

Assuming that the system response function is represented by the DM, the values of B and φ in Equation (9) can be obtained by Equations (2), (7), and (8):

$$B = A\exp\left(\frac{1}{2D_p}\right)\exp\left(-\sqrt{\frac{1}{2D_p}}\sqrt{\sqrt{\omega^2\tau^2 + \left(\frac{1}{4D_p}\right)^2} + \frac{1}{4D_p}}\right), \tag{12}$$

$$\varphi = -\sqrt{\frac{1}{2D_p}}\sqrt{\sqrt{\omega^2\tau^2 + \left(\frac{1}{4D_p}\right)^2} - \frac{1}{4D_p}}, \tag{13}$$

by using τ and D_p as variables. Details of the mathematical derivations of these equations have been provided by Kubota (2000). Kubota (2000) also showed a purely mathematical way to derive fundamental and practical formulas for estimating the MRT of water, as described by Maloszewski et al. (1983).

The rainfall d value was used as the d value for the input function d_{in}. The d values of the spring and stream waters were used as the d values of the output functions d_{out}. The model parameters τ and β (or D_p), which best reproduced each observed fluctuation, were then estimated. The root mean squared error (RMSE) was computed to estimate the goodness of fit of the model. The RMSE is expressed as

$$\text{RMSE} = \sqrt{\frac{\sum_{i=1}^{n}(O_i - X_i)^2}{n}}, \quad (14)$$

where O_i is the ith observed d value, and X_i is the corresponding value calculated from the model. The value of τ was calculated per month; each month was assumed to be 30 days long.

RESULTS AND DISCUSSION

Annual Water Balance in the Whole TEW

Although the annual water loss ($L = P - Q$, where P is annual precipitation, and Q is annual discharge) ranged widely from 526.5 to 998.7 mm during the period from 1979 to 1990 (with 1988 excluded due to lack of data), the long-term average of annual water loss L_{ave} was 787.5 mm (figure 5), which is almost equivalent to the annual evapotranspiration amount in this region. Annual evapotranspiration in 2003 was estimated as 768 mm by the Bowen ratio energy balance method at the TEW (Nobuhiro et al., 2006).

The difference in the daily discharge amount ($Q_{31\ Dec}$ - $Q_{1\ Jan}$ mm/day) between the start day (1 January) and end day (31 December) in each observation year was used to create an index of change in the watershed storage amount ΔS. A strong positive correlation was found between $Q_{31\ Dec}$ - $Q_{1\ Jan}$ and the annual water loss L (figure 6). This shows that water level was not stable between the start and end days of each water year and that change in the watershed storage amount affected the annual water loss.

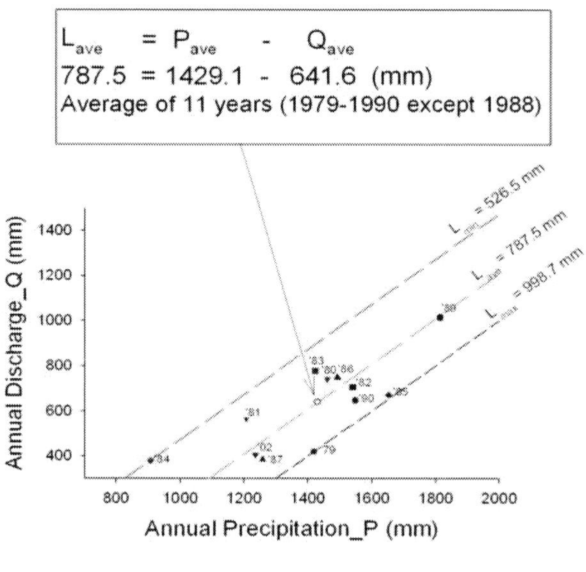

Figure 5. Relationship between annual precipitation (P) and annual discharge (Q).

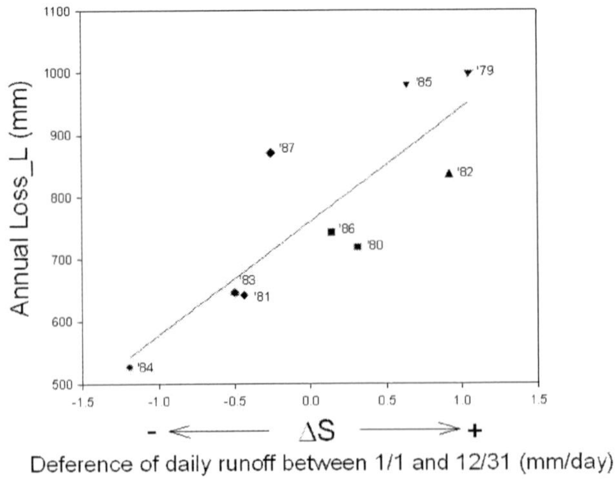

Figure 6. Relationship between the index of change in the watershed storage amount(Q31Dec - Q1Jan) and annual water loss (L).

Rainfall-Discharge Response

Although the discharge amount responded quickly to rain events at O, S, and C, no remarkable response to rain was observed at A and B, except for the responses to heavy rain events in mid-August (figure 7). However, discharge amounts (water level) that rose due to heavy rain in mid-August did not decrease to pre-rain amounts by 31 December of the same year. The watershed showed a slow discharge component that took several months or more to decrease to the amount prior to a rain event. Thus the water level and discharge amount at points within the water year could change widely due to heavy rain and the timing of such rain. Points O, S, and C responded quickly to rain events; these sites are located in areas with comparatively large valley bottoms and the river channel. The valley bottom, river channel, and neighbouring slope features likely contributed to the quick discharge components generated at those sites.

The rainfall-runoff response characteristic of each watershed was investigated for August 2003 (figure 8). A rain event was considered to end when no new rain was recorded for 12 hours. Six rain events (numbered 1–6 in table 1) with total rainfall of 2.0 mm or more occurred during that period. Clear discharge peaks were seen at O, S, and C within 0 to 20 min of the rain peak, but the discharge peaks at A and B were not clear within that time period. The direct runoff ratio for each event was calculated by a level abstraction method using the initial discharge amount of a flood. The direct runoff ratio was high at O (1.8 to 10.7%), S (1.7 to 6.8%), and C (0.1 to 8.2%) and low at A (0.3 to 0.6%) and B (0.1 %). At A and B, no responses were seen for rainfalls #1 and #2, which had relatively small rain amounts (figure 8). However, A and B showed smooth discharge peaks more than 1 day after events #4 and #5, which had relatively large rain amounts. The discharge then decreased successively, taking several months or more to reach the initial discharge amount. This showed that the slow discharge component from the temporal subsurface water storage was generated and expanded only when a certain amount of rainfall was exceeded in TEW.

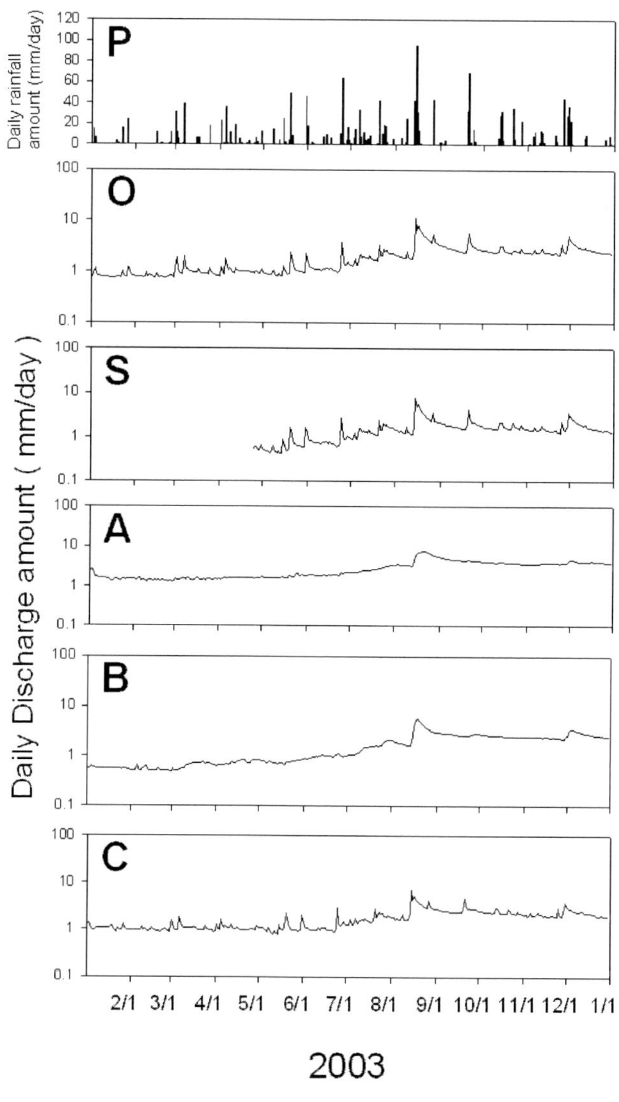

Figure 7. Relationship between daily rainfall and daily discharge amounts in 2003.

Table 1. Characteristics of rainfall events in August 2003

Event no.	Date	Duration (h)	Amount (mm)	Max. intensity (mm/10min)
#1	8/5 16:00~8/6 3:30	11.5	5.7	0.8 (8/5 17:10)
#2	8/8 14:20~8/8 15:50	1.5	2.3	0.9 (8/8 15:50)
#3	8/9 6:10~8/9 19:50	13.7	21.1	4.5 (8/9 8:30)
#4	8/14 1:40~8/15 21:00	43.3	112.6	3.2 (8/15 7:10)
#5	8/16 11:10~8/18 9:00	45.8	35.6	1.7 (8/17 2:10)
#6	8/26 18:20~8/27 10:20	16.0	37.2	9.7 (8/27 1:50)

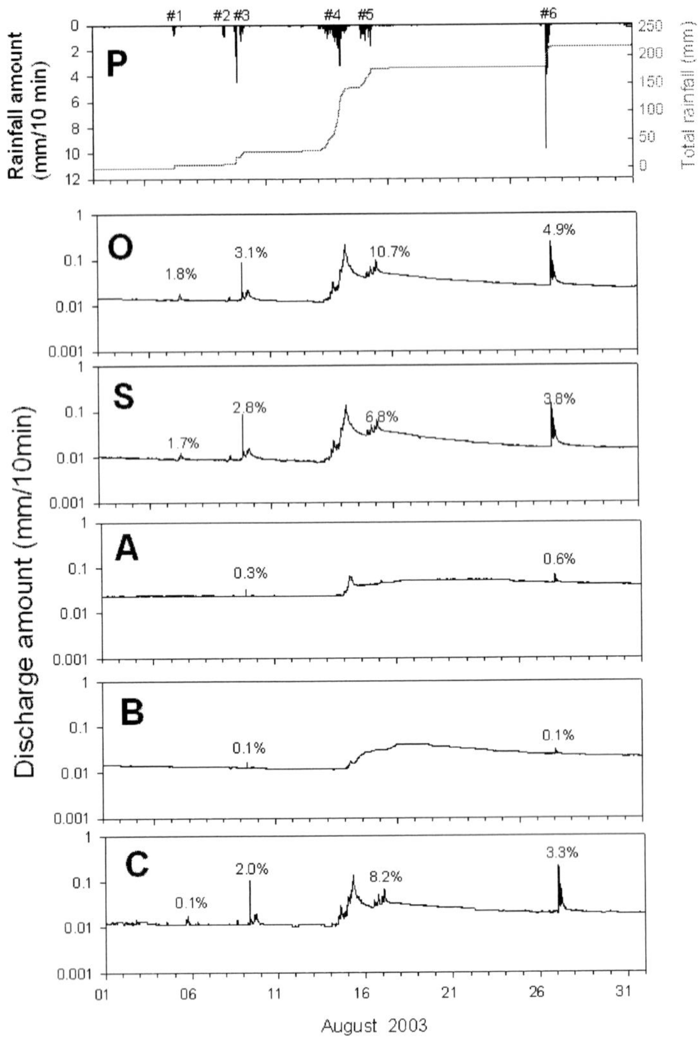

Figure 8. Relationship between 10 minute rainfall and 10 minute discharge amounts in August 2003.

Stable Isotope Compositions in Rainfall, Stream Water, and Spring Water

The stable isotopes of rain are shown in a δ-diagram figure 9. One year is roughly divided into two seasons: summer (April to September) and winter (October to March). The regression lines for the stable isotope compositions in rainfall are as follows:

$$\delta D = 7.50\ \delta^{18}O + 16.03 \quad r^2 = 0.96, \text{ in winter} \tag{15}$$

$$\delta D = 7.93\ \delta^{18}O + 11.14 \quad r^2 = 0.97, \text{ in summer.} \tag{16}$$

Seasonal changes in the *d* value of rain have been reported for many sites in Japan (Waseda and Nakai, 1983; Sanjo, 1990; Taniguchi et al., 2000; Yabusaki and Tase, 2005). These changes have been considered to depend on changes in the origin of water vapour, which mainly comes from the Sea of Japan in winter and Pacific Ocean in summer (figure 1).

The weighted mean values of $\delta^{18}O$, δD, and *d* in rainfall were -7.83, -49.0, and 13.7‰, respectively (table 2). The mean values of stable isotope ratios in stream water was -7.88 in $\delta^{18}O$, -49.6 in δD, and 13.4 ‰ in *d*, respectively (table 2). The distributions of the stable isotope compositions in the stream and spring waters closely followed those of the weighted means of the stable isotope compositions in rainfall (figure 9). The standard deviation and analysis accuracy suggest that the mean values of stable isotope ratios in the stream and spring waters were almost equal to the weighted mean values of stable isotope ratios in rainfall. Thus, isotopic fractionation caused by evaporation on the forest floor can be disregarded in the TEW.

Table 2. Mean values of $\delta^{18}O$, δD, and *d* in rainfall, stream water, and spring waters in the TEW from 27 December 2001 to 27 December 2002

	$\delta^{18}O$		δD		*d*	
	Mean	S.D.	Mean	S.D.	Mean	S.D.
	(per mil)		(per mil)		(per mil)	
Rainfall n=30	-7.83*	3.03	-49.0*	23.2	13.7*	6.1
Spirng A n=31	-7.93	0.04	-50.2	0.6	13.3	0.8
Spirng B n=31	-7.99	0.04	-50.6	0.9	13.3	0.9
Spirng C n=30	-7.77	0.06	-49.3	1.1	13.0	1.2
Stream water O n=32	-7.88	0.05	-49.6	1.1	13.4	1.1

*: weighted mean; S.D.: standard deviation; n: sampling number

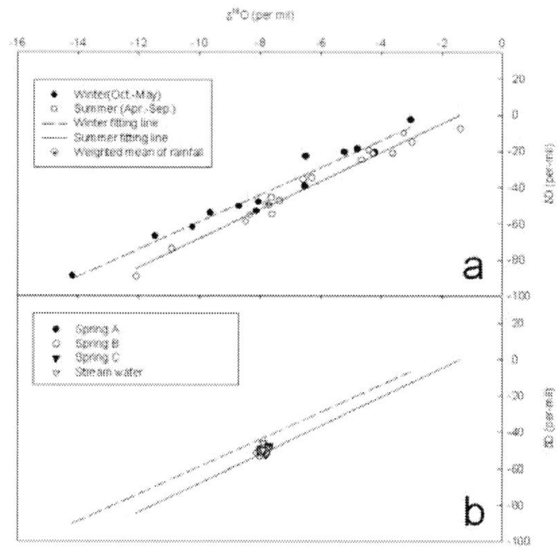

Figure 9. Isotopic compositions of rainfall, spring waters, and stream water.

Seasonal Variation of the *d* Value and MRT Estimations

The *d* value in rainfall showed annual periodic change (figure 10), reaching about 30‰ during January and February but dropping to about 5‰ during June and July. An input function was obtained by fitting equation (8) to the observed data as follows:

$$d_{in} = 13.3 + 7.0 \sin(0.017\,t + 1.048). \tag{17}$$

The *d* values in the stream and spring waters also showed annual periodic changes, but with amplitudes smaller than those in rainfall (figure 11). The estimated MRTs of spring waters ranged from 16.9 to 38.9 months, while the MRTs of stream water were estimated as 15.7 months by the DM and 27.2 months by the EPM.

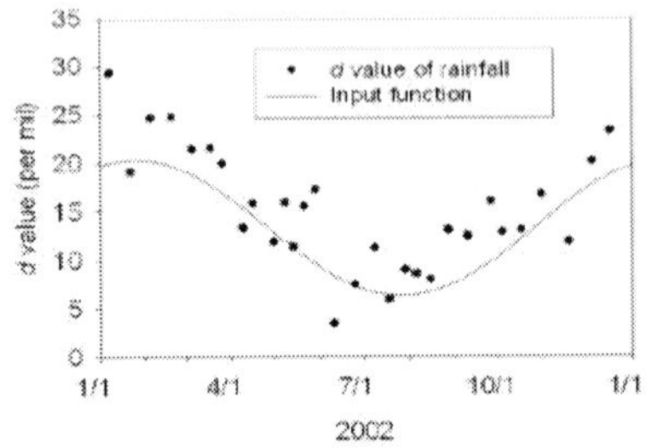

Figure 10. Temporal variation of the *d* value in rainfall and the input function.

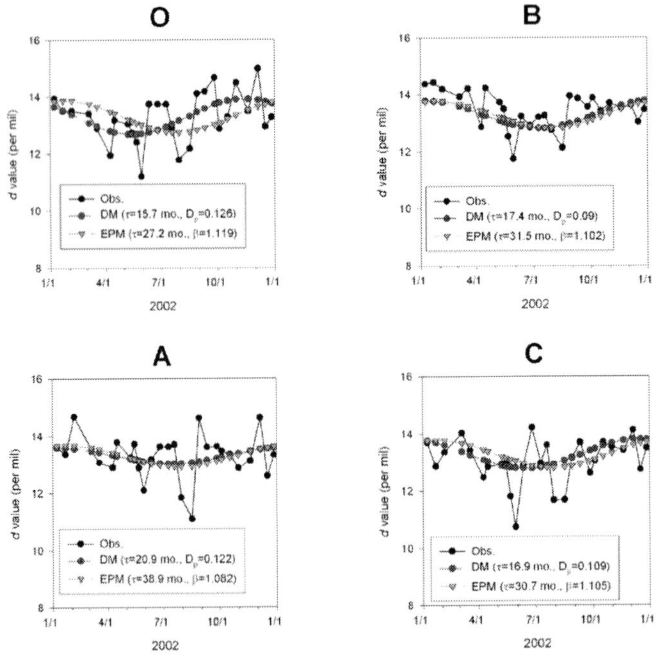

Figure 11. Temporal variations of *d* values in the stream and spring waters and outputs by the exponential piston flow model (EPM) and the dispersion model (DM).

Baseflow Generation and Dominant Flow Pathways in the TEW

The MRT of stream water from the whole catchment (O) was estimated to be 1.3 years by the DM and 2.3 years by the EPM, corresponding to mobile water storage volumes of 839 and 1454 mm, respectively (i.e., $V_m = \tau\ Q_{\mathrm{ave}}$, where Q_{ave} is 641.6 mm). These values are within the limits of the water storage capacities of brown forest soil and heavily weathered gneiss layers (2148 mm; the effective porosities of brown forest soil and of heavily weathered gneiss were set to 0.29 and 0.30, respectively, based on their soil moisture retention curves). In the Matsuzawa catchment of the Kiryu Experimental Watershed, the MRTs of stream water were estimated to be 0.68 years by the EPM and 0.73 years by the DM, corresponding to mobile water storage volumes of 443 mm and 476 mm, respectively (Kabeya et al., 2007). Thus, the mobile water storage volume and residence times of the TEW are two or three times larger than those at the Matsuzawa catchment.

The distribution of stream and spring water transit times in the TEW showed almost no contribution from rain that fell within the previous 90 days (figure 12). This result differs from that at Matsuzawa, which showed a comparatively large contribution of younger water (τ < possibly 90 days). In the Matsuzawa catchment, a hillslope with a thin soil layer (about 30 cm deep) adjoins a valley in which permanent groundwater exists. Kabeya et al. (2007) showed that corresponding to rain events, saturated throughflow occurred along the interface of the soil and bedrock layers in the Matsuzawa catchment; they estimated the MRT of the saturated throughflow to be as short as around 2 to 3 months (Kabeya et al. 2007). Because saturated throughflow with short MRT flows into the permanent groundwater in the valley part, the comparatively young water was considered to make a large contribution in the distribution of the stream water transit time at Matsuzawa. However, the TEW does not have a similar hillslope with a thin soil layer. Rather, rain water has to permeate thick soil and weathered layers before it reaches the groundwater table (figure 13).

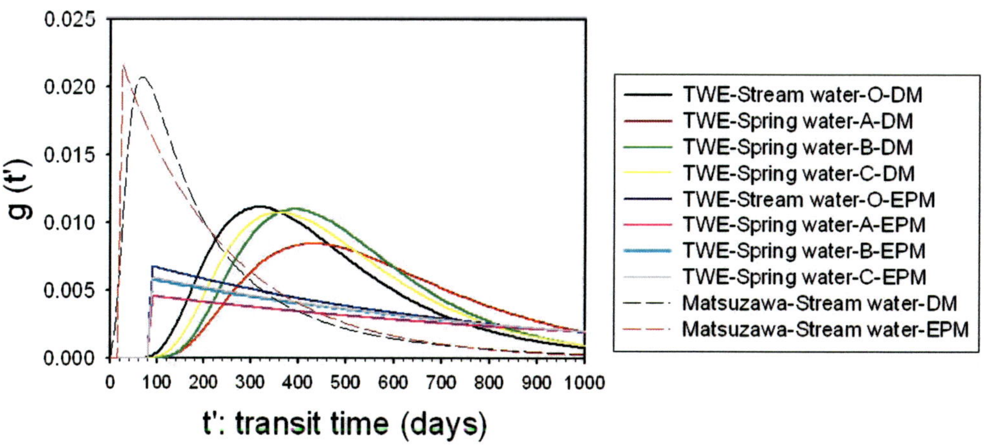

Figure 12. System response functions of stream water and spring waters in the TEW and stream water in the Matsuzawa catchment.

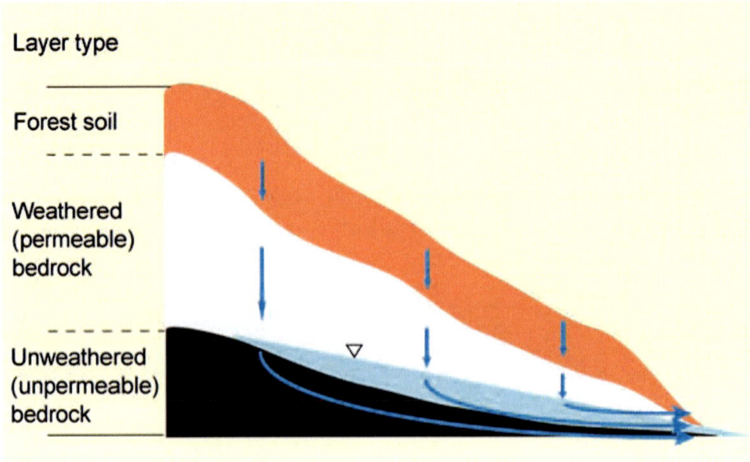

Figure 13. Mechanism of baseflow generation in the TEW.

CONCLUSION

We estimated the MRTs of stream and spring waters based on the seasonal variation in deuterium excess in a small gneiss watershed with a thick weathered layer. The mean residence times of the stream water and spring waters were estimated to be approximately 1 to 3 years. The mean residence time of stream water from the whole catchment was 1.3 years as estimated by the dispersion model (DM) and 2.3 years by the exponential piston flow model (EPM), corresponding to mobile water storage volumes of 839 and 1454 mm, respectively (i.e., $V_m = \tau Q$). These values are within the limits of the storage water capacities of brown forest soil and of heavily weathered gneiss layers (2148 mm). The shapes of system response functions indicated that there was almost no contribution from rain that had fallen within the past 90 days. This result can be explained by the longer time needed by rainwater to permeate the thick soil layer and reach the groundwater table.

This chapter demonstrates that seasonal variations of deuterium excess can be used to estimate the MRT of water within a small gneiss watershed with a thick weathered layer. Recently, ^{17}O has attracted attention as an air-mass-origin parameter that changes in accordance with deuterium excess. Angert et al (2004) showed that the $^{17}\Delta$ value (also called "^{17}O excess" or "^{17}O anomaly") of precipitation was controlled primarily by kinetic effects during the evaporation of the initial vapour and, in contrast to the deuterium excess, was independent of the temperature at evaporation (and condensation) site, suggesting that $^{17}\Delta$ might be superior to deuterium excess as an index of the vapour source. To date, only a few laboratories worldwide perform ^{17}O analysis. However, if a convenient ^{17}O analysis method can be established, the $^{17}\Delta$ value can be expected to provide more detailed information on the residence time of water in areas where the origin of water vapour changes seasonally.

ACKNOWLEDGEMENTS

We thank Associate Professor Nobuhito Ohte and Dr. Yasuhiro Ohnuki for their valuable and encouraging suggestions. We also thank former Chief Tatsurou Kanazashi, present Chief Takashi Yoshitake, Mr. Masaaki Sugiyama, and the other staffs of the Forest Site Section in the Forestry and Forest Products Research Institute for their help in the field. Some of the hydrometric data used in this chapter were measured by several former researchers of the Water Resources Laboratory and Flood Control Laboratory of the Forestry and Forest Products Research Institute.

REFERENCES

Angert, A., Cappa, C. D. and DePaolo, D. J. (2004). Kinetic ^{17}O effects in the hydrologic cycle: Indirect evidence and implications. *Geochimica et Cosmochimica Acta,* 68(17), 3487-3495.

Asano, Y., Uchida, T. and Ohte, N. (2002). Residence times and flow paths of water in steep unchannelled catchments, Tanakami, Japan. *J. Hydrol.,* 261, 173-192.

Band, L., Ogden, F., Butler, J., Goodrich, D., Hooper, R., Kane, D., Lyons, B., McKnight, D., Miller, N., Williams, M., Potter, K., Scanlon, B., Pielke, R., and Peckhow, K. (2002). Hydrologica observatory network, Technical Report #4. Washington, D.C.: Consortium of Universities for the Advancement of Hydrologic Sciences, Inc.

Chappell, N. A., Sherlock, M., Bidin, K., MacDonald, R., Najman, Y. and Davies, G. (2007). Stable isotope studies of rainfall and stream water in forest watersheds in Kampong Thom, Cambodia. In H. Sawada, M. Araki, N. A.Chappell, J.V. LaFrankie, A. Shimizu (Eds.), *Forest Environments in the Mekong River Basin,* pp 3-23. Tokyo: Springer.

Clark, I. and Fritz, P. (1997). Environmental isotopes in hydrogeology, 328. Boca Raton: CRC Press.

Dansgaard, W. (1964). Stable isotopes in precipitation. *Tellus,* 16, 436-468.

DeWalle, D.R., Edwards, P.J., Swistock, B.R., Aravena, R. and Drimmie, R.J. (1997). Seasonal isotope hydrology of three appalachian forest catchments. *Hydrol. Process.,* 11, 1895-1906.

Hooper, R.P. (2004). Designing observatories for the hydrologic sciences. EOS, Transactions of the American Geophysical Union, 85(17), *Jt Assem. Suppl.,* Abstract H24B004.

Kabeya, N., Shimizu, A., Nobuhiro, T., Zhang, J., Kubota, T. and Abe, T. (2005) An exploration of surface structure by using surface wave in Tsukuba Experimental Watershed. *Trans. Kanto Branch Jpn. For. Soc.,* 55, 251-252 (in Japanese).

Kabeya, N., Katsuyama, M., Kawasaki, M., Ohte, N and Sugimoto, A. (2007). Estimation of mean residence times of subsurface waters using seasonal variation in deuterium excess in a small headwater catchment in Japan. *Hydrol. Process.,* 21, 308-322.

Katsuyama, M., Ohte, N. and Kabeya, N. (2005). Effects of bedrock permeability on hillslope and riparian groundwater dynamics in a weathered granite catchment. *Water Resour. Res.,* 41, W01010.

Kawaraya, H., Matsuda, H. and Matsubaya, O. (1999) Origin and mixing of groundwater estimated by oxygen and hydrogen isotope tracers technique in Yachi landslide area,

Akita, Japan. *Journal of the Japan Landslide Society,* 34, 48-55 (in Japanese with English summary).

Kawasaki, I. (1988). The subsurface exploration of the Sawara region by a full automatic underground probe –use of Sato type VIC GR-810 –. Report of investigation about prediction of the landslide generation on the heavy rainfall – a case of the Sawara region -, pp. 93-103, Sammu: Northern Chiba Forest Office.

Kondoh, A. and Shimada, J. (1997) The origin of precipitation in eastern Asia by deuterium excess. *Journal of the Japan Society of Hydrology and Water Resources,* 10, 627-629.

Kubota, T. (2000) On the formulas expressing mean residence time of subsurface water. *Journal of the Japan Society of Hydrology and Water Resources,* 13, 472-475 (in Japanese with English summary).

Lee, K.S., Kim, J.M., Lee, D.R., Kim, Y. and Lee, D. (2007). Analysis of water movement through an unsaturated soil zone in Jeju Island, Korea using stable oxygen and hydrogen isotopes. *J. Hydrol.,* 345, 199-211.

Maloszewski, P. and Zuber, A. (2002). Manual on lumped-parameter models used for the interpretation of environmental tracer data in groundwaters. In Y. Yurtsever (Ed). Use of Isotopes for Analyses of Flow and Transport Dynamics in Groundwater Systems, IAEA-UIAGS / CD no. 02-00131, pp. 1-50. Vienna: IAEA.

Maloszewski, P., Rauert, W., Stichler, W. and Herrmann, A. (1983). Application of flow models in an alpine catchment area using tritium and deuterium data. *J. Hydrol.,* 66: 319-330

McGuire, K.J. and McDonnell, J.J. (2006). A review and evaluation of catchment transit time modeling. *J. Hydrol.,* 330, 543-563.

Nobuhiro, T., Shimizu, A. and Kabeya, N. (2006) Estimation of evapotranspiration in Tsukuba Experimental Watershed in 2003. *Trans. Kanto Branch Jpn. For. Soc.,* 57, 279-281 (in Japanese).

Ohnuki, Y., Yoshinaga, S. and Noguchi, S. (1999). Distribution and physical properties of colluvium and saprolite in unchannelized valleys in Tsukuba Experimental Basin, Japan. J. For. Res., 4, 207-215.

Sanjo, K. (1990). Environmental isotope hydrology of Mt. Tsukuba. Ph. D. dissertation, 101. Ibaraki: University of Tsukuba.

Soulsby, C., Malcolm, R., Helliwell, R., Ferrier, R. C. and Jenlins, A. (2000) Isotope hydrology of the Allt a' Mharcaidh catchment, Cairngorms, Scotland: implications for hyfrological pathways and residence times. *Hydrol. Process.,* 14, 747-762.

Shimizu, A., Kabeya, N., Nobuhiro, T., Zhang, J., Kubota, T. and Abe, T. (2007) Estimation of subsurface sutructure in Tsukuba Forest Experimental Watershed. *Trans. Kanto Branch Jpn. For. Soc.,* 58, 153-156 (in Japanese with English summary).

Taniguchi, M., Nakayama, T., Tase, N. and Shimada, J. (2000) Stable isotope studies of precipitation and river water in the Lake Biwa basin, Japan. *Hydrol. Process.* 14, 539-556.

Uchida, T., Asano, Y., Ohte, N. and Mizuyama, T. (2003) Seepage area and rate of bedrock groundwater discharge at a granitic unchanneled hillslope. *Water Resour. Res.,* 39(1), 1018.

Uchida, T., McDonell, J.J. and Asano, Y. (2006) Functional intercomparison of hillslope and small catchments by examining water source, flowpath and mean residence time. *J. Hydrol.* 327, 627-642.

Vitvar, T. and Balderer, W. (1997) Estimation of water residence times and runoff generation by 18O measurements in a Pre-Alpine catchment (Rietholzbach, Eastern Switzerland). *Applied Geochemistry*, 12, 787-796.

Vitvar, T., Aggarwal, P.K. and McDonnell, J.J. (2005) A review of isotope applications in catchment hydrology. In P.K. Aggarwal, J.R. Gat and K.F.O. Froehlich (Eds.) *Isotopes in the water cycle: past, present and future of a developing science,* pp. 151-169, Netherlands : IAEA.

Vitvar, T., Aggarwal, P.K. and Herczeg, A.L. (2007) Global newrork is launched to monitor isotope in rivers. *EOS,* 88(33), 325-326.

Yabusaki, S. and Tase, N. (2005) Characteristics of the $\delta^{18}O$ and δD of monthly and event precipitation in Tsukuba from 2000 to 2002. *Journal of the Japan Society of Hydrology and Water Resources,* 18, 592-602 (in Japanese with English summary).

Waseda, A. and Nakai, N. (1983) Isotopic compositions of meteoric and surface waters in central and northeast Japan. *Geochemistry,* 17, 83-91 (in Japanese with English summary).

Water Resources Laboratory and Flood Control Laboratory (1993) Statistical reports of hydrological observation at the Tsukuba Experimental Watershed (May, 1978 – December, 1987). *Bull. For. Forest Prod. Res. Inst.,* 364, 125-168 (in Japanese with English summary).

Zhang, J., Shimizu, A., Kabeya, N. and Nobuhiro, T. (2005) Research on suspended sediment of upland small forest watershed in Japan. *Journal of Beijing Forestry University,* 27(6), 12-19.

Zuber, A. (1986). Mathematical models for the interpretation of environmental radioisotopes in groundwater systems. In P. Fritz and J. C. Fontes (Eds.) *Handbook of Environmental Isotope Geochemistry*, vol 2, part B, pp. 1-59. Amsterdam: Elsevier.

Reviewed by Dr. Nick A Chappell, Department of Environmental Science, Lancaster Environment Centre, Lancaster University

In: Lake Pollution Research Progress
Editors: F. R. Miranda, L. M. Bernard, pp. -
ISBN: 978-1-60692-106-7

Chapter 10

An Analysis of Internal Phosphorus Loading in White Lake, Michigan

Alan D. Steinman*[*], *Mary Ogdahl* and *Mark Luttenton

Annis Water Resources Institute;
Grand Valley State University;
Lake Michigan Center;
740 W. Shoreline Dr.;
Muskegon, MI 49441

Abstract

Internal loading can account for a significant percentage of phosphorus entering a lake, and may prevent the recovery of lake water quality even after external loads are reduced. Prior studies in two west Michigan coastal lakes showed that internal phosphorus loading can account for more than 80% of the total phosphorus load,

[*] Alan D. Steinman: steinmaa@gvsu.edu

depending on the time of year. However, these studies occurred on lakes with dense shoreline development and in rapidly urbanizing watersheds. To provide a contrast, we conducted an analysis of internal phosphorus loading in White Lake, MI, which has a less-developed shoreline and a watershed with much lower urban/developed land cover, and greater agricultural land use, than lakes we previously studied.

Sediment cores were removed from 4 sites in White Lake and incubated in the laboratory under aerobic (with oxygen) and anaerobic (without oxygen) conditions. Phosphorus flux from the sediments into the overlaying water column was measured over a 27-day period and compared to rates measured from sediment cores collected previously from Mona Lake and Spring Lake, MI. Internal total phosphorus (TP) loading from White Lake sediments ranged from 1.55 to 7.78 mg TP/m^2/d in anaerobic conditions and from -0.18 to 0.14 mg/m^2/d in aerobic conditions. The negative value suggests that the sediments in some areas of White Lake could act as a sink for TP during aerobic periods. The anaerobic phosphorus release rates were approximately ½ the rates measured during summer months in Mona Lake and ¼ those measured in Spring Lake in previous years.

Internal loading contributed 1.24 tons of total phosphorus based on our laboratory study, with about half coming from the eastern-most basin in White Lake. Compared to an estimated external total phosphorus load of 15.48 tons/yr, internal loading of TP accounted for ~7.4% of the total TP load entering White Lake. This percentage is much lower than what has been measured in Spring Lake (~55-67% of TP load) and Mona Lake (~9-82% of TP load), indicating internal phosphorus loading is less important in White Lake than in the other lakes. These data suggest that management strategies should be currently focused on reducing external phosphorus loads to White Lake.

INTRODUCTION

White Lake is one of several coastal lake systems unique to the West Michigan region. As a drowned river mouth system, White Lake functions similarly to marine estuaries, in that it serves as an interface between upland terrestrial habitats and its ultimate receiving water body, which in this case is Lake Michigan. White Lake is listed as a Great Lakes Area of Concern (AOC), so designated by the International Joint Commission and the US Environmental Protection Agency, because of severe environmental contamination. Its designation as an AOC stems primarily because of contaminated groundwater migrating to the lake from the Occidental Chemical site (formerly Hooker Chemical Company). There are eight other sites of contamination with the potential to affect White Lake, with some of them in varying states of remediation. The White Lake Public Advisory Council (PAC) and the Michigan Department of Environmental Quality have identified eight beneficial use impairments (BUIs) in White Lake, including both eutrophication and the loss of fish and wildlife habitat.

Over the past several decades, land use changes within the watersheds draining to these Great Lakes drowned river mouth systems have resulted in increased nutrient loading. In the White Lake watershed, nutrient loading from the White River was measured as early as the 1960s, when it was estimated that 98% of total phosphate and 96% of total nitrogen entered White Lake via the White River (Freedman et al. 1979). Furthermore, White Lake retained 66–76% of phosphate and between 5–51% of nitrogen received from the White River (Freedman et al. 1979). These data have generally led to the conclusion that most nutrient-based water quality problems in White Lake are due to nutrient loading from the White River.

However, recent changes in White Lake have initiated a re-evaluation of the environmental factors and processes that have primary influence over various biological components within the lake. For example, some improvement in water quality was noted following the diversion of sewage and industrial discharges to the Whitehall treatment system (RAP 1995). In addition, zebra mussels, which reached significant population densities during the early to mid-1990s, have reduced suspended material and increased water clarity. However, lakeshore development has increased and has moved from seasonal residential use to year-round use, resulting in more pressures on the system. Finally, there has been more than a two-fold increase in the maximum biomass of aquatic plants in White Lake between the mid-1970s and 1995 (Luttenton 1996), with much of the increase occurring during the 1990s. In aggregate, these changes suggest that the system has experienced significant modification during the past two decades. These changes, coupled with the fact that White Lake is a Great Lakes Area of Concern, were compelling reasons to reevaluate the nutrient conditions and determine the sources of nutrient loading in the lake.

Attempts to limit or reduce external loading to water bodies have resulted in smart growth and low impact initiatives in many municipalities. By detaining and retaining water on site, there is a greater opportunity for pollutants to be captured before they enter local streams and lakes (Johnston 1991; Steinman and Rosen 2000). These watershed-based practices are founded on the premise that external nutrient loads are the primary source of nutrients entering surface water bodies, and that their control will result in improved water quality. However, in systems where internal nutrient loads are important, nutrient control efforts that are focused solely at the watershed-level will address only part of the problem (Welch and Cooke 1995). Increased development, changing land uses, and intense upstream agricultural activity all contribute to heavier nutrient loads to White Lake; retention of these nutrients in the lake sediments could lead to high rates of internal nutrient loading. Ultimately, an integrated approach is needed that examines source control in a holistic way, especially in urbanizing systems where multiple stressors exist (Paul and Meyer 2001, Walsh et al. 2005, Steinman et al. 2006). In this study, we examine the spatial variability of internal phosphorus loading in White Lake, and evaluate internal P loading in a regional context by comparing measured rates of sediment phosphorus release in White Lake to those measured previously in Mona and Spring Lakes.

METHODS

Site Description: White Lake has a surface area of 9.98 km^2 and a mean and maximum depth of 6.9 and 21.3 m, respectively. White Lake serves as the receiving water body in the White River watershed (figure 1); the watershed covers 1,360 km^2 along the eastern shoreline of Lake Michigan, and contributes approximately 1% of the total phosphorus load and 2% of the total sediment load to Lake Michigan (Robertson 1997). With headwaters in Newaygo County, the White River flows for ~190 river km before discharging to White Lake, which then discharges directly into Lake Michigan through a short navigation channel. Prior to European settlement, the watershed was a continuous system of dense riparian forests, sprawling wetlands and marshes, inland lakes, and flowing streams. The lumber industry changed the landscape dramatically in the 1800s, resulting in reduced forest cover, increased sediment loads to the watershed, and increased development. Agriculture today accounts for 46.5% of the land use in the watershed, primarily in the form of vegetable, livestock, and row crop farms. Other major land use categories include forest at 51.8% and urban (0.7%) (figure 2).

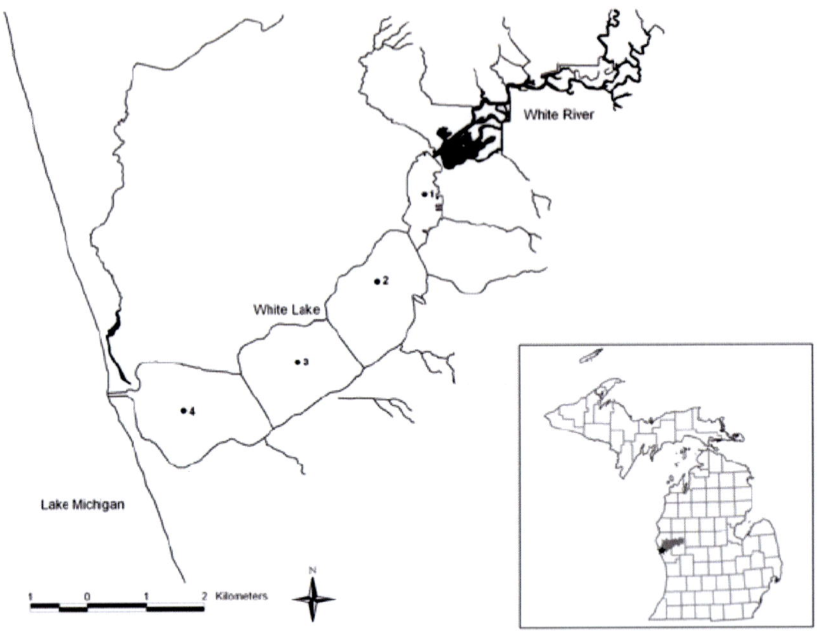

Figure 1. Map of White Lake, showing sampling locations (1-4) of the sediment cores. White Lake is divided into four basins based on depth distributions. The area of each basin was used as a weighting function when summing the total internal phosphorus load in White Lake. Inset: Location map of White Lake watershed (shaded gray).

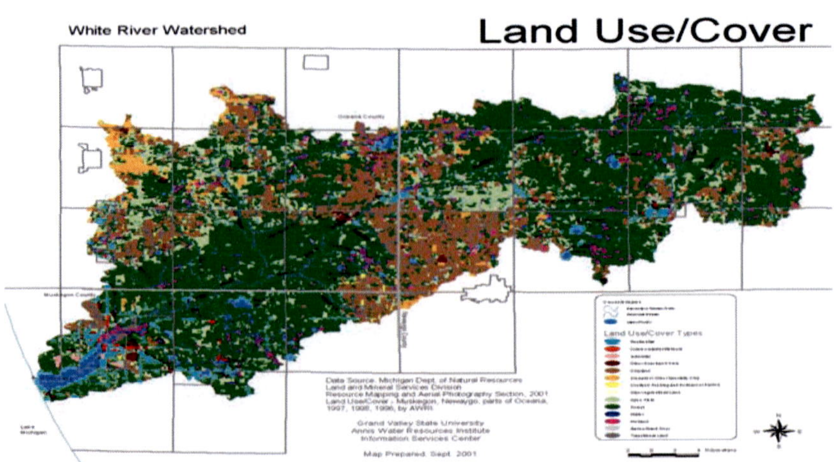

Figure 2. Land use in the White River watershed. Data source: Michigan Department of Natural Resources Land and Mineral Services Division, Resource Mapping and Aerial Photography Section, 2001. Land use/cover by AWRI: Muskegon, Newaygo, parts of Oceana Counties, 1997, 1998, 1996.

Field Methods: We divided White Lake into 4 compartments based on depth distributions (figure 1). One site per compartment was sampled in June, 2006 for environmental conditions and sediment cores. At each site, dissolved oxygen, pH, temperature, specific conductance, total dissolved solids, and chlorophyll *a* were measured at near-surface and near-bottom depths using a Hydrolab DataSonde 4a equipped with a Turner Designs fluorometer. Water samples for phosphorus analysis were collected with a Van Dorn bottle. Water for soluble reactive phosphorus (SRP) analysis was syringe-filtered immediately through 0.45-μm membrane filters into scintillation vials. Samples were stored on ice until transported to the laboratory, always within 5 h of collection. TP samples were stored at 4°C and SRP samples were frozen until analysis. SRP and TP were analyzed on a Bran+Luebbe Autoanalyzer III (US EPA 1983). SRP values below detection were calculated as ½ the detection limit (5 μg/L).

Six sediment cores were collected from each site using a piston corer (Fisher et al. 1992, Steinman et al. 2004). The corer was constructed of a graduated 0.6-m long polycarbonate core tube (7-cm inner diameter) and a polyvinyl chloride (PVC) attachment assembly for coupling to aluminum drive rods. The piston was advanced 20 to 25 cm prior to deployment to maintain a water layer on top of the core during collection. The corer was positioned vertically at the sediment–water interface and pushed downward with the piston cable remaining stationary. After collection, the core was brought to the surface and the bottom was sealed with a rubber stopper prior to removal from the water, resulting in an intact sediment core that was ~20 cm in length, with a 25-cm overlaying water column. The piston was then bolted to the top of the core tube to keep it stationary during transit. Core tubes were placed in a vertical rack and maintained at ambient temperature during transit. An additional core was collected from each site and the top 5 cm removed for sediment analyses in the lab.

Laboratory methods. The 24 cores (6/site) were placed in a Revco environmental growth chamber, with the temperature maintained to match ambient bottom water conditions in White Lake at the time of collection. The water column in 3 of the 6 cores from each site was bubbled with N_2 (with 330 ppm CO_2) to create buffered anaerobic conditions, while the remaining 3 were bubbled with oxygen to create aerobic conditions.

Internal load estimates were made using the methods outlined in Moore et al. (1998), with minor modifications (Steinman et al. 2004). Briefly, a 40-mL water sample was removed by syringe through the sampling port of each core tube at times 0, 2 h, 12 h, 1 d, 2 d, 4 d, 8 d, 12 d, 16 d, 20 d, 24 d, and 27 d. The 40-mL subsample was replaced with filtered water collected (at the same time as the cores were removed) from the corresponding site in the lake; this maintained the original volume in the core tubes. Immediately after removal, a 20-mL subsample was refrigerated for analysis of TP, and a 20-mL subsample was filtered through a 0.45-μm membrane filter and frozen for analysis of soluble reactive P (SRP). P was analyzed as described previously.

Phosphorus release rate calculations were based on the increase in water column TP or SRP using the following equation (Steinman et al. 2004):

$$P_{rr} = (C_t - C_0)\ V/A$$

where, P_{rr} is the net P release rate or retention per unit surface area of sediments, C_t is the TP or SRP concentration in the water column at time t, C_0 is the TP or SRP concentration in the water column at time 0, V is the volume of water in the water column, and A is the planar surface area of the sediment cores. P release rate was calculated from the time when P concentrations stabilized in the water column (~day 1) until the time when an asymptote was reached. For this chapter, only the TP internal loading data are presented; patterns for TP and SRP were similar, although the magnitude of P release was much lower for SRP than TP.

Following the incubations, the top 5 cm of sediment was removed from each core. The sediment was homogenized and subsampled for ash-free dry mass (AFDM). The ashed material was analyzed for TP as described previously. Another subsample (5 g) from the wet sediment was centrifuged to remove excess porewater and sequentially fractionated (Moore and Reddy 1994) to determine the fraction of phosphorus bound to iron and calcium minerals in the sediments. Porewater was filtered, frozen, and analyzed for SRP as described previously. Residual sediment was shaken for 17 h with 0.1M NaOH and centrifuged. The supernatant was drawn off, filtered, and frozen, and subsequently analyzed for SRP. This fraction is referred to Al- and Fe-bound phosphorus and represents a mineral association that can become soluble under anoxic conditions. After this extraction, the sediment was shaken for 24 h with 0.5M HCl, centrifuged, and the supernatant filtered, frozen, and subsequently analyzed for SRP. This fraction is referred to as Ca- and Mg-bound phosphorus and represents a stable mineral association.

Lake-wide internal TP load calculations were made by 1) calculating an average internal load at each site under both anaerobic and aerobic conditions, 2) extrapolating the mean areal rate of the cores to the area of each of the four lake compartments, and 3) summing the fluxes in each of the four compartments. External load calculations were made as part of a separate study (Luttenton et al. 2007), and include loads from the White River (the major tributary to White Lake), four minor tributaries, and groundwater inflow directly to White Lake.

Total P release rates were analyzed with a 2-way ANOVA. The main effects were site ($n = 4$) and oxidation state ($n = 2$). Data were ln-transformed where necessary. Statistical tests were undertaken with SigmaStat 3.1.

RESULTS

There was a distinct depth gradient in the sampling sites. Site 1, located farthest to the east and nearest the mouth of the White River (figure 1), was the shallowest and also had the lowest secchi depth (i.e., least transparent). Sites became progressively deeper and more transparent as one moved west, toward the channel to Lake Michigan (table 1). Stratification of the water column, based on temperature and dissolved oxygen concentration, was evident at sites 3 and 4, incipient at site 2, and not evident at site 1, consistent with its shallow depth (table 1). Elevated total phosphorus concentrations were measured near the lake bottom at sites 3 and 4, where hypoxic conditions were present (table 1). At sites 1 and 2, where near-bottom DO concentrations were still relatively high, TP concentrations were very similar at the near-surface and near-bottom.

In the anaerobic treatments, TP concentrations peaked between days 16 and 27 (figure 3), and averaged 480, 180, 280, and 350 µg/L at sites 1-4, respectively. In the aerobic treatments, TP concentrations stayed very low, and declined slightly over time in several of the core tubes (figure 3). The mean TP concentrations on day 27 at sites 1-4 were 30, 30, 30, and 35 µg/L,

Table 1. Selected limnological characteristics of sampling sites in White Lake. ND = no data

Parameter		Site			
		1	2	3	4
Depth (m)		3.2	10.2	13.5	16.1
Secchi Depth (m)		1.6	2.2	2.25	2.6
Chl a ($\mu g\ L^{-1}$)	Surface	4.5	8.1	7.8	6.4
DO (mg L^{-1})	Surface	7.40	7.84	8.09	8.09
	Bottom	6.72	2.07	0.89	1.11
Temperature (°C)	Surface	21.89	23.14	23.3	23.29
	Bottom	20.62	18.22	15.24	14.47
TP ($\mu g\ L^{-1}$)	Surface	30	30	20	20
	Bottom	30	20	420	80
SRP ($\mu g\ L^{-1}$)	Surface	9	6	16	5
	Bottom	9	7	ND	21

respectively. TP release rates were significantly greater in the anaerobic treatments than in the aerobic treatments (2-way ANOVA: $F = 70.324$; $P < 0.001$), but site had no significant effect on TP flux. The interaction term between redox and site was significant ($F = 4.021$; $P < 0.026$), as the influence of redox was more evident at sites 1, 3, and 4 than at site 2 (figure 3).

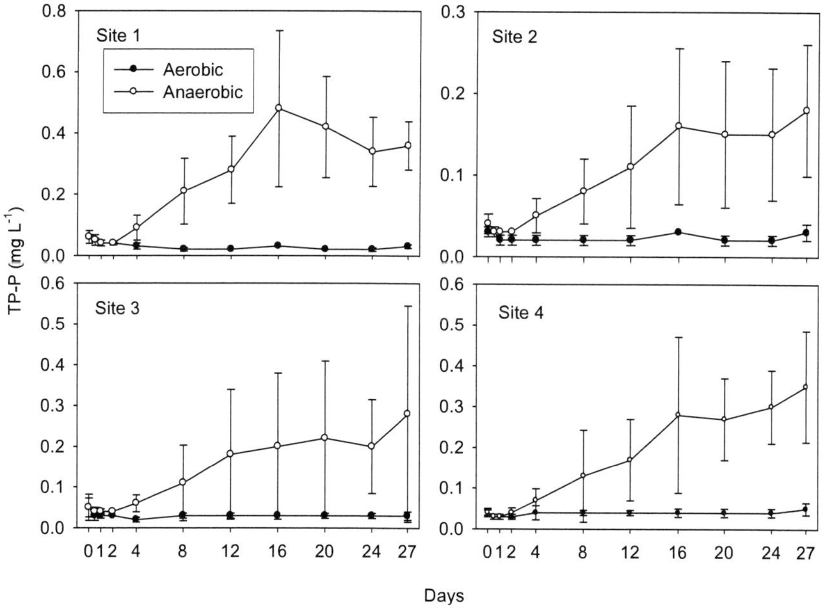

Figure 3. TP concentrations in the water column overlying sediment cores from 4 sites in White Lake sampled in summer, 2006. Note different y-axis scales for the sites.

Mean release rates of TP ranged from 1.55 to 7.78 mg TP/m^2/d in anaerobic conditions and from -0.18 to 0.14 mg/m^2/d in aerobic conditions (table 2). The negative value at site 1 suggests that the sediments could act as a sink for TP during aerobic periods. The anaerobic TP release rates were ~½ the rates measured in Mona Lake and ~¼ those measured in Spring Lake (table 2).

Table 2. Mean summer flux rates of TP from sediment cores collected from White Lake (this study), Mona Lake (2004 and 2005; Steinman et al. 2008), and Spring Lake (Steinman et al. 2004)

TP Release Rate (mg P/m^2/d)		
Site	Anaerobic	Aerobic
	White Lake (2006)	
1	7.78	-0.18
2	1.55	0.07
3	2.46	0.03
4	3.21	0.14
	Mona Lake (2004)	
1	6.46	-1.84
2	5.38	-2.41
3	13.63	0.99
4	12.82	0.30
	Mona Lake (2005)	
1	4.16	0.22
2	5.57	0.21
3	4.85	0.52
4	3.19	0.19
	Spring Lake (2003)	
1	26.71	0.40
2	16.02	-2.00
3	9.04	0.16
4	10.64	-1.04

The initial TP concentration in the sediment cores (prior to incubation) ranged from 899 (Site 1) to 1280 mg/kg (Site 2; figure 4). No inferential statistics were applied to these data as replicate cores were not sampled at each site. These numbers are similar to mean concentrations measured in Spring Lake (1282 mg/kg) and Mona Lake (1394 mg/kg). At the end of the incubation, site-specific differences in TP in White Lake were still evident and highly significant (2-way ANOVA: $F = 12.852$, $P < 0.001$), irrespective of redox status; mean TP concentration was significantly lower at Site 1 compared to all other sites (figure 4). TP concentration in the sediment was not statistically different between the aerobic and anaerobic cores at the end of the laboratory incubation ($F = 0.291$; $P = 0.597$). The interaction between redox and site was not statistically significant ($F = 0.530$, $P = 0.668$).

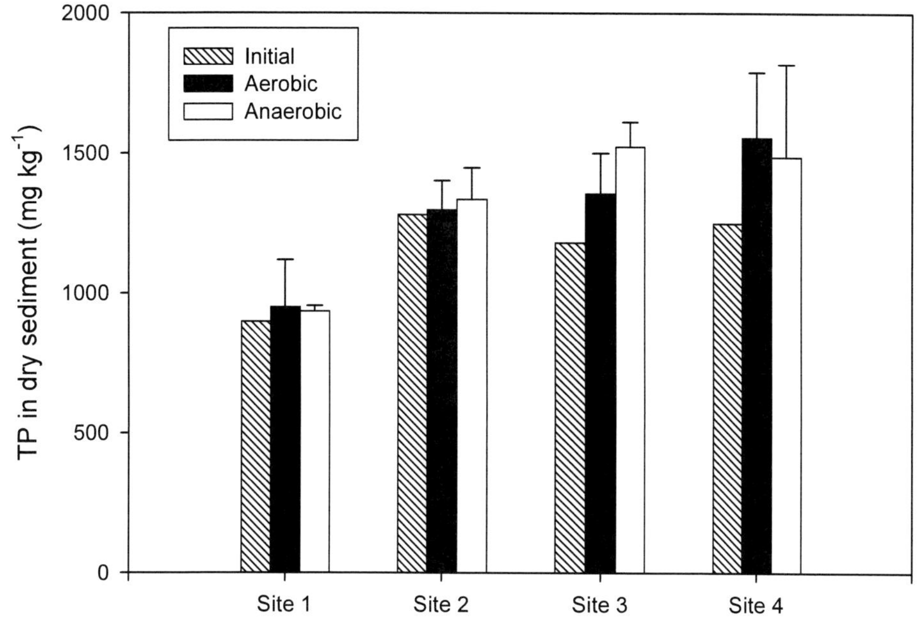

Figure 4. TP concentration in dry sediment (mg/kg) from summer 2006 sediment cores.

SRP in the porewater at the end of the incubations ranged from below detection to an anomalously high concentration of 0.167 mg/L in one core from Site 3 (figure 5). Overall, site had no statistically significant effect on SRP porewater concentration; the high variance in the data at Site 3 precluded statistically significant differences, despite the apparently higher mean concentrations at that site. Porewater SRP was statistically higher under anaerobic than aerobic conditions (2-way ANOVA: $F = 44.817$, $P < 0.001$; figure 5). The interaction term between redox and site was not statistically significant ($F = 1.818$, $P = 0.184$).

There were clear differences between the NaOH-extractable and HCl-extractable fractions of SRP, regardless of site (figure 5), with the SRP from the NaOH extraction significantly less than the SRP from the HCl extraction (3-way ANOVA: $F = 363.652$, $P < 0.001$). The mean NaOH-extractable SRP fraction was 3.14 mg/L, while the mean HCl-extractable SRP fraction was 10.2 mg/L. Site also was statistically significant (3-way ANOVA: $F = 4.470$, $P = 0.01$), with Sites 1 and 4 having significantly more extractable SRP than Site 2 (figure 5). The interaction term between site and extraction type was marginally significant ($F = 2.377$, $P = 0.088$), as the effect of site was influenced by the type of extraction; the significant effect of site appeared to be much more prominent in the NaOH fraction than the HCl fraction (figure 5). Redox had no significant effect on extractable SRP ($F = 0.869$, $P = 0.358$), and the redox × extract and site × redox × extract interactions terms were non-significant, as well.

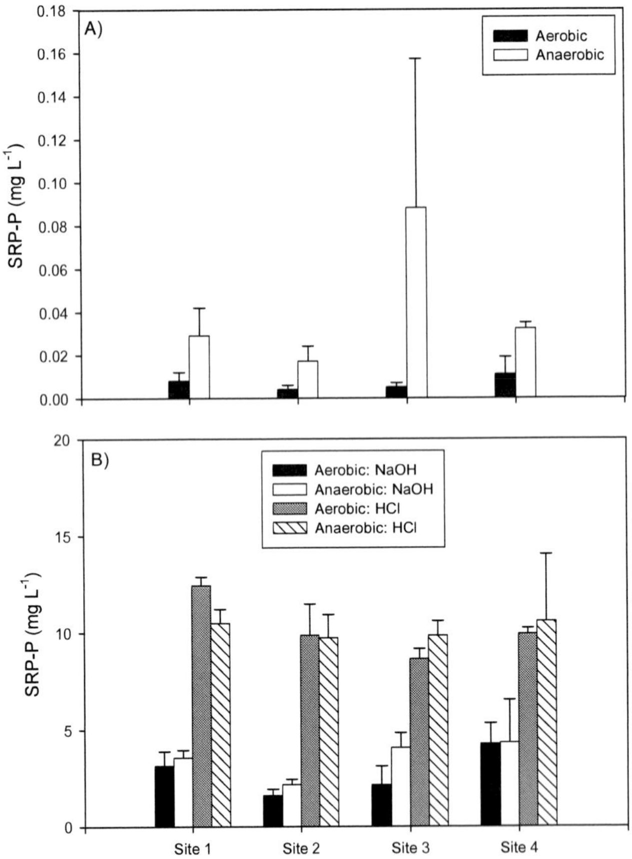

Figure 5. Top panel = Porewater SRP concentrations in sediment cores. Bottom panel = NaOH-extractable P (Fe/Al-bound P) and HCl-extractable P (Ca/Mg-bound P). Data are mean (± SD) values at the end of the laboratory incubations.

Lake-Wide Internal Loading

Sediment cores from the eastern-most lake compartment (#1) contributed potentially 2.5 to 5X more TP than any of the other compartments (table 3) based on the anaerobic laboratory results. However, in nature, the shallow depth in this region results in a well-mixed water column, keeping the sediments frequently aerobic, which likely minimizes the release of phosphorus that would otherwise be disassociated from reduced iron. Therefore, the estimates in table 3 should be viewed as maximum potential rates, and may be overestimates compared to natural conditions. More dissolved oxygen data are needed, on both spatial and temporal (seasonal and diel) bases, to reduce the uncertainty in these flux estimates. The internal loads from compartment #4 were the second highest, but it is unclear if the released P actually reaches the epilimnion of the lake given the depth at this site (and presumably deeper point of stratification of the column). Internal TP loads were generally very low under aerobic conditions (table 3), as expected when the phosphorus dynamics are being driven by iron biogeochemistry.

Table 3. Potential mean internal TP load estimates (tons) by site for summer, 2006 in White Lake. See methods for description of calculation

Lake Compartment	Anaerobic	Aerobic
1	0.640	-0.014
2	0.128	0.006
3	0.203	0.003
4	0.264	0.011
Sum	1.235	0.006

Internal vs. External Loading Analysis

Luttenton et al. (2007) estimated external TP loads to White Lake from the White River and minor tributaries flowing directly into White Lake (13.05 tons/yr), as well as groundwater (based on well data from both south and north sides of White Lake: 2.43 tons/yr). Results from this chapter revealed a potential internal TP load of 1.24 tons/yr. Summing all these sources resulted in a total (i.e., internal and external) potential TP load of 16.72 tons/yr. Based on these data, internal loading of TP accounts for up to ~7.4% of the total TP load entering White Lake.

Discussion

Internal P loading can be a significant source of nutrients in shallow, eutrophic lakes, and can result in serious impairment to water quality (Welch and Cooke 1995, 1999; Steinman et al. 1999, 2004; Søndergaard et al. 2001; Nürnberg and LaZerte 2004). This process has both ecological and societal implications; internal loading rates can be sufficiently great that reductions in external loading fail to improve water quality. If these reductions in external loading require the investment of money and create expectations of success, the resulting disappointment (or worse) from stakeholders at the failure to improve lake conditions can set back future restoration activities and harm the reputation and credibility of natural resource managers. Hence, it is critical to assess the relative importance of internal nutrient loading in eutrophic lakes as part of a lake management plan.

Water column TP concentrations were lower in White Lake compared to other lakes that have been assessed in the region. Near-surface TP levels in White Lake were 20 to 30 μg/L, compared to summer surface concentrations of 30 to 60 μg/L in Mona Lake (Steinman et al. 2008), and 60 to 120 μg/L in Spring Lake (Steinman et al. 2004). However, near-bottom TP concentrations in White Lake were similar to those measured in the other lakes; levels ranged from 20 to 420 μg/L in White Lake (although only one of the four sites had severely elevated concentrations), whereas near-bottom TP levels were 30 to 420 μg/L in Mona Lake and 40 to 80 μg/L in Spring Lake.

The summer TP release rates from White Lake under anaerobic conditions (1.6-7.8 mg P/m^2/d) were in the same general range as those measured in mesotrophic systems (~3 to 7 mg/m^2/d; Nürnberg and LaZerte 2004). These release rates are lower than what we have measured in other phosphorus-impacted lakes (table 4). For example, median summer TP release rates from Spring Lake and Mona Lake were ~2 to 5X greater than the measured rates in White Lake. Despite its considerably larger surface area (table 5), the overall amount of TP potentially released from White Lake sediments during the summer (1.24 tons) is very similar to the summer estimates in Mona Lake (1.09 tons), and is considerably less than the estimate in Spring Lake (2.7 to 6.4 tons). The Spring Lake estimate is inflated slightly because it also includes spring internal loading, but those fluxes were very small compared to the summer (Steinman et al. 2004).

Table 4. Comparison of mean TP release rates (mg P/m^2/d) from coastal west Michigan lakes under anaerobic conditions

Lake	Date of Core Collection	Minimum Release Rate	Maximum Release Rate	Median Release Rate
White Lake	June, 2006	1.55	7.78	2.84
Spring Lake	June-July, 2003	1.64	29.54	12.50
Mona Lake	July, 2004	5.38	13.63	9.64
Mona Lake	June, 2005	3.19	5.57	4.51

Table 5. Selected morphometric characteristics of coastal West Michigan lakes

Lake	Surface Area (km^2)	Mean Depth (m)	Max Depth (m)	Watershed Area (km^2)
White Lake	9.98	6.9	21.5	1,360
Spring Lake	5.25	6.0	13.0	134
Mona Lake	2.65	6.1	8.3	200

The lower sediment TP release rates in White Lake also result in internal loading accounting for a much lower percentage of total TP load in White Lake compared to either Mona or Spring Lakes. Internal loading during summer months was estimated to account for 55 – 67% of the total TP load in Spring Lake (Steinman et al. 2004) and 67 – 85% of the total TP load in Mona Lake (Steinman et al. 2008), whereas it accounted for only ~7% of the total TP load in White Lake. Furthermore, the 7% value may be an overestimate, as external load estimates included minimal storm event sampling (Luttenton et al. 2007), when most of the nutrients are transported downstream.

Several reasons could account for the relatively low TP release rates in White Lake. First, the total phosphorus sediment concentrations are lower than in Mona and Spring Lakes. This suggests that there may be less phosphorus available for internal loading to the water column in White Lake. However, TP can be a poor predictor of internal loading because it includes both highly mobile and immobile phosphorus fractions. Analyzing the sediment by different fractions allows one to determine the relative abundance of mobile vs immobile phosphorus. The NaOH-extractable SRP fraction provides an estimate of the Al- and Fe-bound phosphorus, which are relatively mobile and can become soluble under anoxic conditions. In contrast, the HCl-extractable SRP fraction provides an estimate of Ca- and Mg-bound phosphorus, which is a relatively stable phosphorus fraction.

The mean NaOH-extractable SRP fraction in White Lake sediment was ~½ the amount in Spring Lake (3.14 mg/L vs 6.75 mg/L, respectively) and about the same as Sites 1, 2, and 4 in

Mona Lake (2.9 mg/L) but considerably lower than the amount at Site 3 in Mona Lake (49 mg/L). The mean concentration of the less mobile HCl-extractable SRP fraction in White Lake sediment was ~¾ the amount in Spring Lake (10.2 vs 13.9 mg/L, respectively) and a bit greater than at Sites 1, 2, and 4 in Mona Lake (8.9 mg/L) but again, considerably less than at Mona Lake Site 3 (63 mg/L). These data suggest that the lower levels of mobile P in White Lake sediments account, at least in part, for the lower internal P loading rates in this system.

Second, the morphometry of White Lake is very different than Mona or Spring Lakes (table 5); White Lake is deeper than the other two, so even if P flux was occurring from the sediments to the water column during periods of low DO, the fate of the released P is unclear. More information is needed on the hydrodynamics of White Lake, to determine if water currents in the hypolimnion may transport P-rich bottom water from the deep sites in the western basins toward shallow sites in the eastern basin, where it is more likely to become available to algae, or if this P-rich water either stays in the hypolimnion or flows out into Lake Michigan. If the released P stays in the hypolimnion, then it may not cause ecological impairments, such as algal blooms.

Our analyses included several assumptions. It was assumed that release rates from the sediment cores were representative of sediments and conditions in White Lake. Our sampling strategy was designed to cover as much of the geographic range in White Lake as possible, given the study's constraints. However, there is likely considerable sediment variation within each lake compartment; sampling only one site per compartment does not allow us to estimate the importance of this variation. In addition, internal loading varies on an annual basis (Steinman et al. 2008); our analysis in summer 2006 should be viewed as a snapshot, and not as definitive.

The second assumption was that the incubation conditions were representative of natural conditions. Although the laboratory conditions mimicked the ambient temperature and light regime, clearly the hydrodynamics were altered. It is likely that the laboratory set-up for the anaerobic water column represented an optimal situation for release of P (constant anaerobic conditions) compared to natural conditions. Hence, our estimates of internal loading are likely higher than what is occurring in nature. However, because our measurements cover only one season, the estimated internal load of 1.24 tons in White Lake probably underestimates the overall annual load. The P release rates measured under anaerobic conditions indicate that internal loading is a distinct and measurable phenomenon in White Lake, albeit a relatively small source overall of P to White Lake.

Conclusion

Internal phosphorus loading in White Lake is of relatively low importance to the overall phosphorus budget, especially compared to other regional drowned-river mouth lakes, such as Mona and Spring Lakes. These lakes have much greater developmental pressures along their shorelines, which may account for historically greater P loads. Our results provide useful guidance for future research and management activities. Although overall internal P loading is low in White Lake, it may be an important source of P in certain regions and at specific times of the year in White Lake. More detailed hydrodynamic data are needed to determine if P release in the hypolimnion is reaching the epilimnion or near-shore areas, thereby becoming available to primary producers. With respect to lake management strategies, the relatively low internal P loading rates in White Lake suggest that the current focus should be on reducing external loads. It is critical that external load reduction complement any in-lake management strategies (Hansson et al., 1998). Minimization of stormwater discharge, regular maintenance of septic systems, use of low-P fertilizer, and implementation of other best management practices (e.g. riparian buffer strips, on-farm comprehensive nutrient management plans, and constructed wetlands/detention areas) should be emphasized, along with the provision of appropriate incentives in the White Lake watershed; these actions are likely to have the greatest influence on reducing phosphorus concentrations in White Lake in the short-term.

ACKNOWLEDGEMENTS

We are grateful to Brian Scull, Jennifer Cymbola, Kelly Wessell, Matt Cooper, and Aaron Parker for their help with field sampling, Jennifer Cymbola who also assisted with laboratory sampling, Rod Denning for GIS assistance, and Gail Smythe for assistance with laboratory analysis. We are grateful to Jim Duncan, Dave Farhat, and the President's Office at GVSU for helping to fund this project.

REFERENCES

Fisher, M.M., Brenner, M. and Reddy, K.R. (1992). A simple, inexpensive piston corer for collecting undisturbed sediment/water interface profiles. *J. Paleolimnol.* 7, 157-161.

Freedman, P.L., Canale, R.P. and Auer, M.T. (1979). Applicability of land treatment of wastewater in the Great Lakes area basin: Impact of wastewater diversion, spray irrigation on water quality in the Muskegon County, Michigan, lakes. U.S. EPA Report, EPA-905/9-79-006-A.

Hansson, L-A., Annadotter, H., Bergman, E., Hamrin, S.F., Jeppesen, E. Kairesalo, T., Luokkanen, E., Nilsson, P-Å., Søndergaard, M. and Strand, J. (1998). Biomanipulation as an application of food chain theory: constraints, synthesis and recommendations for temperate lakes. *Ecosystems* 1, 558-574.

Johnston, C.A. (1991). Sediment and nutrient retention by freshwater wetlands: effects on surface water quality. *Critical Reviews in Environmental Control* 21, 495-565.

Luttenton, M. (1996). Distribution and abundance of aquatic macrophytes in White Lake. Project Report submitted to Muskegon County Soil Conservation District.

Luttenton, M., Steinman, A.D. and Rediske, R. (2007). Summary report of the White Lake water quality assessment. Annis Water Resources Institute, Muskegon, MI.

Moore, P.A. and Reddy, K.R. (1994). Role of Eh and pH on phosphorus geochemistry in sediments of Lake Okeechobee, Florida. *J. Environ. Qual.* 23, 955-964.

Moore, P.A., Reddy, K.R. and Fisher, M.M. (1998). Phosphorus flux between sediment and overlying water in Lake Okeechobee, Florida: spatial and temporal variations. *J. Environ. Qual.* 27, 1428-1439.

Nürnberg, G.K. and LaZerte, B.D. (2004). Modeling the effect of development on internal phosphorus load in nutrient-poor lakes. *Water Resour. Res.* 40:W01105, doi:10.1029/2003WR002410.

Paul, M.J. and Meyer, J.L. (2001). Streams in the urban landscape. *Annu. Rev. Ecol. Syst.* 32, 333-365.

RAP (1995). White Lake Area of Concern Remedial Action Plan: 1995 Update. Public Sector Consultants.

Robertson, D.M. (1997). Regionalized loads of sediment and phosphorus to Lakes Michigan and Superior-high flow and long-term average. *J. Great Lakes Res.* 23, 416-439.

Søndergaard, M., Jensen, J.P. and Jeppesen, E. (2001). Retention and internal loading of phosphorus in shallow, eutrophic lakes. *TheScientificWorld* 1, 427–442.

Steinman, A.D., Chu, X. and Ogdahl, M. (2008). Spatial and temporal variability of internal and external phosphorus loads in an urbanized watershed. *Aquatic Ecology* doi: 10.1007/s10452-007-9147-6.

Steinman, A.D., Havens, K.E., Aumen, N.G., James, R.T., Jin, K.-R., Zhang, J. and Rosen, B. (1999). Phosphorus in Lake Okeechobee: sources, sinks, and strategies. In: Reddy, K.R., O'Connor, G.A., Schelske, C.L. editors. *Phosphorus Biogeochemistry of Subtropical Ecosystems: Florida as a case example* (pp. 527-544). New York, NY: CRC/Lewis Publ.

Steinman, A.D., Rediske, R. and Reddy, K.R. (2004). The reduction of internal phosphorus loading using alum in Spring Lake, Michigan. *J. Environ. Qual.* 33, 2040-2048.

Steinman, A.D., Rediske, R., Chu, X., Denning, R., Nemeth, L., Uzarski, D., Biddanda, B. and Luttenton, M. (2006). An environmental assessment of an impacted, urbanized watershed: the Mona Lake Watershed, Michigan. *Arch. Hydrobiol.* 166, 117-144.

Steinman, A.D. and Rosen, B.H. (2000). Lotic-lentic linkages associated with Lake Okeechobee, Florida. *J. N. Am. Benthol. Soc.* 19, 733-741.

U.S. EPA. (United States Environmental Protection Agency). 1983. Methods for Chemical Analysis of Water and Wastes. Washington, D.C., USA.

Walsh, C.J., Roy, A.H., Feminella, J.W., Cottingham, P.D., Groffman, P.M. and Morgan, R.P. II. (2005). The urban stream syndrome: current knowledge and the search for a cure. *J. N. Am. Benthol. Soc.* 24, 706-723.

Welch, E.B. and Cooke, G.D. (1995). Internal phosphorus loading in shallow lakes: importance and control. *Lake Reserv. Manage.* 11, 273-281.

Welch, E.B. and Cooke, G.D. (1999). Effectiveness and longevity of phosphorus inactivation with alum. *Lake Reserv. Manage.* 15, 5-27.

In: Lake Pollution Research Progress
Editors: F. R. Miranda and L. M. Bernard

ISBN: 978-1-60692-106-7

Chapter 11

Hydraulic Characterization of Deep Aquifer(s) in the Arsenic Affected Meghna Floodplain, Southeastern Bangladesh

***Anwar Zahid*[*1], *M. Qumrul Hassan*[2], *Jeff L. Imes*[3], *David W. Clark*[3], *Satish C. Das*[4] and *M. Zainal Abdin*[4]**

[1] Department of Geology, University of Dhaka, Dhaka, Bangladesh and Bangladesh Water Development Board, Dhaka, Bangladesh
[2] Department of Geology, University of Dhaka, Dhaka, Bangladesh
[3] United States Geological Survey, National Drilling Company-USGS Ground-Water Research Program, P.O. Box: 15287, Al Ain, UAE
[4] Bangladesh Water Development Board, Dhaka, Bangladesh.

Abstract

Because of arsenic contamination in shallow groundwater, characterization of deeper aquifers and assess their hydraulic connectivity is now an important issue in Bangladesh.

[*] Anwar Zahid: E-mail: anwarzahid_b@yahoo.com, Tel-+880-2-7287176; +880-1819 105 871

To determine the hydraulic characteristics of aquifers and development potential of deep aquifer for sustainable long-term use, study was undertaken by assessing water levels of different aquifers and conducting pumping test in deep aquifer under Meghna floodplain area of southeastern Bangladesh. Study shows that groundwater pumping for irrigation and other uses cause large seasonal water level fluctuations that is between 2 and 4.5m, 6.5 and 11m and 6.5m in the shallow, main and deep aquifer formations, respectively. The trend of groundwater level fluctuations supports the hydraulic connectivity of these aquifers. Aquitards separating aquifers are not continuous regionally. This implies that uncontrolled development of deeper aquifers may cause leakage of arsenic from contaminated shallow depths to aquifers below. Water levels dropping below sea level for over withdrawal may eventually cause saline water intrusion as well. However, during the 98.5 hours constant-discharge pumping test for deep aquifer, water levels in observation wells open to the shallow and main aquifers showed no noticeable effect from pumping in the deep aquifer i.e. under conditions of moderate groundwater use for public supply, arsenic-rich groundwater in the shallow aquifer are not likely to be drawn into the deep aquifer. The transmissivity values of the aquifer is generally favorable for groundwater development and ranged from about 1,070 m^2/day using Jacob's straight line method to 2,948 m^2/day using Chow's method at a distance of 44 m from the pumped well. Transmissivity ranged between 1,570 m^2/day using Jacob's straight line method and 2,956 m^2/day using Hantush-Jacob solution at a distance of 120m. Transmissivity was calculated as 2,385 m^2/day using recovery data. Estimated storage coefficient values ranged between 0.0000375 and 0.00268, indicates that the aquifer is confined to leaky-confined or semi-confined in nature.

Keywords: Groundwater level, irrigation abstraction, fluctuation, arsenic leakage, aquifer test, transmissivity, storativity.

INTRODUCTION

Groundwater is the main source for drinking and irrigation water in the lower floodplain areas of south-eastern Bangladesh and mainly withdrawn from shallow aquifer (BWDB-UNDP 1982). The upper aquifer system of the area can yield large quantities of water, however, is not completely suitable for sustainable development because of quality problems. The arsenic contamination of shallow (generally up to 50m depth) groundwater has changed the potentiality of it's use. Besides, high concentration of iron, manganese and salinity at different depth levels of main aquifer makes it unsuitable for drinking use. Considering the increasing demand for municipal and rural water supplies, agricultural, industrial and other uses and quality problems in shallow and main aquifers, development of deep aquifer has already been started in some areas. However, before large scale withdrawal of groundwater from deep aquifers, understanding the natural distribution of groundwater for long term sustainability is very important. Sustainable yield from aquifers, effective use of the water stored in aquifers, preservation of water quality, and movement of water between different aquifer formations into a comprehensive management system needs to be studied carefully. Rapid and uncontrolled development of the deep aquifer could severely limit the usefulness and the productive duration of the aquifer.

A pumping test is one of the most useful means of determining hydraulic properties of water-bearing layers and confining beds, and important for a proper understanding of some groundwater system. Groundwater levels of different aquifers were monitored throughout Kachua upazilla (sub-district) area under Chandpur district and an aquifer test was conducted for the deep aquifer at Sreerampur village of Kachua to determine the aquifer hydraulic characteristics, potability of the aquifer system, and the response of the deep aquifer pumping to upper aquifers and to development stresses. In particular, it is essential to understand whether pumping stresses can induce arsenic contaminated water from the shallow aquifer or high salinity water from depth or the shallow aquifer to migrate into the aquifer. The specific conductivity of water pumped from the production well was also monitored during the test to determine if higher salinity water was being captured by the well during the test. The long-term constant-discharge pumping test was performed as the aquifer properties determined from the analysis of constant-discharge pumping test represent the regional hydraulic properties (Nury et al, 1998) in alluvial aquifers like Bengal Delta.

Methods and Design of Aquifer Test

Study Area

Most of Bangladesh including the study area lies within Bengal basin which began forming during the Late Mesozoic as the continental landmass of Gondwana fragmented and continued to form during the Tertiary when the Indian plate collided with the Eurasian plate resulted in the formation of the Himalayan ranges. The Bengal basin contains a maximum of 15-km to 22-km thick sequence of Cretaceous to Recent sediments and occupies some 100,000 km^2 of lowland floodplain and delta (DPHE-BGS 2001). Alluvial deposit carried by the Ganges-Bramaputra-Meghna (GBM) river systems have gradually built up the delta and Meghna estuary (Brammer 1996). Physiographically, the study area lies within Meghna estuarine floodplain under Tippera surface (Morgan et al. 1959) that is bounded by the Meghna river in the west, Lalmai hills in the east and Old Meghna estuary at its south. The total discharge of the lower Meghna river was contributed from the eastern territory and the huge discharge hit the western bank and caused erosion in the western bank and subsequent deposition in the eastern bank towards the study area (Hussain and Huq 1998). The surface of the sequence, composed of silt, silty clay, silty loam and grayish clay, has been assigned to the Chandina Formation (Bakr 1977), while the status of the underlying medium sand aquifer is not well known. The Geological Survey of Bangladesh (GSB) has dated the surface of the Chandina deltaic plain as between 3,000 and 6,000 years. The underlying sands locally contain brackish water and may belong to the Dupi Tila.

Based on elevation and morphological features the area is divided into three units (BWDB-USGS 2005). Southeastern part of the upazilla has been defined as Chandina Alluvium High that always remains above normal flood level. The sediments mainly consist of clayey silt, silty clay and very fine sand. Below this oxidized zone grey colored fresh sandy silt and fine sand are present. The northwestern and southwestern part have been defined as Chandina Alluvium Medium. This part normally remains under water for about 3 to 4 months. Maximum and minimum elevation of this unit is 6.0 and 4.6 m above mean sea level (AMSL) respectively. The northeastern and central-western part of the area is defined as

Chandina Alluvium Low. The surface remains under water for longer periods than the other units.

In part of the southeast Bangladesh, multi-layered aquifer conditions exist. On a regional basis BWDB-UNDP (1982) described three aquifers between Holocene and Plio-Pleistocene formations. In Kachua area BWDB-UNDP aquifers may be classified as follows (figure 1):

The shallow (1st) aquifer extends down to 40 to 80m, below the 3 to 6 m thick upper clay and silt unit. The aquifer sediments are composed of sand with lenses of clay. Water of this unit is severely contaminated by arsenic.

The main (2nd) aquifer extends down to 250 to 350m and is generally underlain and overlain by silty clay bed, and composed mainly of fine to medium sand, grey to light brown in color, occasionally inter-bedded with clay lenses. It is either semi-confined/leaky or consists of stratified interconnected, unconfined water-bearing zones. Irrigation water in drawn predominantly from these strata.

The deep (3rd) aquifer has been encountered to depths of 400m below a 10 to 15m thick silty clay bed. This aquifer is composed mainly of grey to dark grey fine to medium sand that in places alternates with thin sandy shale/clay lenses. This deeper water bearing unit is separated from the overlying main aquifer by one or more clay layers of varied thickness.

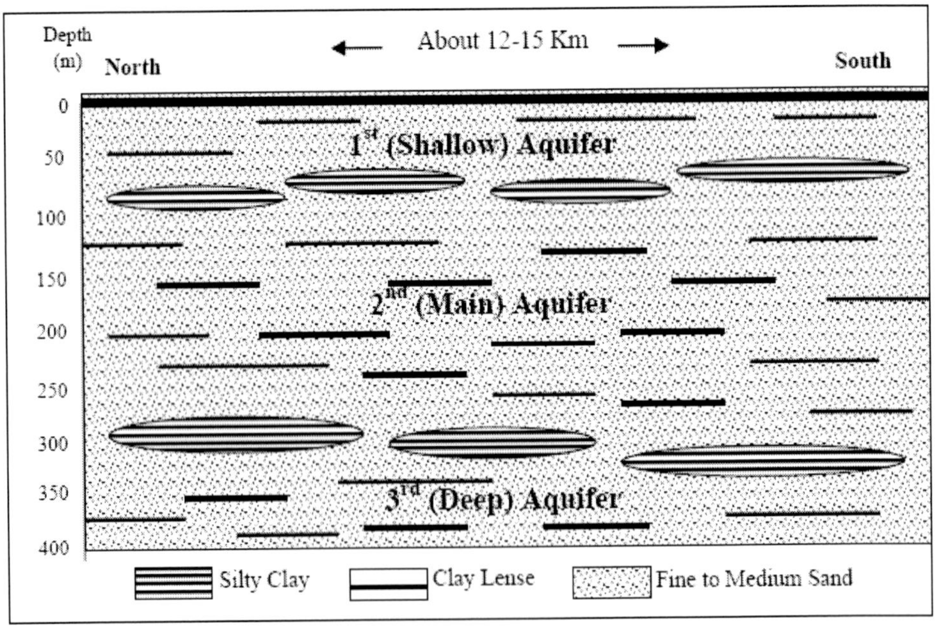

Figure 1. Aquifer system under Kachua Upazila area (Zahid et al 2007).

Observation of Water Levels in Different Aquifers

1998 to 2003 groundwater level data of 4 Bangladesh Water Development Board (BWDB) monitoring wells (20-30m deep) installed in the shallow aquifer (table 1), 2004-05 to 2005-06 data of 13 selected Department of Public Health Engineering (DPHE) hand tubewells (180-225m deep) installed in the main aquifer (table 2) and 2003 to 2004 data of 2 BWDB deep monitoring wells (336-352m deep) installed in the deep aquifer (table 3) were used to asses the response of groundwater level above or below mean sea-level (MSL) considering recharge and withdrawal for different uses. Locations of monitoring wells are shown in figure 2.

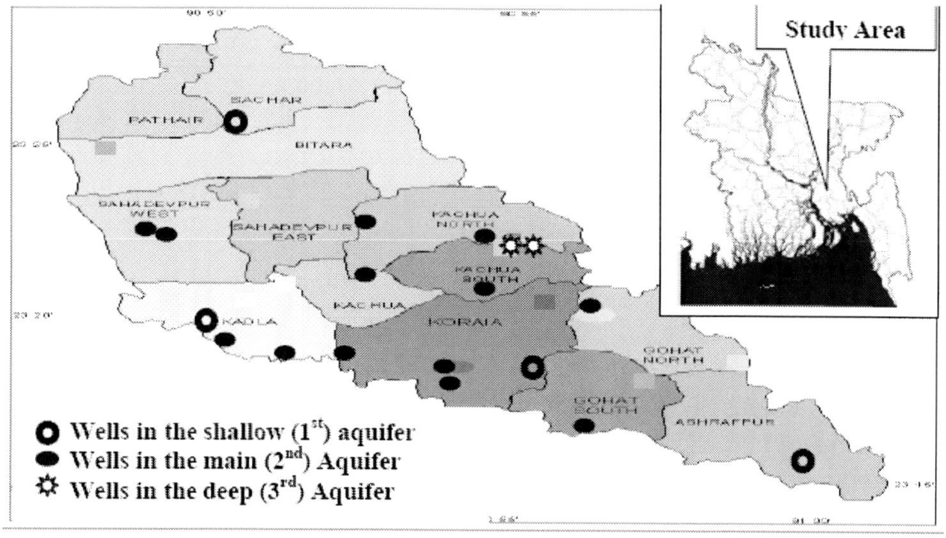

Figure 2. Location of groundwater level monitoring wells.

Table 1. Information on BWDB observation wells screened in shallow aquifer

Location	Latitude (North)	Longitude (East)	Depth (m)	measuring point (m) Above MSL
Jagatpur	231555.0	910000.0	19.20	12.70
Machimpur	231897.7	905532.1	27.44	6.48
Sacher	232649.6	905051.2	27.44	5.36
Muradpur	232000.0	905000.0	28.66	6.53

Table 2. Tube wells selected for monitoring groundwater movement in main aquifer

Location	Latitude (North)	Longitude (Rast)	Depth (m)	measuring point (m) Above MSL
Nalua, Karaia	231827.3	905440.7	215	6.288
Sahedapur, Karaia	231833.6	905456.4	215	6.376
North Dumuria, Karaia	231906.1	905240.9	220	5.848
Hossainpur, Kachua South	232123.7	905515.9	220	7.016
Ghagra, Kachua South	232134.5	905222.8	218	6.335
Tetaia, Kachua North	232249.4	905331.4	213	6.317
Ujani, Kachua North	232219.0	905522.0	211	6.28
Hasimpur, Gohat North	232006.9	905651.1	191	6.342
Gobindapur, Gohat South	231724.1	905638.8	218	6.649
Singua, Sahadevpur West	232303.3	904953.9	223	5.992
Khilmeheb, Sahadevpur West	232342.0	904942.8	220	6.236
Kadla, Kadla	231852.7	905128.9	210	5.987
Chaumuhari, Kadla	231946.9	904946.6	225	7.188

Setting of Pumping Well and Observation Wells

For a well performance test, yield and drawdown are recorded to calculate the specific capacity of the well. The aquifer test at Sreerampur village was conducted mainly to provide data from which the principal factors of aquifer performance [transmissivity (T) and storage coefficient (s)] can be calculated. An aquifer test with the setup of a 365-m deep pumping well with a constant discharge of 708 gpm (gallons per minute), and 5 (five) observation wells installed at different depth levels from 25 to 352 m (figure 3, table 3), were performed to obtain a picture of the general hydraulic properties of the aquifer and also to predict the effect of withdrawals on the aquifer system, the drawdown in the tested well with time, and different discharges and the radius of the zone of influence for individual or multiple wells. In a uniform and homogeneous aquifer the piezometer should be installed at about the same depth as the middle of the well screen in the pumped well. Amongst observation wells, two have been installed in the pumped aquifer and others are above the confining layer that separates the two aquifer systems. The test continued for 98.5 hours and water levels were measured in both the pumped well and five piezometers. The constant-discharge pumping test was followed by recovery test. Important water quality parameters were also monitored and samples were collected at different time intervals during the test.

Table 3. Major features of pumping and observation wells

Wel l ID	Aquifer	Depth (m)	Screen length (m)		Diameter of well (m)	Distance from PW (m)	measuring point (m) Above MSL
			From	To			
PW	Deep	365	326	362	0.35	0	6.84
P-1	Main	281	269	278	0.076	14.5	6.3
P-2	Deep	336	324	333	0.076	44.1	6.94
P-3	Deep	352	333	349	0.076	105	6.28
P-4	Shallow	25	20	23	0.076	7.5	6.72
P-5	Main	183	171	180	0.076	9.25	6.54

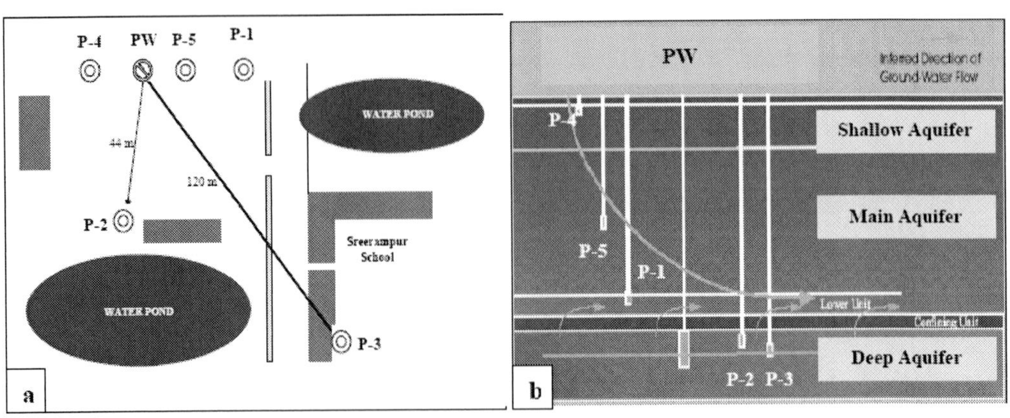

Figure 3. (a) Location and (b) position of pumping well and observation well screens.

The flow rate of 3,860 m^3/day was determined using standard tables for the discharge pipe diameter considering a water height of 8.33 m in a manometer attached to the side of the pumping-well discharge tube. The rate at which the discharge water stream dropped from the horizontal was also measured. A flow rate of 3880 m^3/day was determined using standard tables that relate the rate of fall to discharge. The two methods were in good agreement and an average value of 3870 m^3/day flow was used to analyze the aquifer-test data.

The constant-rate test for the time duration of 98.5 hours was conducted at Sreerampur. The pumping rate was estimated and monitored using two different methods. The circular orifice weir was used to measure and monitor the discharge rate of the pump. The height of the water column, the static water level just before the test started, time since the pump started, pumping rate, dynamic groundwater levels at various intervals during the pumping period, and time the pump stopped were recorded. Measurement of water levels after the pump stopped (recovery data) were also measured as these are extremely valuable to verify the aquifer storage coefficient calculated during the pumping phase of the test. The test was started on November 13, 2003 at 12:00 noon. Before the test began, water level transducers were placed in the pumping and observation wells. Drawdown was also recorded manually by measuring tapes. To obtain better and more reliable results pumping continued till the depression cone had reached a stabilized position. The depression cone continues to expand until the recharge of the aquifer equals the pumping rate.

Best Fit Analytical Method

Amongst different analytical methods, it is important to select a numerical solution which is more appropriate to actual field conditions. Two analytical solutions were studied in detail to determine the most appropriate solution to the deep observation well aquifer-test data. The Papadopolous-Cooper (1967) solution for a confined aquifer with well-bore storage and the Hantush-Jacob (1955) solution for a leaky confined aquifer were determined to be the best choices. Besides, Jacob (1950) straight line plot was drawn manually and Jacob (1950), Chow (1952) and Theis recovery methods were applied using computer programs developed by Abdin (2004) for manually measured drawdown data.

The following assumptions as well as parameters were considered for analyzing aquifer-test data using different methods valid for confined or leaky-confined aquifer.

All geologic formations are horizontal and the aquifer has a seemingly infinite areal extent. In reality, hydrogeologic settings rarely have aquifers that can be considered of infinite areal extent, rather they change laterally in grain size, shape or lithology that affect the shape of a time-drawdown curve. Although such aquifers do not exist in lower delta, many aquifers are of such wide extent that all practical purposes they can be considered infinite.

The aquifer is homogeneous, isotropic and of uniform thickness over the area influenced by the pumping test. Homogenous aquifers seldom occur in lower delta of Bengal basin and most aquifers are stratified to some degree. As a result of this stratification the drawdown observed at a certain distance from the pumped well may be different at various depths within the aquifer, because of differences in hydraulic conductivity in vertical and horizontal direction. However, these differences in drawdown diminish with increased pumping time.

Groundwater has a constant density and viscosity.

Prior to pumping, the piezometric surface were nearly horizontal over the area influenced by the pumping test.

The aquifer was pumped at a constant discharge rate.

All changes in the position of the potentiometric surface were due to the effect of the pumping well alone,

Pumping well was 100% efficient.

Another important assumption is that the pumped well fully penetrations the entire aquifer and thus receives water from the entire thickness of the aquifer by horizontal flow. If the well only partially penetrates the aquifer, like Sreerampur study, the flow paths have a vertical component to them. The flow paths are, therefore, longer and converge on a shorter well screen, resulting in an increase in head loss (Driscill 1986). However, observation wells for pumping tests were placed far enough away from the pumping well to avoid partial penetration effects. If the observation well is partially penetrating and more than $1.5b(K_h / K_y)0.5$ away from the pumping well, the effects are negligible (Hantush 1964) (b is the saturated thickness, K_h and K_v are the horizontal and vertical hydraulic conductivities). This condition is valid for Sreerampur aquifer. If this condition is not satisfied, there will be an upward inflection in the response, similar to that obtained in the leaky method or for some sort of recharge boundary (Domenico and Schwartz 1997).

Steady or equilibrium flow occurs when there is equilibrium between the discharge of the pumped well and the recharge of the aquifer by an outside source. Unsteady-state or non-equilibrium flow occurs from the moment pumping starts till the steady state is reached.

RESULTS AND DISCUSSION

Water Level Fluctuations in Different Aquifers

Long-term groundwater hydrograph of four of BWDB wells, installed in the shallow aquifer, show maximum depth to water table in dry season and in the monsoon it regains (figure 4). No permanent declining is observed. The average maximum and minimum water level was observed as 6 and 0.5m above MSL respectively (table 4). Seasonal groundwater table fluctuation ranges in between 2 and 4.5m in the Upazila area. Generally, groundwater withdrawal from shallow aquifer for domestic purposes and during dry period by shallow irrigation wells is balanced with the vertical percolation of rain water and inflow from surrounding aquifers in monsoon. There is a clear difference between the amount of water that can potentially recharge the aquifer system and the actual quantity. Actual recharge is the quantity of water when the groundwater table has risen to the ground surface and no more water can enter the aquifer system. Its quantity depends on the degree of depletion of this reservoir during the dry season. Potential recharge is the maximum amount of recharge that can occur if enough storage reservoir in the aquifer system is available. Any surplus rainfall is then rejected and contributes to surface flooding. The ultimate limit on groundwater development is controlled by the long-term average amount of potential recharge. Assessment done by WARPO (2000) indicates that when the clay is relatively thin, but exhibits low vertical permeability, the piezometric level may drop below the base of the clay. In this case the total gross quantity of groundwater abstracted from the aquifer is balanced by infiltration through the upper clay. When the upper clay varies in thickness, the piezometric level in the aquifer may locally drop below the base of the clay, creating local unconfined conditions within the aquifer and groundwater abstraction is partially balanced by leakage through the clay and partially by unconfined storage change in the aquifer. The recharge to the aquifer, which is the leakage during the period of no abstraction, may be less than the potential recharge, particularly when the vertical permeability is low.

During the peak irrigation season in March and April, the hydrograph is fairly smooth from year to year and steepest rise in the hydrograph is immediately after the irrigation pumps are switched off (mid-April to mid-May) rather than at the start of the monsoon (late June) as might be expected. During the monsoon, the hydrographs rise steadily until the levels is within about 1-2m of the surface which point a dynamic equilibrium is established between the water table, deep rooted vegetation and surface water bodies (Ravenscroft 2003). The aquifer full condition i.e. no more storage capacity is attributed before the end of June.

Table 4. Average groundwater levels above or below mean sea-level in different aquifers

Aquifer Formation	Nos. of Wells	Depth of Wells (m)	Groundwater Level (m)		Fluctuation (m)	
			Maximum	Minimum	Maximum	Minimum
Shallow (1st)	04	20-30	6	0.5	4.5	2
Main (2nd)	13	180-225	4	-8	11	6.5
Deep (3rd)	02	336-352	4.5	-2	6.5	6.5

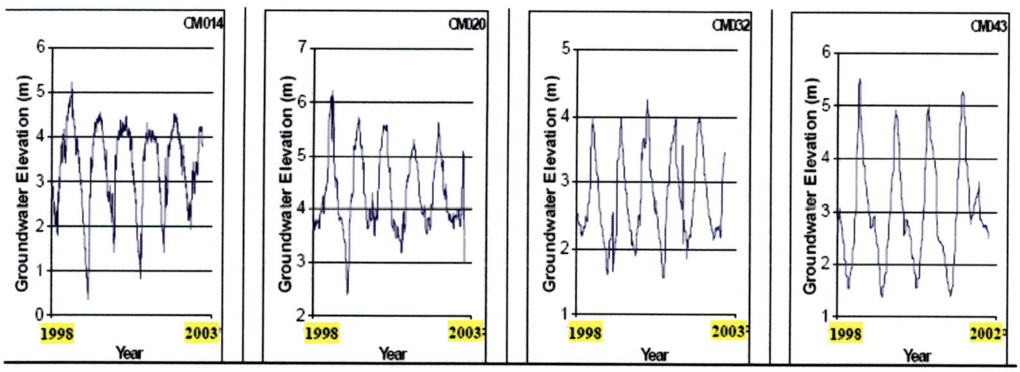

Figure 4. Groundwater level hydrograph of four BWDB wells screened in shallow aquifer at Kachua sub-district.

A shallow aquitard is common in this region at depth about 75m. Below this aquitard all deep irrigation wells are installed in the upper part while DPHE domestic and community hand tubewells are screened in the deeper part of main aquifer. Groundwater pumping for irrigation by deep irrigation wells causes large seasonal water level fluctuations (figure 5), ranges from 6.5 to maximum of 11m (table 4). The average maximum and minimum water level was observed as 4 and -8m above and below MSL respectively. Withdrawal by shallow irrigation wells may also influence on huge fluctuation of water level in the main aquifer. However, these levels in the deeper part get recovered rapidly when seasonal pumping stops. This rapid recovery, parallel to water levels in shallow aquifer, reveals that shallow and main aquifers are hydraulically connected and rainfall also contributes recharge to main aquifer through shallow aquifer. However, with intensive abstraction, groundwater levels may fall towards a permanent new equilibrium state.

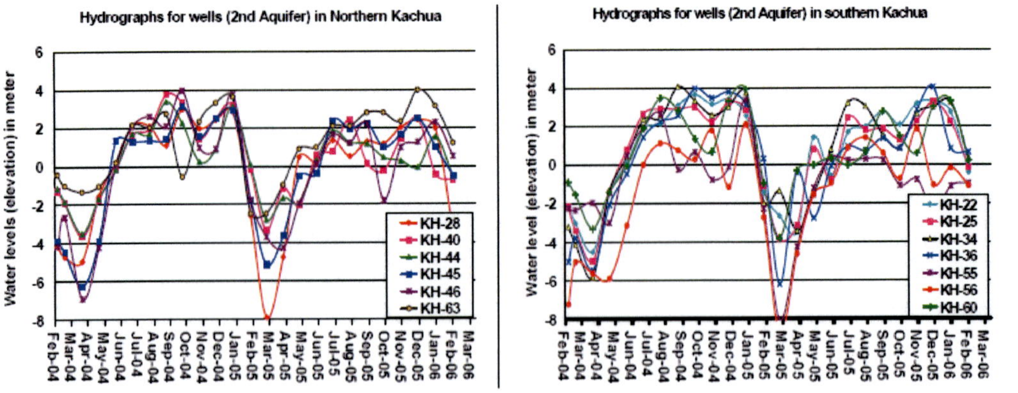

Figure 5. Groundwater level hydrographs of hand tubewells screened in the deeper part of main aquifer.

At Sreerampur village a deep confining layer is encountered at 302 to 320m depth, which is also common in surrounding areas of Meghna floodplain. Only two monitoring wells are installed at entire Kachua area in this deep aquifer formation. The average maximum and minimum water level of one hydrologic year was observed as 4.5 and -2m above and below MSL respectively for two nearby wells (figure 6a). Hence the seasonal fluctuation is 6.5m that is less compare to average water level fluctuation of main aquifer (figure 6b), but still high as there is only one low discharge production well installed in deep aquifer in Kachua upazilla area. This huge fluctuation indicates the influence of irrigation withdrawal in upper aquifers.

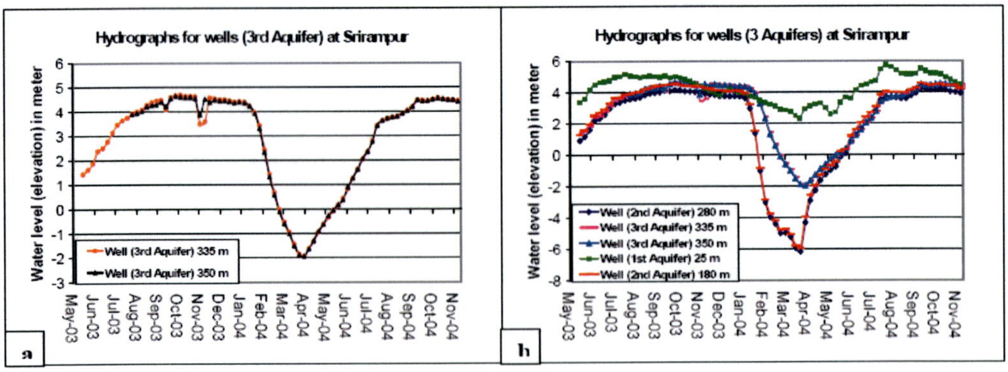

Figure 6. (a) Groundwater level hydrograph of observation wells screened in the deep aquifer; (b) Response of irrigation pumping on water level of different aquifers.

In Figure 6b, water level hydrographs of five wells installed in all three aquifers within $500m^2$ area are drawn for comparison study. It is clear that maximum water level i.e. static water level is almost same (about 4.5m above MSL) for three aquifers and during irrigation period, all five wells response similarly with different declining intensity. Because of continuous percolation of rain water through unsaturated zone, lowering of water level in shallow aquifer is very low (about 1.0m) that is about 10.5m for main aquifer and about 6.5

for deep aquifer. As most of the irrigation wells are installed in the main aquifer, lowering of water level is highest in this aquifer. However, the trend of water level fluctuations in different aquifers support the hydraulic connectivity of these aquifer i.e. the aquitards separating aquifers are not continuous regionally rather locally extended. This implies that uncontrolled development of deep aquifers may cause both qualitative and quantitative degradation of groundwater. Water levels dropping below sea level for over withdrawal in dry season may eventually cause saline water intrusion as well as leakage of arsenic from shallow aquifer to upper part of main aquifer.

Response of Ground-Water Levels to Pumping

The water level in the pumping (deep) aquifer declined abruptly during the first 3 minutes of pumping, then stabilized at about 12.5 to 13.0m drawdown for the duration of the test (figure 7a). The pumping phase of the aquifer test was terminated after 5,911 minutes (98.5 hours), and data was collected during the recovery phase of the test until 7,168 minutes (119.5 hours) after the test had begun. During the pumping test, water levels in observation wells open to the shallow and main aquifers showed no noticeable effect from pumping in the deep aquifer (figure 7b). The overall fluctuation of water levels (as caused by factors other than long-term declines or barometric pressure changes) measured in P-1, P-4, and P-5 was less than about 0.04m during the test. This indicates that the 12 to15m confining unit that separates the main and deep aquifer retards the movement of ground water between the aquifers. Therefore, under conditions of moderate groundwater use for public supply, arsenic-rich, iron-rich, and saline ground water in the shallow aquifer are not likely to be drawn into the deep aquifer.

Groundwater levels in the deep aquifer observation wells responded to the withdrawal of water from the pumping well. Water levels in P-2 (44 m from the pumping well) declined about 1.1 m in response to pumping (figure 7a). Water levels in P-3 (120m from the pumping well) declined about 0.68m in response to pumping. Small fluctuations in the measured water levels may be partly caused by variations in the pumping rate during the test as voltage in the power supply fluctuated.

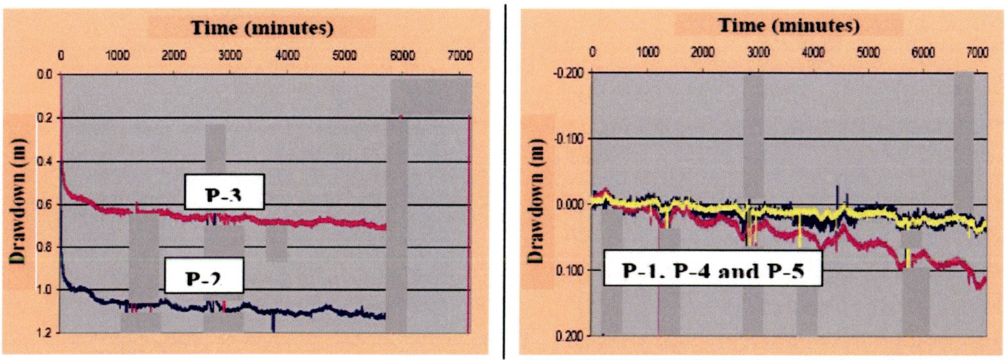

Figure 7. Monitoring of water level (drawdown) during aquifer test (a) Observation wells in deep (3rd) pumping aquifer below aquitard; (b) Observation wells in upper aquifers.

Water levels in both observation wells rose rapidly at the end of the aquifer test, and had returned to within 0.05m of the pre-test water levels when data collection was stopped. Comparison of water level altitudes in the shallow and deep observation wells shows that water levels in the deeper aquifer were lower than water levels in the shallower aquifer in an area about 100m radius centered on the pumping well after about 3,600 (60 hours) minutes into the test.

Analysis of Drawdown Data

Gunther Theim (1962) published the first formula on the response equation for steady radial flow (Ferris et al. 1962), based on the work of Darcy and Dupuit, and computed the hydraulic characteristics of a water-bearing formation by pumping a well and observing the effect of this pumping in a number of other wells. Now, most of the more or less complicated flow problems can be solved by applying proper mathematical methods. Aquifer test analyses may provide unrealistic estimates of hydraulic properties (Halford and Kuniansky. 2002). However, it makes no difference in performing a test whether response curves are obtained as analytical solutions or by other methods, e.g. flow nets, digital computers, etc (Stallman 1976).

The investigated deep aquifer under Sreerampur village may be considered as a confined aquifer as it is overlain by a 10 m to 12 m thick nearly-impervious silty clay layer, 295m below the surface. The lower boundary of the aquifer was not encountered above the investigated depth of 362 m; however, from the depositional history of the region it may be assumed that another impermeable formation can exist beneath the tested depth. A confined aquifer is a completely saturated aquifer whose upper and lower boundaries are impervious layers. However, completely impervious layers (aquatard) rarely exist in nature. A semi-confined or leaky aquifer, is a completely saturated aquifer that is bounded above by a semi-pervious layer and below by a layer that is either impervious or semi-pervious. Physically, there is induced recharge during an aquifer test through the semi-pervious layer. The rate of leakance is determined by the hydraulic conductivity of the aquitard and head difference across the aquitard. In a semi-confining scenario, confining units may pinch out laterally and part of the aquifer may become unconfined. This may true for the studied aquifer, too.

Theis Method for Predicting Drawdown in a Confined Aquifer

A major advance was made by Theis (1935), who was the first to develop a nonsteady-state formula which introduces the time factor and the storage coefficient. Theis developed the analytical solution for flow to a well in a confined aquifer. He noted that when a well penetrating an extensive confined aquifer, is pumped at a constant rate, the influence of the discharge extends outward with time. The rate of decline of head, multiplied by the storage coefficient and summed over the area of influence, equals the discharge. Because the water must come from a reduction of storage within the aquifer, the head will continue to decline as long as the aquifer is effectively infinite. Since all sources of recharge are moving through a semi-confining unit, the flatness of the recharge effect on a Theis curve will continue in a horizontal manner (Lohman 1979). However, when the semi-confining layer is saturated and has a head higher than the head in the aquifer being pumped, this head differential may cause the aquitard to release water from storage in an attempt to reach equilibrium (Weight and Sonderegger 2000). As in the semi-confining scenario, the initial drawdown tends to follow the confined Theis curve. As the aquifer becomes stresses and the head lowers in the pumping well, the head change stimulates leakage from above and below to act as recharge to the system. The result is a flattening of the curve rather than following the Theis curve.

Besides the assumptions mentioned above the following assumptions are also considered for Theis solution.

- Aquifer is fully confined and discharge is derived exclusively from storage in the aquifer.
- The flow to the well is in unsteady state, i.e. the drawdown differences with time are not negligible nor is the hydraulic gradient constant with time.
- The water removed from storage is discharged instantaneously with decline of head.
- The storage in the well can be neglected.

The nonsteady-state or Theis equation which was derived from the analogy between the flow of groundwater and the conduction of heat for predicting drawdown (s) at the well is as follows:

$$s = \frac{Q}{4\pi T}\int_u^{\alpha}\frac{e^{-y}dy}{y} = \frac{Q}{4\pi T}W(u) \text{ and, consequently } T = \frac{Q}{4\pi s}W(u)$$

The well function $W(u)$ is the infinite series part of the analytical solution to the nonsteady, radial groundwater flow equation that is approximated by:

$$W(u) = -0.577216 - \ln u + u - \frac{u^2}{2\bullet 2!} + \frac{u^3}{3\bullet 3!} - \frac{u^4}{4\bullet 4!} + \ldots$$

where, $u = \frac{r^2 S}{4Tt}$ and, consequently $S = \frac{4Ttu}{r^2}$

u determines the radius of a cone of depression (Theis 1940). The radius not only increases with increasing time but, for a given time, is larger for decreasing values of storativity and increasing values of transmissivity. Theis' type curve u versus $W(u)$ is unique only because it pertains to a particular set of conditions at the pumped well and in the aquifer (Stallman 1976).

Hantush-Jacob Leaky Aquifer Model

For all practical purposes, the time-drawdown behavior before the inflection represents a withdrawal of water from storage from the aquifer, no part of which was contributed from other sources. This part of the curve, along with its straight line extension, may be analyzed with methods discussed for time-drawdown behavior. There are several ways in which this condition may be compromised in field conditions: direct recharge from streams, recharge across bounding low permeable materials etc. The problem of leakage has been extensively investigated by Hantush and Jacob (1955) and Hantush (1956, 1960, 1964). In a leaky aquifer, the drawdown curve will initially follow the nonleaky curve. However after a finite time interval, the lowered hydraulic head in the aquifer will induce leakage from the confining layer. If there is storage in the confining layer, the rate of drawdown will be slower than that in the case where the there is no storage in the confining layer (Hantush 1960).

In many aquifer tests, water is contributed from the less permeable confining units, in addition to the aquifer that is pumped (Halford and Kuniansky 2002). Hantush and Jacob (1955) presented a solution for drawdown in a pumped aquifer that has an impermeable base and a leaky confining unit above. Conceptually, this would be a four layer system, from top to bottom, a water table aquifer, a leaky confining unit, a confined aquifer, and an extremely low permeability bed rock. During the early time of pumpage, water is coming out of storage from the pumped aquifer and the leaky confining unit. Eventually, the discharge comes into equilibrium with the leakage through the confining unit from the unstressed water-table aquifer and the system is at steady-state.

The additional assumptions for the analysis are:

- Aquifer is leaky, horizontal flow stressed aquifer, vertical flow through confining unit.
- Drawdown in the water table or unstressed aquifer is negligible.
- Well storage can be neglected.
- Water instantaneously comes out of storage in the aquifer.
- Confining unit storage is negligible.

The equation is based on the drawdown of a well pumped at a constant discharge rate in a leaky aquifer (Hantush and Jacob 1955).

$$s = \frac{Q}{4\pi T} W(u, r/B)$$

where, $u = \dfrac{r^2 S}{4Tt}$ is dimensionless time,

$$\frac{1}{B} = \sqrt{\frac{K_Z / b'}{T}}$$

K_Z/b' is the leakance *(1/T)*, where K_Z is vertical hydraulic conductivity of the confining unit *(L/T)* and b' is thickness of the confining unit *(L)*.

Early drawdown data from the field curve match the non-leaky part of the curve, but they soon deviate and follow one of the leaky *r/B* curves. The match point method yields values of *w(u, r/B), 1/u, t* and *s*. In addition, the *r/B* curve followed by the field data is noted. *T* and *S* are readily determined from

$$T = \frac{Q}{4\pi s} W(u, \frac{r}{B}) \text{ and } S = \frac{4uTt}{r^2}$$

Unaware of the work done many years earlier by De Glee (1930, 1951), Hantush and Jacob (1955) also derived the same equation, which express the steady-state distribution of drawdown in the vicinity of a pumped well in a semi-confined aquifer in which leakage takes place in proportion to the drawdown. Hantush (1956, 1964) noted that if r/L is small ($r/L \leq 0.05$), the equation may, for practical purposes, be approximated by

$$s_m \approx \frac{2.30Q}{2\pi T} (\log 1.12 \frac{L}{r})$$

Thus a plot of s_m against r on semi-logarithmic paper, with r on the logarithmic scale, will show a straight-line relationship in the range where r/l is small. The slope of the straight portion of the curve, i.e. the drawdown difference Δs_m per log cycle of r, is expressed by

$$\Delta s_m = \frac{2.30Q}{2\pi T}, \text{ consequently } T = \frac{2.30Q}{2\pi \Delta s_m}$$

The Hantush-Jacob leaky aquifer solution was chosen to analyze the aquifer test data because of the possibility that there was some induced leakage of ground water across the confining unit during the test. With the advent of computer programs to analyze aquifer test data, we no longer have to manually plot graphs and calculate aquifer parameters by hand. This makes it much easier to analyze the data, but makes us much less aware of the assumptions behind the analytical solutions. Drawdown data for the individual deep observation wells can be matched well to about 500 minutes using the Hantush-Jacob solution. Drawdown from about 200 to 500 minutes becomes nearly constant at about 1 m for P-2 and nearly constant at about 0.6 m for P-3. This response may be caused by equilibrium between leakage across the confining unit and drawdown from well pumpage. After about 500 minutes, drawdown in both observation wells begins to increase in an approximately linear manner, at a rate of about 0.00001 m/minute in P-2 and about 0.00002 m/minute in P-3. This may indicate decreased leakage through the confining unit, draining of stored water in a permeable sand lens, or the presence of a lateral barrier. Of the three possibilities, a lateral barrier seems unlikely considering the geological setting.

The best fit curve match (figure 8a) to P-2 yielded an aquifer storage of 0.0013, aquifer transmissivity of 2,300 m^2/day, vertical to lateral aquifer hydraulic conductivity ratio of 0.0044 (*Kz/Kr*), and a confining unit vertical hydraulic conductivity of 0.35 m/day ($r/B = 0.1533$). The best fit curve match (figure 8b) to P-3 yielded an aquifer storage of 0.0022, aquifer transmissivity of 2,956 m^2/day, vertical to lateral aquifer hydraulic conductivity ratio of 0.0044, and a confining unit vertical hydraulic conductivity of 0.43 m/day ($r/B = 0.365$).

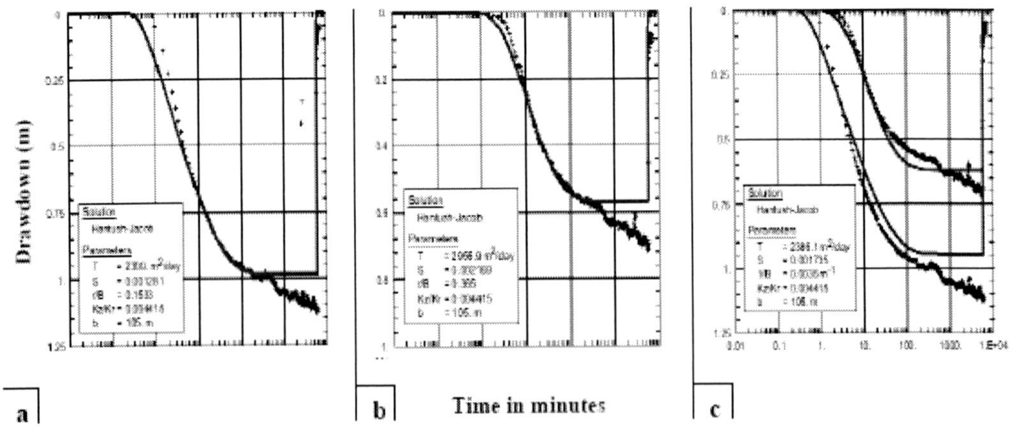

Figure 8. Best fit Hantush-Jacob solution to drawdown data in semilog scale collected from (a) P-2; (b) P-3; (c) Average drawdown data for P-2 and P-3.

Because the aquifer values are similar for the individual observation well data curve matches, both sets of data can be approximately fit to a single set of values (figure 8c). The best fit curve match to both observation wells taken together yielded an aquifer storage of 0.0017, aquifer transmissivity of 2,386 m^2/day, vertical to lateral aquifer hydraulic conductivity ratio of 0.0044, and a confining unit vertical hydraulic conductivity of 0.35 m/day ($1/B = 0.0035$).

The parameter $1/B$ in the Hantush-Jacob solution is related to the confining unit vertical hydraulic conductivity by the relation $Kv' = (T*b')/2B$, where T is the deep aquifer transmissivity and b' is the confining unit thickness. The lateral hydraulic conductivity of the deep aquifer is estimated as 23 m/day using $Kh=T/b$, where b is the aquifer thickness. The vertical hydraulic conductivity of the deep aquifer as determined from the solution derived ratio of the aquifer vertical hydraulic conductivity to aquifer lateral hydraulic conductivity ($Kz/Kr = 0.0044$) is about 0.10 m/day, about 200 times smaller than the lateral hydraulic conductivity of the deep aquifer. This may indicate that the cumulative hydraulic effect of disseminated clay and silt in the aquifer is similar to the clay-rich confining unit.

During the long-term monitoring period from May 2003 to November 2003 water levels in P-2 (deep aquifer) were from 0.15 m to 0.60 m higher than water levels in P-1. Therefore, under natural flow conditions experienced during the summer months, the groundwater gradient is upward from the deep aquifer to the shallow aquifer, and the specific discharge through the confining unit is estimated to be from about 0.004 m/day to about 0.017 m/day. Assuming a porosity of 0.30, the average velocity of water through the 12.2 m thick confining unit is 0.014 m/day to 0.057 m/day. The time for groundwater to move through the confining unit under these gradients at the estimated average velocities range from about 200 days to about 850 days.

Papadopulos-Cooper Solution to Drawdown Data

A solution for drawdown in large diameter wells that takes into consideration the storage within the well, which is assumed to be negligible in the Theis method, has been presented by

Papadopulous and Cooper (1967). Besides the general assumptions mentioned earlier, the added conditions are,

- Storage in the well cannot be neglected as well diameter is relatively large.
- The aquifer is confined.
- The well losses are negligible.

The general flow equation inside a large diameter well is (Kruseman and Ridder, 1983),

$$s_w = \frac{Q}{4\pi T} F(u_{w,} \beta)$$

where $F(u_{w,} \beta)$ is a function for which numerical values are given.

$$u_w = \frac{r_w{}^2 S}{4Tt} \text{ and } \beta = \frac{r_w{}^2 S}{r_c{}^2}$$

The index w stands for 'at the pumped well' and r_c is the radius of the unscreened part of the well.

The best visual fit to the complete data sets, using each observation well individually, was the Papadopolous-Cooper solution (figures 9a and 9b). However, the best match for each set of observation well data required aquifer properties and well bore diameters that were not reasonable. Aquifer storage values were extremely small (on the order of 10-9). The size of the well bore needed to match the delayed drawdown response was much larger than the actual bore of the pumping well for both data sets. Both observation well P-2 and P-3 have an actual casing radius of 0.175 m. The predicted casing radius for P-2 was 1.735 m, and the predicted casing radius for P-3 was 3.372 m. Also, the transmissivity value predicted from the data for P-2 (6921 m^2/day) differed considerably from the value predicted from the data for P-3 (8401 m^2/day). Because the transmissivity and well-bore radius values predicted from the individual observation well curve matches differed considerably, the drawdown data cannot be matched simultaneously for both observation well P-2 and P-3 using average transmissivity values (figure 9c).

Jacob's Method

The Cooper-Jacob (1946) method of time-drawdown and Jacob (1950) method of distance-drawdown are a modification of the Theis equation and assume that u is small. However, the conditions for its application are more restricted than for the Theis and Chow method. Cooper-Jacob and Jacob methods are valid where the time (t) is sufficiently large or radial distance (r) is sufficiently small. The Cooper-Jacob time-drawdown approach is modified to plot drawdowns at various radial distances from the pumping well. By graphing drawdown on a linear scale versus distance on a logarithmic scale, a straight line fit results (Jacob 1950). Estimates of the hydraulic properties can also be made, if the assumption of u

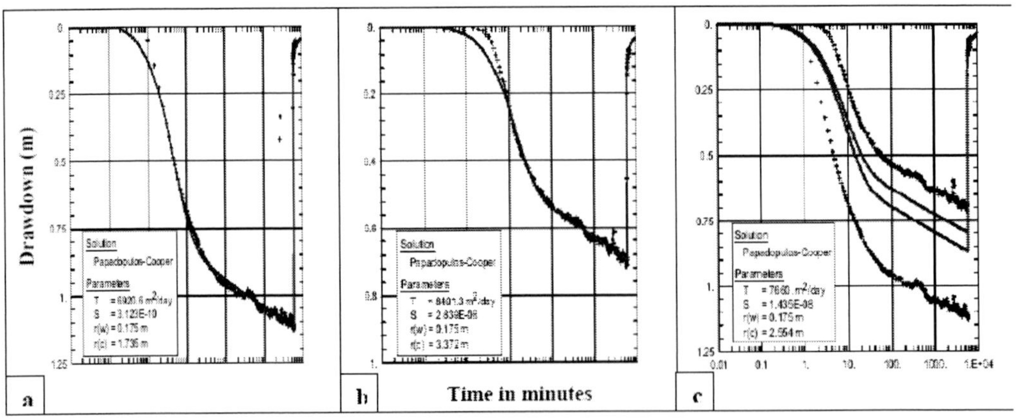

Figure 9. Best fit Papadopulos-Cooper solution to drawdown data in semilog scale collected from (a) P-2; (b) P-3; (c) Average drawdown data collected from both P-2 and P-3.

(Jacob 1950). Estimates of the hydraulic properties can also be made, if the assumption of u being sufficiently small (0.02) is valid. The intercept where the straight-line fit crosses the zero drawdown represents the range of influence of the cone of depression (Weight and Sonderegger 2000).

The same assumptions apply to the Cooper-Jacob analytical solution as the Theis solution, but the well function $W(u)$ is calculated for $u<0.01$ in order to neglect all but the first two terms of the infinite series of the well function. A straight-line approximation of $W(u)$ is adequate for most applications even where u is as great as 0.1 (Halford and Kuniansky 2002).

In the Theis formula, the exponential integral can be expanded in a convergent series, so that the drawdown (s) may be written as (Kruseman and Ridder 1983)

$$s = \frac{Q}{4\pi T}(-0.5772 - \ln u - \frac{u^2}{2.2!} + \frac{u^2}{3.3!} \cdots)$$

From $u = \frac{r^2 S}{4Tt}$ it will be seen that u decreases as the time of pumping t increases. Accordingly, for large values of t and/or small values of r the terms beyond $\ln u$ in the series of the equation become negligible. So for small values of u (<0.01) the drawdown can be expressed as

$$s = \frac{Q}{4\pi T}(-0.5772 - \ln \frac{r^2 S}{4Tt})$$

After rewriting and changing into decimal logarithms this equation reduces to

$$s = \frac{2.30Q}{4\pi T}\log\frac{2.25Tt}{r^2 S}$$

Therefore, a plot of drawdown versus the logarithm of t forms a straight line. If this line is extended till it intercepts time-axis where $s=0$, the interception point will have the coordinates $s=0$ and $t=t_0$. Substitution of these values into the drawdown equation gives

$$0 = \frac{2.30Q}{4\pi T}\log\frac{2.25Tt_0}{r^2S},$$

and because $\frac{2.30Q}{4\pi T} \neq 0$, it follows that $\frac{2.25Tt_0}{r^2S} = 1$ or $S = \frac{2.25Tt_0}{r^2}$

If $t/t_0=10$ and hence $\log t/t_0 = 1$, s can be replaced by Δs, i.e. by the drawdown difference per log cycle of time, and it follows that

$$T = \frac{2.30Q}{4\pi\Delta s}.$$

Jacob straight-line plot was applied to interpret the time-drawdown data for Sreerampur test as time was sufficiently large and radial distance is small. Drawdown recorded manually at selected intervals was used to create the hand-drawn data graph in figure 10. In this method, the field data are plotted on semi-logarithmic paper and a straight line is drawn through the field data points and extended backward to the zero drawdown axis. This is the distance at which the well is not affecting the water level. It may intercept the axis at some positive value of time (t_0). The value of drawdown per log cycle (ΔS) is obtained from the slope of the graph.

Drawdown of the observation wells installed in the deep aquifer were used to calculate transmissivity and storativity values of the deeper formation. The rate of drawdown after 1000 minutes is negligible because the water table was nearly in the steady condition.

For P-2 two straight lines were drawn and transmissivity and storativity values were calculated as 1,070.784 m^2/day and 0.00095 (first cycle) and 2,078.496 m^2/day and 0.00268 (third cycle) respectively. The average values are 1,574.64 m^2/day and 0.001815 respectively for transmissivity and storativity. These values are more appropriate as this well (P-2) is within a reasonable distance from the pumped well. Two straight lines were also drawn for P-3 and transmissivity and storativity values were calculated as 1,570.464 m^2/day and 0.00068 (second cycle) and 2,944.656 m^2/day and 0.00010 (third cycle) respectively. The average values are 2,257.56 m^2/day and 0.00039 respectively for transmissivity and storativity. Using computer program, transmissivity and storativity were estimated as 2,468 m^2/day and 0.000546 respectively (figure 11).

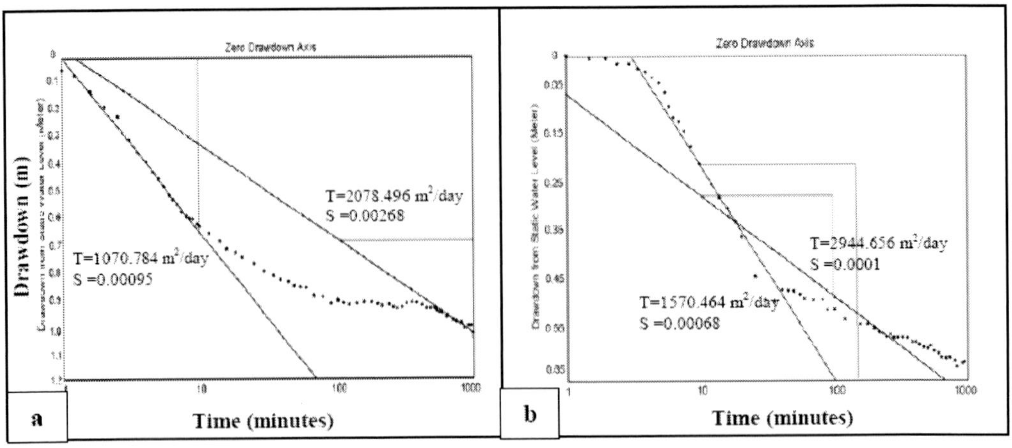

Figure 10. Jacob straight-line time-drawdown method for a fully confined aquifer plotted on semilog papers for (a) P-2 and (b) P-3.

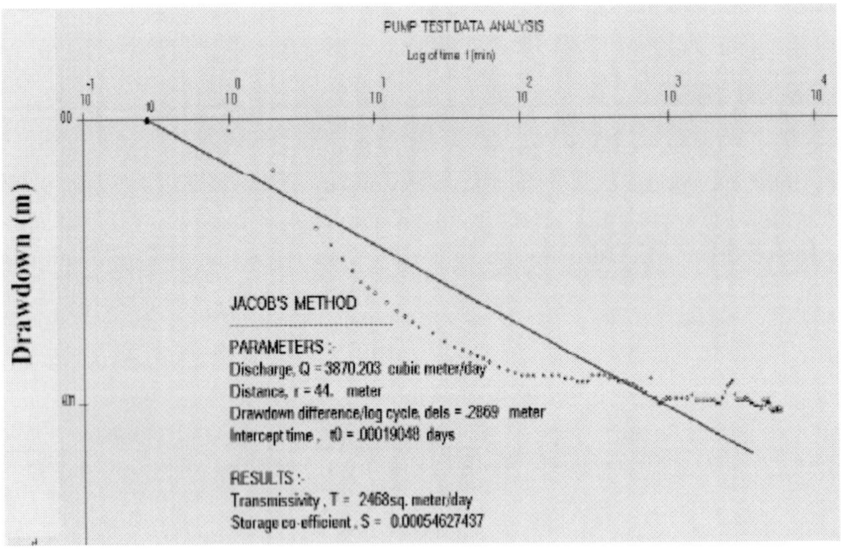

Figure 11. Jacob straight-line time-drawdown method for a fully confined aquifer using computer program for P-2 Chow's method.

Chow (1952) developed a method which has the advantage of avoiding the curve fitting of the Theis method and not being restricted to small values of r and large values of t as is the Jacob method (Kruseman and Ridder 1983). The same assumptions and conditions are generally satisfied as for the Theis method because this method is directly based on the Theis equation.

$$s = \frac{Q}{4\pi T} W(u)$$

To find the values of *W(u)* and *u* corresponding with the drawdown *s* measured at a certain moment *t*, Chow introduced the function

$$F(u) = \frac{W(u)e^{u}}{2.30}$$

F(u) can be calculated from the drawdown (*s*) versus the corresponding time (*t*) on single logarithmic paper (*t* on logarithmic scale).

Using a computer program, transmissivity and storativity were estimated as 2,948 m^2/day and 0.0000357 respectively (figure 12).

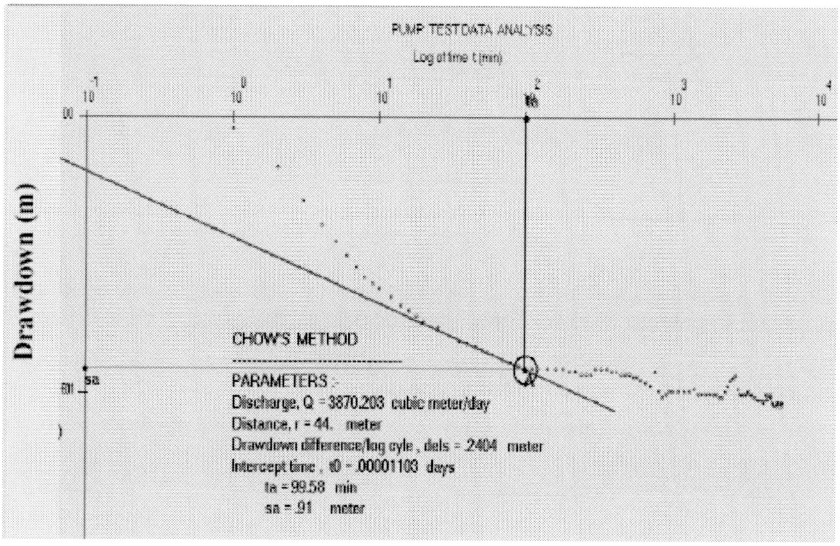

Figure 12. Analysis of data with the Chow method for P-2.

Theis Recovery Data Analysis for Confined Aquifer

The analysis of recovery data involves the measurement of the rise in water levels, also referred to as residual drawdowns, following the cessation of a period of pumping at a constant rate. The field procedure requires a drawdown measurement at the end of the pumping period (*t*) and recovery measurements during the recovery period (*t'*). The residual drawdown is plotted on a linear axis and the value of *t/t'* on the logarithmic axis. The analytical method is based on the Theis theory and applies to confined aquifers with fully penetrating wells. The method relies on the theory of superposition in that the water level rise after the test is assumed to be the combined response to an imaginary well recharging the aquifer and continued pumping. Imaginary recharge occurs at an identical rate to the constant discharge during the pumping test. The equation for residual drawdown after a pumping test with constant discharge is:

$$s' = \frac{Q}{4\pi T}\left\{W(u) - W(u')\right\}$$

where,

$$u = \frac{r^2 S}{4Tt} \text{ and } u' = \frac{r^2 S}{4Tt'}$$

If u and u' are small, less than 0.01, then the above equation can be simplified to:

$$s' = \frac{2.3Q}{4\pi T}\log_{10}\left(\frac{t}{t'}\right)$$

A semilog plot of s' versus t/t' will yield a straight line. The slope of which is:

$$\Delta s' = \frac{2.3Q}{4\pi T}$$

where, $\Delta s'$ is the change in residual drawdown in one log cycle of t/t'. The same assumptions as for the Cooper-Jacob, straight-line method must be met, and the flow to the well is in an unsteady state when $t' > (25\ r^2S)T$ and $u < 0.01$.

Using computer program, transmissivity was estimated as 2,385 m^2/day (figure 13).

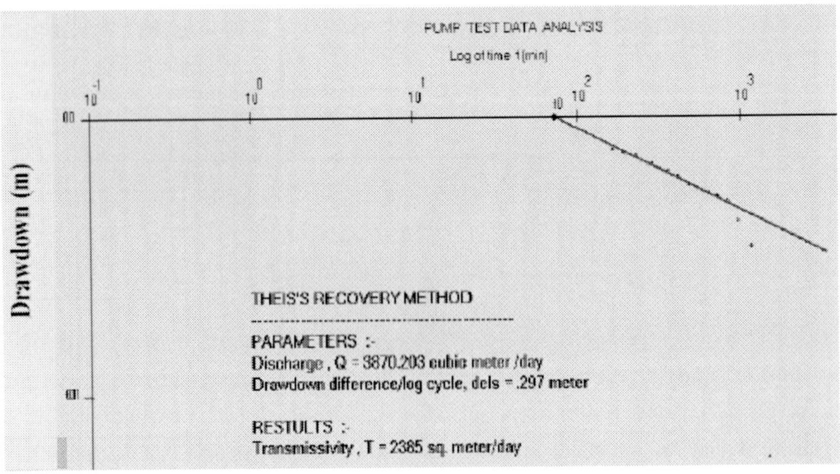

Figure 13. Analysis of recovery data with the Theis recovery method for P-2.

Comparison of Results for Different Methods

The results of the pumping test and recovery data analysis using different methods are presented in table 5. The estimated transmissivity of the aquifer using drawdown data collected at P-2 varied between 1,070 m^2/day for first cycle analysis using Jacob's straight line method and 2,948 m^2/day Chow's method from constant-discharge test, to 6921 m^2/day for Papadopulous-Cooper method. For P-3, the estimated transmissivity ranged between 1,570 m^2/day using Jacob's straight line method from second cycle analysis and 2,956 m^2/day using Hantush-Jacob solution, to 8,401 m^2/day using Papadopulous-Cooper method. Because of solution match difficulties, the Papadopolous-Cooper solution was not considered as the appropriate solution to the aquifer-test data. Transmissivity was calculated as 2,385 m^2/day using recovery data analyzed by the Theis recovery method. As these are graphical methods of solution, there is often slight variation in the results, depending upon the accuracy of the graph construction and subjective judgments in matching field data to type curves (Fetter 1994). The accuracy of the numerical values of the hydraulic characteristics of water-bearing layers and less permeable strata determined during the graphical analyses and the accuracy of the assumed boundary conditions play an important role in the reliability of the results obtained by these methods.

When the transmissivity estimates from the Theis curve method and the Cooper-Jacob method generally are comparable (Weight and Sonderegger 2000), and the resulting answers are almost the same. Well losses and partial penetration have a minimal effect on transmissivity values that are estimated using the Cooper-Jacob method. Additional drawdown at later times is due to declining heads in the aquifer and the rate of decline is controlled mostly by the transmissivity of the aquifer. Analyzing the change in drawdown at later times negates the effect of a fixed offset due to well losses and partial-penetration on the determination of transmissiviy (Halford and Kuniansky 2002).

Estimated storage coefficient values ranged from 0.0000375 to 0.00268 for P-2 and from 0.00010 to 0.0022 for P-3. The estimated storage coefficients indicate that the aquifer is confined to leaky-confined or semi-confined. Bouwer (1978) and Fetter (1994) suggest that storage coefficients for confined aquifer can vary from 0.00001 to 0.001, and Weight and Sonderegger (2000) suggest that storage coefficients for leaky-confined or semi-confined aquifers can vary from 0.001 to 0.03.

Table 5. Aquifer properties determined from constant-discharge pumping test data using different methods

Method	Transmissivity (m^2/day)		Storage Coefficient	
	P-2	P-3	P-2	P-3
Papadopulous-Cooper	6,921	8,401	-	-
Hantush-Jacob	2,300	2,956	0.0013	0.0022
Jacob (Hand Drawn)				
1st Cycle	1,071	-	0.00095	-
2nd Cycle	-	1,570	-	0.00068
3rd Cycle	2,078	2,945	0.00268	0.00010
Jacob	2,468	-	0.000546	-
Chow	2,948	-	0.0000357	-
Theis Recovery	2,385	-	-	-

The sorting of unconsolidated sediments largely controls the expected range of hydraulic conductivity. Well-sorted sediment has a much larger hydraulic conductivity than poorly-sorted sediment, because finer material fills the voids between coarser grains in poorly-sorted sediment. The hydraulic conductivity of unconsolidated sediment can be estimated empirically from the grain-size distribution (Vukovic and Soro 1992). Hydraulic conductivity estimates from grain-size distributions typically have a greater uncertainty than estimates from aquifer tests (Halford and Kuniansky 2002). Using the Hantush-Jacob solution, the vertical to lateral aquifer hydraulic conductivity ratio was estimated at 0.0044, and the confining unit vertical hydraulic conductivity was estimated at 0.35 m/day. Drawdown in both observation wells becomes nearly constant in the interval from about 200 to about 500 minutes.

During the aquifer test, water levels in P-1 were about 0.4 m higher than water levels in P-2. Therefore, under pumping stresses similar to those induced during the aquifer test, the ground-water gradient is downward from the shallow aquifer to the deep aquifer, and the groundwater flux through the confining unit is estimated at about 0.011 m/day. Assuming a porosity of 0.30, the average velocity of water through the confining unit is 0.037 m/day. The time for ground water to move through the confining unit under this gradient at the estimated average velocity is about 330 days.

CONCLUSION

Three aquifer formations have been classified under Kachua sub-district area till investigated depth of about 400m between Holocene and Plio-Pleistocene sediments. The static water level is almost same for these three formations and during irrigation period, groundwater levels response similarly. Lowering of water level in shallow aquifer is very low (about 1.0m) because of continuous percolation of rain water through unsaturated zone, which is about 10.5m for main aquifer and about 6.5m for deep aquifer. The trend of water level fluctuations and lowering of water levels in all three aquifer formations during irrigation withdrawal, mainly from upper part of main aquifer, support that the aquitards separating aquifers are extended locally, not continuous regionally. During the pumping test, water levels in observation wells open to the shallow and main aquifers showed no noticeable effect from pumping in the deep aquifer that indicates at least local hydraulic separation of aquifers and under conditions of moderate groundwater use for domestic and municipal supply, arsenic and/or chloride-rich groundwater in the upper aquifers are not likely to be drawn into the deep aquifer.

Aquifer test results as well as aquifer properties and characteristics of the deep aquifer formation show that the deep aquifer can yield significant amounts of potable water. Though different hydrogeologic factors control the aquifer parameters, solution results can also vary because the graphical methods of solution depend upon the accuracy of the graph construction and subjective judgments in matching field data to type curves. The transmissivity values are ranged from about 1,070 m^2/day using Jacob's straight line method to 2,948 m^2/day using Chow's method at a distance of 44 m from the pumped well. At a distance of 120m, transmissivity ranged between 1,570 m^2/day using Jacob's straight line method and 2,956 m^2/day using Hantush-Jacob solution. Transmissivity was calculated as 2,385 m^2/day using recovery data. Estimated storage coefficient values ranged between 0.0000375 and 0.00268, indicates that the aquifer is confined to leaky-confined or semi-confined in nature. The Hantush-Jacob solution for a leaky confined aquifer was chosen as the most representative of the physical situation and this gives better results considering field condition in the deltaic floodplain aquifers. However, slight deviations are not prohibitive to the application of different methods. When greater deviations from the above assumptions occur, special flow problems may raise. So, it is useful to evaluate data using as many methods as possible. Each may help to provide a different perspective and aid in a better interpretation.

NOTATION

c D'/K': hydraulic resistance of the semi-pervious layer (day)
L $\sqrt{Tc}$: leakage factor (meter)
Q constant discharge rate at the well ($meter^3/day$)
r radial distance from the pumping well (meter)
s drawdown at the well (meter)
S aquifer storage coefficient
S' is the residual drawdown (meter)
s_m maximum drawdown (meter) in a piezometer at distance r (meter) from the pumped well
Δs the drawdown per log cycle of time (meter)
T transmissivity of the aquifer ($meter^2/day$)
t time from the start of pumping (minutes)
t' time from the cessation of pumping (minutes)
t_0 time, where the straight line intersects the zero-drawdown axis (minutes)

ACKNOWLEDGEMENT

The authors are grateful to the respective authorities of Ground Water Hydrology (GWH) of Bangladesh Water Development Board (BWDB) and Bangladesh Arsenic Mitigation Water Supply Project (BAMWSP) for undertaking such an extensive aquifer pumping test component at Sreerampur. Gratitude goes to Mr. Alamgir Hossain and MA Karim, former Directors, GWH, BWDB for supporting authors to participate during the pump test, and to all field personnel of BWDB participated in performing pump test. Thanks are due to the World Bank for providing fund under BAMWSP.

REFERENCES

Bouwer, Herman (1978) Groundwater Hydrology: McGraw-Hill, New York, 480 p.
Brammer H (1996) The Geography of the soils of Bangladesh. University Press Limited.
BWDB-UNDP (1982) Groundwater Survey: the Hydrogeological conditions of Bangladesh. UNDP Technical Report DP/UN/BGD-74-009/1, 113p.
BWDB-USGS (2005) Report on Deep Aquifer Characterization and Mapping Project, Phase-I, Ground Water Hydrology Division-I, BWDB

Chow VT, (1952) On the determination of transmissivity and storage coefficients from pumping test data. American Geophysical Union Transactions, v.33, 397-404

Cooper HH and Jacob CE (1946) A generalized graphical method for evaluating formation constants and summarizing well field history, American Geophysical Union Transactions, v.27,526-534

De Glee GJ (1930) Over grondwaterstromingen bij wateronttrekking door middle van putten. Thesis, J. Waltman, Delft (The Netherlands): 175p

De Glee GJ (1951) Berekeningsmethoden voor de winning van groundwater. In: Drinkwatervoorziening, 3e Vacantie cursus: 38-80. Moorman's periodieke pers. The Hague (The Netherlands)

Domenico PA, Schwartz FW (1997) Physical and Chemical Hydrogeology, 2nd edition, John Wiley and Sons Inc., New York, 506p

DPHE-BGS (2001) Arsenic contamination of groundwater in Bangladesh, British Geological Survey and Department of Public Health Engineering, Govt. of Bangladesh; rapid investigation phase, Final Report

Driscoll FG (1995) Groundwater and Wells, US Filter/Johnson Screens, St. Paul, MN 55112, 1089p

Ferris JG, Knowles DB, Brown RH, Stallman RW (1962) Theory of aquifer tests: U.S. Geological Survey Water Supply Paper 1536-E, P. 69-174

Fetter CW (1994) Applied Hydrogeology, Third Edition: Macmillan, New York, 691 p

Halford KJ and Kuniansky EL (2002) Documentation of spreadsheets for the analysis of aquifer test and slug test data: U.S. Geological Survey, open-file report 02-197, 51 p

Hantush MS (1956) Analysis of data from pumping tests in leaky aquifers: Transactions of the American Geophysical Union, vl.37, 702-714

Hantush MS (1964) Drawdown around wells of variable discharge. *Journal of Geophysical Resources,* vol, 69:4221-4235

Hantush MS (1960) Modification of the theory of leaky aquifer. *Journal of Geophysical Resources,* vol, 65:3713-25

Hantush MS and Jacob CE (1955) Non-steady flow in an infinite leaky aquifer: *Transactions of the American Geophysical Union,* vl.36, 95-100

Hussain SI and Huq NE (1998) Late Quaternary morphostratigraphy and paleogeography of Chandpur-Shariatpur area, south-central Bangladesh: *Bangladesh Journal of Geology,* Vol. 17, P 1-9

Jacob CE (1947) Drawdown test to determine the effective radius of artesian well: Transaction of the American Society of Civil Engineers, Paper 2321, v.112, 1047-1064 p

Jacob CE (1950) Flow of ground-water. In *Engineering Hydraulics.* H. Rouse (ed.), John Wiley, New York, pp 321-386

Kruseman GP and Ridder NAD (1983) Analysis and evaluation of pumping test data: International Institute for Land Reclamation and Improvement/ILRI, The Netherlands, 200p

Lohman SW (1979) Ground-water Hydraulics, U.S. *Geological Survey professional paper* 708, 70pp

Morgan JP and McIntire WG (1959) Quaternary geology of the Bengal Basin, East Pakistan and India, *Geological Society of America Bulletin,* Vol. 70, pp. 319-342

Nury SN, Bashar K, Chowdhury KR (1998) Determination of transmissivity and storage coefficient from step-drawdown pumping test data of Rajshahi and Dhaka using Birsoy and Summers's method: *Bangladesh Journal of Geology,* Vol. 17, P 43-54

Papadopulos IS and Cooper HH (1967) Drawdown in a well of large diameter. Water Resources Research, v. 3, p. 241-244

Ravenscroft P (2003) Overview of the Hydrogeology of Bangladesh. In. Groundwater Resources Development in Bangladesh: Background to the Arsenic Crisis, Agricultural Potential and the Environment. Editors: A Atiq Rahman and Peter Revenscroft. The University Press Limited, Bangladesh. P 43-86.

Stallman RW (1976) Aquifer test design, observation and data analysis, Techniques of water-resources investigations of the U.S. Geological Survey, Book 3, Chapter B1, 26p

Theim G (1906) Hydrologic methods: Leipzig, J.K. Gebhardt, 56p

Theis CV (1935) The relation between the lowering of the piezometric surface and the rate and duration of discharge of a well using ground water storage: Transaction of American Geophysical Union, v. 16, 519-524 p

Theis CV (1940) The source of water derived from wells-essential factors controlling the response of an aquifer to development. Civil Eng., American Society of Civil Engineers, p. 277-280

Vukovic M and Soro (1992) Determination of hydraulic conductivity of porous media from grain-size composition: Water Resources Publications, Littleton, Colorado, 83 p

WARPO (2000) Draft Development Strategy (DDS), Estimation of groundwater resources, Annex-C, Appendix 6, National Water Management Plan, WARPO, Dhaka, Bangladesh, 2000.

Weight WD and Sonderegger (2000) *Manual of Applied Field Hydrogeology,* McGraw-Hill, New York, 608p

Zahid A, Hassan MQ, Balke K-D, Flegr M, Clark DW (2007) Groundwater Chemistry and Occurrence of Arsenic in the Meghna Floodplain Aquifer, Southeastern Bangladesh. *Journal of Environmental Geology,* DOI 10.1007/s00254-007-0907-3

In: Lake Pollution Research Progress
Editors: F. R. Miranda and L. M. Bernard

ISBN: 978-1-60692-106-7

Chapter 12

WEIGHT-OF-EVIDENCE ASSESSMENT OF IMPACTS FROM AN ABANDONED MINE SITE TO THE DASSERAT LAKE WATERSHED, QUEBEC, CANADA

Richard R. Goulet[1] and Yves Couillard[2]
[1] Department of Earth Sciences, University of Ottawa, Marion Hall, Ottawa, Canada, K1N 6N5
[2] Environment Canada, Existing Substances Division,
200 Sacré-Coeur Bd. 7th Fl., Gatineau, Québec, Canada K1A 0H3

ABSTRACT

This paper presents the approach used to determine the impacts of metal contamination from the Aldermac abandoned mine to Dasserat Lake. Rather than relying on a single risk quotient approach, we built a weight of evidence assessment of impacts at Dasserat Lake. As lines of evidence of impacts, we used regional surface water pH, water quality criteria, acute biotic ligand models, predicted no effect concentrations for three groups of species (i.e. algae, invertebrates and fish), transplanted bivalve experiments, indigenous bivalve population survey, invertebrate community surveys, fish condition indicators and finally, fish demographic information. pH levels were acidic in Lake and Bay Arnoux while close to neutrality in Dasserat Lake. Cadmium levels in most parts of Dasserat Lake were above the Canadian water quality guideline for the protection of aquatic life. Correspondingly, transplanted bivalves did not survive in Dasserat Lake 1 km downstream of Arnoux Bay and the natural population of bivalves was confined to the north-west and south east portion of Dasserat Lake. All these lines of evidence indicated that along with Arnoux Lake, who is currently completely acidified, almost 75% of Dasserat Lake was negatively impacted by drainage from the Aldermac abandoned mine site. These results were used to evaluate if the current risks at Dasserat Lake were unreasonable.

Introduction

Determining whether an ecological risk is unreasonable or when contamination becomes pollution (Chapman 2007) is a challenging task. In a society where sustainable development is valued as a driver for the future economy, some degree of risk has to be accepted in order to sustain some economical growth. The International Atomic Energy Association (1999) defines reasonable risk as equating no effects or negligible effects. By negligible effects, IAEA (1999) refers to one affecting the population of a specific group of organisms at a localized area/or over a short period of time (one generation or less), but not affecting other trophic levels or the integrity of any population as a whole. This definition of risk considers five lines of evidence: the hazard itself vs environmental contamination, the level of organization at which the effect is observed (organism vs population), the scale of effects (local vs regional), the timeframe of effects (one generation vs several generations) and the link of the impacted population to other communities in the food chain (structure and function of an ecosystem). A risk assessment should therefore address at least all five lines of evidence in order to determine if the ecological risk is unreasonable or if there are significant adverse effects.

It is common that determination of impacts is based on deterministic risk quotients (e.g. Doyle et al. 2003) or when more data are available, on Bayesian/Monte Carlo probabilistic assessments (HERA 1996). Hull and Swanson (2006) recommend the use of weight of evidence, or the generation of multiple lines of evidence, to support decision-making such that no single line of evidence, alone, drives decision making (Suter 1996). Weight of evidence can be defined as the process by which measurement endpoints are related to an assessment endpoint to evaluate whether a significant risk of harm [an unacceptable risk outcome] is posed to the environment (Menzie et al. 1996; Hull and Swanson 2006). Assessment endpoints are explicit expressions of an actual environmental value to be protected (e.g., maintenance of resident communities); measurement endpoints are the lines of evidence that evaluate the assessment endpoints (e.g., survival, growth, or reproduction in the laboratory, species diversity measurements).

In recent years, there have been numerous improvements in the methods to evaluate the impacts of metals to the aquatic ecosystems (Chapman et al. 2003, Chapman 2008). It is generally accepted that it is the free metal ion that is the best predictor of metal accumulation to aquatic organisms (Campbell et al. 1995). Acute effects of metals can also be predicted by the free metal ion, with due consideration that dissolved organic carbon and other inorganic ligands can bind the free ion, decreasing its bioavailability or that other cations can be taken up instead of the free ion, decreasing, again, its uptake into the organisms (DiToro et al. 2001; Santore et al. 2001). The so-called Biotic Ligand Model (BLM) have been used extensively to predict acute metal effects to invertebrates and fish (Paquin et al. 2002). Similarly, chronic effects of Cu (De Schamphelaere and Janssen 2004) and Zn (De Schamphelaere et al. 2005) have also been predicted using the BLM, although the importance of cationic competition is not as important (Schwartz and Vigneault 2007). These acute and chronic biotic ligand models have been mostly derived from laboratory toxicity tests and their application to field conditions, although very promising, may not fully be applicable.

It may be desirable to couple these models with field evidence of impacts. The best approaches to investigate field impacts are before and after/ control and impact (BACI) designs and the Reference Condition approach (Reynoldson et al. 1997). The Canadian Environmental Effects Monitoring program recommends investigating the size, length and fish, benthic community's indices and fish tissue levels for human consumption, upstream and downstream of mining facilities as lines of evidence of impacts. All of these indicators of impacts can be used as lines of evidence to indicate if ecological risks are likely to occur or have happened in a particular ecosystem. However, this determination of risk in a weight-of-evidence approach becomes much more complicated compared to conventional assessment relying solely on risk quotients.

In this assessment, we used studies conducted on Dasserat Lake, located in the Rouyn-Noranda mining area, Québec, Canada, to assess whether or not the lake was impacted by acid mine drainage loading that has been occurring for over 5 decades now. We also aimed at determining if the impacts occurring justified remediation or a *status quo*. Although Dasserat Lake is subject to a polymetallic contamination including Cd, Cu and Zn, a number of elements indicate that Cd is the main contaminant of concern in this aquatic ecosystem. First, cadmium is roughly one hundred times more acutely toxic to aquatic invertebrates than are Cu or Zn (e.g., Borgmann et al. 2005). Furthermore, a recent study has demonstrated that metals toxic effects observed in the indigenous benthic communities of the Rouyn-Noranda area are probably caused mainly by Cd (Borgmann et al. 2004). In addition, populations of the bivalve *Pyganodon grandis* inhabiting lakes of the region produce metallothionein concentrations (MT) primarily as a defence against Cd intoxication. As nutrients for the bivalve, Cu and Zn are efficiently regulated (Zn more than Cu), and MT levels are not related at all to the internal levels of these metals (Couillard et al. 1993; Perceval et al. 2006). Therefore, the ecotoxicological part of this assessment has focussed exclusively on Cd.

METHODS

Site Description

The Aldermac abandoned mine is situated west of the city of Rouyn-Noranda, approximately 600 km north-west of Montréal, Canada (Figure 1). The mine operated from 1932 to 1943 during which time 1.5 million tonnes of mining residues was spread over 76 hectares. Since then, the abandoned mine tailings released acid mine drainage to the Arnoux River and Arnoux Lake, which resulted in a wipe out of the fish community (Provost 2008). Arnoux Lake discharges into Arnoux Bay which then discharges into Dasserat Lake. Seven sampling stations were established in the littoral zone of the latter lake (named DS-1 to DS-7: Figure 1).

Surface Water Quality

Water samples were obtained twice in 1998 using an *in situ* dialysis technique. A dialysis cell consisted of a 250 mL jar (Nalgene: total volume ~ 300 mL) filled with ultra-pure deionized water (<18 Mohms) and capped with a custom-made closure designed to hold a filter membrane (0.2 μm polysulfone, Pall). The filter membrane allows for the free diffusion

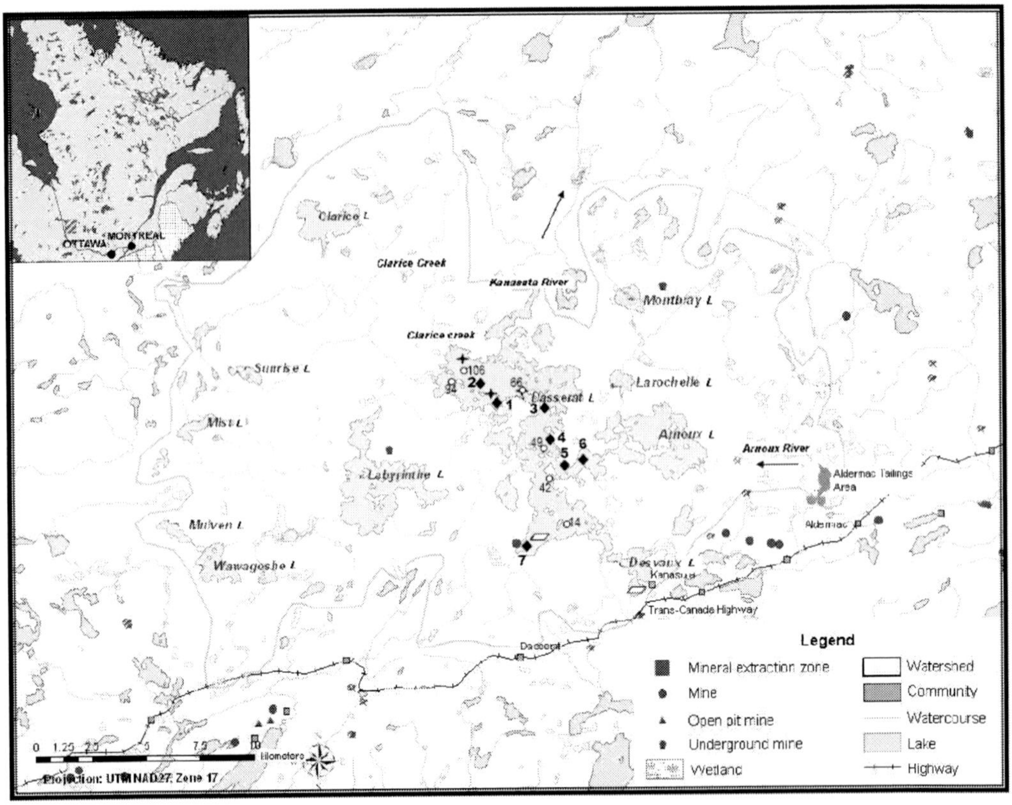

Figure 1. Map of Dasserat Lake and Arnoux Lake watershed situated downstream of the Aldermac Abandoned mine site, near Arntfield, Québec, Canada. Meaning of the symbols in Dasserat Lake: ◆, sampling sites of this study; ✦, sampling sites for the benthos study of Borgmann et al. (2004); ○, fishing sites for the walleye study of Nadeau and Gaudreau (2006); □, fishing areas for the perch study of Kovecses et al.(2005). Numbers on the map refer to sampling stations.

of dissolved species from the water column into the ultra-pure water until equilibrium is reached. The cells were placed by SCUBA-equipped divers in the epilimnion of stations DS-1, DS-4, DS-6, and DS-7 at depths between 2 and 5 meters and fixed to a plastic rod approximately 15 cm above the sediment surface. They were left to equilibrate with the surrounding waters for about 14 days. Tests conducted *in situ* indicated that this deployment time was long enough for reaching steady-state (i.e., 95% SS) between water inside the cell and ambient water.

Once in the laboratory, the sample containers were opened in a clean Class 100 laminar flow hood and two sub-samples were allocated for the determination of major cations (Ca, Mg, Na, K) and total dissolved trace metal concentrations (Al, Fe, Mn, Cd, Cu, Zn). Depending on the metal to be analyzed and the concentration range, a variety of analytical techniques were used for these determinations: inductively-coupled atomic emission spectrometry (ICP-AES, AtomScan 25 spectrophotometer, Thermo Jarrell Ash), flame atomic absorption spectroscopy (FAAS, Spectra AA-20 spectrophotometer, Varian), graphite furnace atomic absorption spectroscopy (GFAAS, SIMAA 6000, Perkin-Elmer). A third sub-sample was used for the determination of major anions (Cl, NO_3, SO_4, PO_4) by ion chromatography (DX-300 Gradient Chromatography Systems; Dionex). The last sub-sample was analyzed for dissolved organic and inorganic carbon using a Total organic carbon analyzer (TOC-5000A, Shimadzu). Quality assurance/quality control (QA/QC) methodology included field blanks, laboratory blanks, and certified reference standards.

Surface Sediment Sampling

In September 1997, divers collected sediment cores with plexiglass tubes (9 cm internal diameter) at each of the seven sampling stations of Dasserat Lake. The cores were brought to shore with minimum perturbation and were extruded on shore to collect only the top 0.5 cm oxic layer. Cd, Cu and Zn concentrations were measured in the samples using procedures for metal extractions and analyses described in Couillard et al. (1993).

Benthic Invertebrate Investigation

Borgmann et al. (2004) collected benthic invertebrates in ten lakes of the Rouyn-Noranda region in August 2000. A minibox corer was used to this effect and was subsampled using five plastic core tubes (6.5 cm diameter, 10 cm height), which were then pooled. Three distinct corer samples were taken at different sites in the lakes, all at similar depths (10 m). For Dasserat Lake, samples were obtained at three locations in the north-western part (Figure 1). Sediment and invertebrates were fixed in 5% formalin for 48–72 h, and then rinsed with water and preserved in 70% ethanol. The fixed invertebrate samples were sieved through a 250 μm mesh screen. All organisms retained were counted and identified to the genus level. Further details on sediment collections and analyses are given in the paper.

During summer of 1998, we conducted an assessment of *P. grandis* presence at the stations 1 to 7 of Dasserat Lake. At each station, SCUBA divers searched carefully ~ 10 000 m^2 of bottom area for 30 minutes.

At the beginning of July 1999, indigenous specimens of the freshwater bivalve *Pyganodon grandis* were transplanted in lakes of the Rouyn-Noranda area for a period of 400 days (Perceval et al. 2006). Five sites were used for this transplant experiment with one site in each of five lakes Opasatica (48◦04'N; 79◦17'W), Joannès (48◦10'N; 78◦41'W), Vaudray (48◦05'N; 78◦40'W), Dasserat (site DS-4: 48◦15'N; 79◦24'W) and Dufault (48◦18'N; 78◦59'W). Based on a preliminary study, these lakes were selected in order to have the widest Cd concentration gradient possible (calculated free Cd^{2+} ion concentrations ranging from ~ 0.05 to 1 nM) from a set of lakes presenting water bodies with comparable trophic status. Opasatica Lake, which acted as the source of experimental animals, is a headwater lake with no point sources of metal pollution and is therefore classified as the reference lake. Lakes Joannès and Vaudray have been polluted by metals via atmospheric transport and are classified here as intermediate-contaminated lakes. Lakes Dasserat and Dufault are subject to point-source and atmospheric metal inputs and are classified as highly contaminated lakes.

The transplant sites were located in the littoral zone of the lakes, on the leeside, at a distance of ~ 40m from the lakeshore. Water depth at all sites was approximately 3 m. All sites were characterized by a gentle substrate slope (< 5%) and homogenous sediments, and by the absence of dense beds of macrophytes. All sites contained a few resident *P. grandis*, except for the sites located in Lakes Dasserat and Dufault.

At the beginning of the experiment, divers placed enclosures at each of the experimental sites, each consisting of plastic borders arranged in circles of 95 cm diameter, inserted 10 cm in the sediments (leaving a 10 cm wall projecting above the sediment surface). On day 400 (September 2000), bivalves were withdrawn from the sites, transported to the laboratory, and gills were manually isolated from each specimen using a scalpel. Gill sub-samples were used for metallothionein quantification, for determining bioaccumulated Cd concentrations, and for subcellular Cd partitioning. Perceval et al. (2006) provide further details on transplant sites, bivalve collection, processing and deployment, and on bivalve retrieval and analysis.

Fish Health

Fish populations of 23 lakes of the Abitibi-Témiscamingue region (western Québec) were sampled in the falls of 1989, 1997 and 1998 (Nadeau and Gaudreau 2006). A standard method for collecting fish was used where gill nets were set at depths ranging from 5 to 15 meters. In Dasserat Lake, precautions were taken over the years so that fish were collected at the same station upstream (station 14), close to the Arnoux Bay outlet (stations 42 and 49) and downstream of Arnoux Bay (stations 66, 94 and 106) (Figure 1). Additional details on experimental fishing and on measurements performed on fish (e.g., determinations of age, sexual maturity, and abundance) can be found in the report.

Perch Stomach Content

Kovecses et al. (2005) used stomach contents and the mean weight of prey in the gut contents to characterize the diet of yellow perch (*Perca flavescens*) populations of six lakes of the Rouyn-Noranda area including Dasserat (Figure 1). Fish and invertebrate sampling was performed in June 2000. To measure the taxonomic richness of the perch population's diet of the study lakes, the authors measured the percent occurrence of prey items (percentage of fish/age category/lake with a particular prey item in their stomach contents). In addition, for each fish the authors determined the total number of individuals of each prey item in the gut contents and the average dry weight of prey/fish (total weight of gut contents divided by the number of diet items) in the stomach contents. Additional information on perch and invertebrate sampling and analyses are provided in Kovecses et al. (2005).

Evaluation of Effects

Cadmium concentrations in surface water were first compared with the Canadian Water Quality Guideline for Cd (CWQG) for the protection of aquatic life (CCME 1999). If Cd levels in water were above the CWQG, predicted no effects concentrations (PNECs) for algae, invertebrates and fish were derived using chronic toxicity data retrieved from the literature. The derivation of PNEC values was done using species sensitivity distributions populated with the chronic data collected and computed using the ETX 2.0 software (http://www.rivm.nl/rvs/overige/risbeoor/Modellen/ETX.jsp).

Acute Biotic Ligand Model (BLM) simulations were performed using the Biotic Ligand model software (http://www.hydroqual.com/wr_blm.html). This acute BLM predicts Cd toxicity to *Daphnia pulex*, *D. magna*, *Cerodaphnia dubia*, Rainbow trout, and Fathead minnow with due consideration of the Cd^{2+} ion. The ions H^{+}, Ca, Mg, K, Na are considered competing for the biotic ligand, and anions and dissolved organic carbon are considered binding the potential toxic cation, rendering it less available for binding with the biotic ligand. All dissolved parameters used in the BLM modeling are presented in Table 1.

Speciation of Cd in the dissolved phase was determined with the help of the Windermere Humic Aqueous Model (WHAM Model 6.0.7: Tipping 2002). Input parameters included total dissolved Cd, major anions and cations, as well as other metals potentially competing for dissolved organic matter binding sites (Mn, Fe and Al). Thermodynamic constants for metal-inorganic ligand interactions were obtained from the NIST Standard Reference database 46 (Smith and Martell 2004).

RESULTS

Water Quality

Cd levels in surface waters are above the CWQG of 0.03 µg/L (adjusted for lake water hardness) at all sampling stations (1 to 7; Figure 1) in Dasserat Lake, which confirms that this lake may be potentially impacted by Cd (Table 1). Additional calculations of predicted no effects concentrations for algae (3 µg/L), invertebrates (0.2 µg/L) and fish (5 µg/L) indicates that chronic cadmium effects to invertebrates are to be expected in the Arnoux Bay only, where dissolved cadmium is at 0.76 µg/L. Furthermore, acute biotic ligand modelling for indicated no acute cadmium effects on *Ceriodaphnia dubia* (67-112 µg/L), *Daphnia magna* (0.9-224 µg/L), Fathead minnow (6.3 – 103 µg/L), and Rainbow trout (2.8-45 µg/L). There is currently no chronic BLM available for Cd.

Sediment Quality

Sediment levels in Arnoux Bay were in the range of 5µg/g dry wt at stations 4, 6 and 7 (Figure 1) and 1.54 µg/g at station 1, situated the farthest downstream of Arnoux Bay. Levels of cadmium in uncontaminated lakes are usually close to 1 µg/g (Couillard et al. 2004). Hence, the sediments downstream and upstream of Arnoux Bay appeared to be contaminated with Cadmium.

Table 1. Water chemistry parameters and trace elements and calculated free ions (in parentheses) concentrations at four littoral sampling stations within Dasserat Lake

station	pH[b]	[Ca] (mg L^{-1})	[K] (mg L^{-1})	[Mg] (mg L^{-1})	$[Cd]_d$ (µg L^{-1})	[Na] (mg L^{-1})	[Cl] (µM)	[SO4] (µM)	[DIC][a] (mg L^{-1})	[DOC][a] (mg L^{-1})	Cd(sediment) (µg/g dry wt)
1	7.76	8.4	0.5	2.1	0.03 (0.03)	1.0	6.0	130	5.0	6.5	1.54
4	7.14	7.3	0.5	2.0	0.09 (0.04)	1.0	6.2	165	2.8	5.6	5.10
6	4.52	9.0	0.5	2.7	0.76 (0.55)	1.2	4.2	434	0.9	3.8	5.38
7	7.41	7.8	0.5	2.1	0.06 (0.04)	1.0	6.7	176	3.0	6.9	5.99
CWQG	6-9				0.03						
PNEC algae					3 (0.2-12.8)						
PNEC invert					0.2 (0.02-0.8)						
PNEC Fish					5 (0.5-24)						

[a] DIC = dissolved inorganic carbon; DOC = dissolved organic carbon.

[b] Mean determined on $[H^+]$, not pH.

Benthic Invertebrate Community

A total number of 13 taxa were reported in the north western part of Dasserat Lake (table 2). This is lower than the 22 total number of taxa collected in Opasatica Lake, which served as control in the work of Borgmann et al (2004). The densities per species were also generally lower in Dasserat Lake compared to Opasatica Lake. For instance, sensitive species such as sphaeriids, Ephemeroptera and Tanytarsini were present in lower numbers in Dasserat Lake compared to Opasatica Lake (Table 2). In contrast, Chironomidae and Chaoboridae were present in higher densities in Dasserat Lake than in Opasatica Lake (Table 2) There were still *Hexagenia* mayfly larva reported, which indicate that young perch can potentially grow well in this lake but the number of large perch may be limited compared with Opasatica Lake (Sherwood et al. 2002, Campbell et al. 2003, Rasmussen et al. 2008).

Table 2. Mean (S.D., n=3) numbers of benthic invertebrates per benthic cores (33.18 cm^2), total number of taxa observed

Lake	Sphaeriidae	Ephemeroptera	Tanytarsini	Chironomidae	Chaoborus	Total number of taxa
Opasatica	4.1(2.3)	0.5(0.9)	3.3(4.1)	3.1(1.7)	0.2(0.3)	22
Dasserat	0.7(0.4)	0.2(0.4)	1.73(0.6)	11.2(12.8)	14.5(3.0)	13

Freshwater Bivalves

The survival of transplanted bivalves in Dasserat Lake was severely affected at station 4 (Figure 1). In fact, only 43% of the bivalves survived at this site compared to good survival at lakes Joannès, Vaudray and Opasatica (Table 3). Bivalves had very poor survival at Lake Dufault, which was also historically contaminated with Cd (Couillard et al. 2004). There was no indigenous bivalve at station 4. In fact, indigenous bivalves were only found at the northwest part and south-east part of the lake. The natural bivalve population hence seems to be segmented by the Cd loading coming from the Arnoux Bay.

Table 3. Dissolved Cd, Survival of transplanted bivalves and natural population densities in five lakes of the Rouyn-Noranda region

Lake	Opasatica	Joannes	Vaudray	Dasserat	Dufault
Dissolved Cd (µg/l)	0.006	0.039	0.076	0.12	0.38
Mortality after 400 days	13.8	15.7	6.2	57	63
Indogeneous density (ind/m^2)	1.12	0.77	0.18	0	0

* modified from Perceval et al. (2006).

Fish Health

The stomach content of perch caught upstream of Bay Arnoux were generally bigger than perch collected in Opasatica Lake. In addition, the mean weight of their gut contents was also higher in Dasserat Lake than in Opasatica Lake (Kovecses et al 2001). These results suggest that perch were in good condition in Dasserat Lake.

The total number of fish collected went from 200 in 1989 to about 250 in 1997-98. There was no notable difference in the number of fish species caught upstream, as measured by catch per unit effort, near or downstream of the Arnoux Bay. Nadeau and Gaudreau (1996) indicated that the population of *S. vitreum* in Dasserat Lake was similar to other lakes in the region (Table 4). However, the number of *S. canadense* was much lower than other lakes in the region (Table 4).

Table 4. Catch per unit effort (CPUE) for walleye species in Dasserat Lake compared to other lakes in the region

Lake	Stizostedion canadense	Stizostedion vitreum
Achepabanca	--	6,7
Cuvillier	--	10,7
Dasserat	0,8	15,1
Duparquet	7,7	13,3
Faillon	1,3	6,7
Garde (La)	--	12,8
Kipawa	--	15,0
Malartic	7,2	12,1
Opasatica	4,0	16,1
Pommeroy	--	23,5
Preissac	7,6	21,3
Quinze (Des)	5,5	8,3
Temiscamingue	11,7	14,9
Valets	--	5,2
Wetetnagami	--	7,2

DISCUSSION

The objective of this paper was to collect several lines of evidence to indicate if environmental impacts were of concerns in Dasserat Lake.

First, Cd levels in sediment were high at most stations suggesting long term loading of Cd to Dasserat Lake. Surface water levels were above the Cd CWQG for the protection of aquatic life at almost all sampling stations in Dasserat Lake. Further investigation of Cd impacts using derived predicted no effects concentrations or the acute BLM indicated that this contamination level should only be of concern in the Arnoux Bay where the pH was 4.0 and levels of Cd were high. However, transplanted bivalves did not survive downstream of Arnoux Bay (at station 4), which suggest that Cd loading from Arnoux Bay are deleterious to part of the Dasserat Lake ecosystem. This limited survival was confirmed by the presence of native bivalve populations only upstream of Arnoux Bay (e.g., Lake Desvaux and Berthemet) and far downstream in the North-western part of Dasserat Lake, which indicates that the contamination spread extensively across this lake. Bivalves are sessile organisms that are constantly exposed to drainage from Arnoux Bay and unlike other species such as fish, cannot avoid exposure from point sources. Hence, they are excellent bioindicators of impacts in

Dasserat Lake. Bivalves are also a good source of food for aquatic mammals such as Mink, Otter and Muskrats, which provide evidence that small mammals could also be affected in that part of the lake. Overall, there is a habitat fragmentation of the bivalve population in Dasserat Lake due to the Cd loading from Arnoux Bay. Similarly, sensitive species such as Hephemeroptera, Sphaeriids and Tanytarsini were found in lower numbers in Dasserat Lake compared to Opasatica Lake, which served as control in the Borgmann et al. (2004) study. Interestingly, species known to be metal tolerant such as Chironomidae and Chaoboridae (Warren et al. 1998) were found in higher numbers in Dasserat Lake compared to Opasatica Lake. The total number of benthic invertebrates was also lower in Dasserat Lake than in Opasatica Lake. This suggests that the quantity of food to fish species may be limited if they were constrained to forage for food in this area of the lake.

Interestingly enough, the impact on the fish community is much less supported by our different lines of evidence. Total fish capture since 1989 have increased (Gaudreau 2007) suggesting that fish populations appear to resist metal contamination coming from the Aldermac site. Populations of *S .canadense* are significantly lower than any other lakes in the region though. However, it is difficult to assign this difference to metal loading from Arnoux Bay since the difference in sensitivity to metals amongst the two walleye species have never been investigated to our knowledge. Further research on this topic would be of interest in explaining the low number of black walleye in Dasserat Lake.

Despite these two anomalies in the fish population, the number of walleye caught in Dasserat Lake are comparable to other lakes in the region (Nadeau and Gaudreau 2006; Table 4) and the size of perch and their gut content is comparable to control lakes (Kovecses et al. 2001). However, perch caught by Kovecses et al. (2001) were all upstream of Arnoux Bay (Figure 1). The limited evidence of impacts to the fish community is somewhat not very surprising. First, our predicted no effects concentration is much higher for fish than algae or invertebrates, which suggests fish are less sensitive to cadmium. Also, the fish in Dasserat Lake can avoid contamination by spending most of their time upstream of Arnoux Bay or far downstream of the Bay. Correspondingly, the northwest and southeast parts of the lakes are the ones mostly visited from local fisherman (Goulet R., personal observation). The home range size of fish provided by Minns (1995) supports the evidence that the fish in Dasserat Lake can avoid the point source contamination. Goulet et al. (2008) also noted that Hg levels in fish were not a good indicator of Hg releases from point sources because they can simply move around, unlike sessile invertebrate species such as freshwater bivalves or benthic insect larva. A number of studies indicate that yellow perch is able to survive and reproduce successfully in areas where water and sediment metal concentrations are toxic to many invertebrates (Rasmussen et al. 2008).

Additional information missing from this weight of evidence analysis is the trend in metal loading to Dasserat Lake. An increasing or stable trend in metal loading to an already impacted ecosystem warrants remediation to occur while decreasing trends, could justify the adoption of natural attenuation as a remediation option provided that natural recovery could occur within a timeframe that would not irreversibly threat the structure and function of an ecosystem. Such trend could be obtained by seasonal monitoring of metal loading from Arnoux Bay into Dasserat Lake over several years. However, such an approach may be quite labor intensive and costly. Rather, a sediment core profile should be taken at the deepest part of North Dasserat Lake. Measurement of metal profiles at a very fine scale would indicate if the sediments are recovering towards background levels. The core should then be dated by

^{210}Pb. The capability of the Pb-210 dating method is well known for the establishment of precise chronologies in sediment columns collected in places where no abrupt changes in the dynamic conditions of sedimentation are expected (Appleby and Olfield, 1988).

Whether a risk is unreasonable or not depends on the hazard itself, if it is affecting at the organismal or population levels, are these effects only localised or spread on a regional scale, do they last for more than one generation and does it affect other population in the ecosystem or in the same food chain. We believe the different lines of evidence indicate that there is unreasonable risk to Dasserat Lake. First, the size of water body affected (~12km^2) compared to the unimpacted areas at the northwestern and southeastern portion of the lake (~15km^2) along with the consideration that upstream Arnoux Bay (~4km^2), Arnoux Lake (~16km^2) and Arnoux River (~2km^2) are already severely impacted by the Abandoned mine site (Provost 2008) are strong indications of impacts. This is particularly convincing when considering the size of the spread of abandoned tailings (0.75km^2) compared to the total impacted area (~30km^2). Second, the fact that metal loading from the Aldermac site has occurred for more than 5 decades is also strong evidence to conclude that the impacts to Dasserat Lake are unreasonable. Moreover, the impacts on the native benthic invertebrates, coupled with the impact on the bivalve population provided by both the onsite survey and the bivalve transplantation experiments suggest that the benthic community is affected and that indirect impacts are also to be expected to other organisms in the food chain such as aquatic mammals. Direct and indirect effects of metal exposure to fish (e.g., Rasmussen et al. 2008, Sherwood et al. 2002) is currently assumed to be minimal but if Cd loading increases in the next couple of years or if the assimilation capacity of the lake is depleted, impacts on fish may eventually occur. A good example of the likely impacts to the fish community in Dasserat Lake is the state of Arnoux Lake where fish were wiped out because it was directly exposed to the acid mine drainage from the Aldermac site (Provost 2008). Hence, it becomes apparent that the ecological risks to the Dasserat Lake ecosystem are unreasonable and that remediation of the Aldermac site is necessary.

CONCLUSION

As reported by Provost (2008), it was generally believed that metal loading to Dasserat Lake appeared to have minimal impacts to Dasserat Lake. This study provides more accurate update of impacts in Dasserat Lake and concludes that there is unreasonable risk to this lake. In addition, a local artist, Véronique Doucet, painted trees of the Aldermac site and showed its paintings to the local community. Some of them were used as postal cards and sent to the provincial government by thousands of local residents (Provost 2008). The evidence from this risk assessment combined with the political pressure created by Véronique Doucet and others provided the provincial government with adequate information to support decision-making regarding the remediation of the Aldermac abandoned mine site. As a result, the Québec government announced the remediation of the Aldermac site starting in the fall of 2008 (Québec 2008). Hence, efficient risk communication is also essential for weight-of-evidence risk assessment to lead to the improvement of the environment.

Acknowledgments

The authors would like to thank Patrick Doyle (Environment Canada), Peter Campbell (INRS-ETE, Université du Québec), Beverley Hale (University of Guelph) for continuous support during R.G. internship within the Metals-in-the-Environment Research Network. The authors would also like to thank the Conseil Régional de l'Environnement en Abitibi-Témiscamingue (CREAT) and his director general Maribelle Provost for their constant support with our risk communications. Also, a special thank is due to Véronique Doucet for providing a unique way to communicate environmental risk through art.

References

Appleby, P.G., and Olfield, F., (1988) The assessment of Pb-210 data from sites with varying sediment accumulation rates. *Hydrobiol.* 103, 29-35.

Borgmann, U.; Couillard, Y.; Doyle, P;. Dixon, G. (2005) Toxicity of sixty-three metals and metalloids to *Hyalella azteca* at two levels of water hardness. *Environ. Toxicol. Chem.*, 24, 641–652.

Borgmann, U.; Nowierski, M.; Grapentine, L.C;. Dixon, D.G. (2004) Assessing the cause of impacts on benthic organisms near Rouyn-Noranda, Quebec. *Environ. Pollut.*,129(1), 39-48.

[CCME] Canadian Council of Ministers of the Environment (1999) Canadian water quality guidelines for the protection of aquatic life: Cadmium. In: Canadian environmental quality guidelines, 1999, Canadian Council of Ministers of the Environment, Winnipeg (MA).

Campbell, P.G.C. (1995) Interactions between trace metals and aquatic organisms: a critique of the free-ion activity model: Metal Speciation and Bioavailability in Aquatic Systems Tessier, A., Turner, D.R. (Edition). JohnWiley and Sons Ltd., pp. 45–102.

Campbell, P.G.C.; Hontela, A.; Rasmussen, J.B.; Giguere, A.; Gravel, A.; Kraemer, L.; Kovesces, J.; Lacroix, A.; Levesque, H.; Sherwood, G. (2003) Differentiating between direct (physiological) and food-chain mediated (bioenergetic) effects on fish in metal-impacted lakes. *Hum. Ecol. Risk. Assess.*, 9(4), 847-866.

Chapman, P.M.; Wang, F.Y.; Janssen, C.R.; Goulet, R.R.; Kamunde, C.N. (2003) Conducting ecological risk assessments of inorganic metals and metalloids: Current status. *Hum. Ecol. Risk. Assess.* 9(4), 641-697.

Chapman, P.M. (2007) Determining when contamination is pollution - Weight of evidence determinations for sediments and effluents. Environ. Int., 33(4), 492-501.

Chapman, P.M. (2008) Environmental risks of inorganic metals and metalloids: A continuing, evolving scientific odyssey. *Hum. Ecol. Risk. Assess.*, 14(1),5-40.

Couillard, Y.; Courcelles, M.; Cattaneo, A.; Wunsam, S. (2004) A test of the integrity of metal records in sediment cores based on the documented history of metal contamination in Lac Dufault (Québec, Canada). *J. Paleolimnol.*, 32, 149-162.

Couillard, Y.; Campbell, P.G.C.; Tessier, A. (1993) Response of metallothionein concentrations in a freshwater bivalve (*Anodonta grandis*) along an environmental cadmium gradient. *Limnol. Oceanogr.*, 38, 299-313.

De Schamphelaere, K.A.C; Janssen, C.R. (2004) Development and field validation of a biotic ligand model predicting chronic copper toxicity to *Daphnia magna*. *Env.Toxicol.Chem.*, 23(6), 1365-1375.

De Schamphelaere, K.A.C.; Lofts, S.; Janssen, C.R. (2005) Bioavailability models for predicting acute and chronic toxicity of zinc to algae, daphnids, and fish in natural surface waters. *Env.Toxicol.Chem.*, 24(5), 1190-1197.

Di Toro, D.M.; Allen, H.E. ; Bergman, H.L. ; Meyer, J.S. ; Paquin, P.R.; Santore, R.C. (2001) Biotic ligand model of the acute toxicity of metals. 1. Technical basis. *Env. Toxicol. Chem.*, 20(10), 2383-2396.

Doyle, P.J.; Gutzman, .D.W.; Sheppard, M.I.; Sheppard, S.C.; Bird, G.A.; Hrebenyk, D. (2003) An ecological risk assessment of air emissions of trace metals from copper and zinc production facilities. *Hum. Ecol. Risk. Assess*. 9(2), 607-636.

Gaudreau, A. (2007) Personal communication. Ministère des Richesses Naturelles et de la Faune, Direction de l'aménagement de la faune, 180 boul. Rideau, 1st Fl., Rouyn-Noranda, QC, J9X 1N9.

Goulet, R.R.; Lalonde, J.D.; Chapleau, F.; Findlay, S.C.; Lean, D.R.S. (2008) Temporal trends and spatial variability of mercury in four fish species in the Ontario segment of the St. Lawrence River, Canada. *Arch.Environ.Contam.Tocixol.*, 54(4), 716-729.

Hare, L.; Tessier, A.; Croteau, M.N. (2008) A biomonitor for tracking changes in the availability of lakewater cadmium over space and time. *Hum.Ecol.Risk.Assess.*, 14(2), 229-242.

[HERA] Human and Ecological Risk Assessment. (1996) Special issue: commemoration of the 50th anniversary of Monte Carlo. *Hum. Ecol. Risk Assess.*, 2. 627–1034.

Hull, R.N. and Swanson S. (2006) Sequential analysis of lines of evidence – an advanced weight-of-evidence approach for ecological risk assessment. *Int. Environ. Assessm. Manag.*, 2(4): 302-311.

International Atomic Energy Association (1999) Guidelines for Integrated Risk Assessment and Management in Large Industrial Areas, IAEA TECDOC Series No. 994, Vienna , Austria.

Kovecses, J.; Sherwood, G.D.; Rasmussen, J.B. (2005) Impacts of altered benthic invertebrate communities on the feeding ecology of yellow perch (*Perca flavescens*) in metal-contaminanted lakes. *Can. J. Fish. Aquat, Sci.*, 62, 153-162.

Menzie, C.; Henning, M.H.; Cura, J.; Finkelstein, K.; Gentile, J.; Maughan, J.; Mitchell, D.; Petron, S.; Potocki, B.; Svirsky, S.; Tyler, P. (1996) A weight-of-evidence approach for evaluating ecological risks: report of the Massachusetts Weight-of- Evidence Workgroup. *Hum. Ecol. Risk Assess*. 2, 277–304.

Minns, C.K. (1995) Allometry of home-range size in lake and river fishes. *Can.J.Aquat.Sci.*, 52(7), 1499-1508.

Nadeau, D ; Gaudreau, A. (2006) Bilan de sept années « 1997-2003 » de suivi des populations de doré en Abitibi-Témiscamingue. Ministère des Ressources naturelles et de la Faune, Secteur Faune Québec, Direction de l'aménagement de la faune, Rouyn-Noranda. Québec. 68 p.

Pagenkopf, G.K. (1983) Gill surface interaction model for trace metal toxicity to fishes: role of complexation, pH and water hardness. *Environ. Sci. Technol*, 17, 342–347.

Paquin, P.R.; Gorsuch, J.W.; Apte, S.; Batley, G.E.; Bowles, K.C.; Campbell, P.G.C.; Delos, C.G.; Di Toro, D.M.; Dwyer, R.L.; Galvez, F.; Gensemer, R.W.; Goss, G.G.; Hogstrand, C.; Janssen, C.R.; McGeer, J.C.; Naddy, R.B.; Playle, R.C.; Santore, R.C.; Schneider, U.; Stubblefield, W.A.; Wood, C.M.; Wu, K.B. (2002) The biotic ligand model: a historical overview. *Comp. Biochem. Physiol. C*, 133(1-2), 3-35.

Perceval, O.; Couillard, Y.; Pinel-Alloul, B.; Campbell, P.G.C. (2006) Linking changes in subcellular cadmium distribution to growth and mortality rates in transplanted freshwater bivalves (*Pyganodon grandis*). *Aquat. Toxicol.*, 79(1), 87 -98.

Provost, M. (2008) La restauration du parc à résidus miniers abandonné Aldermac. Franc Vert 5(1) : 9 p. Available from : http://www.francvert.org/pages/51article larestaur ationdu parcminier.asp

Québec, Government. 2008. Le gouvernement du Québec annonce le début des travaux de restauration du site minier Aldermac. Press release of 8 September 2008. [Cited Sept 2008]. Available from :http://communiques.gouv.qc.ca/gouvqc/communiques/GPQF/Septembre2008/08/c2570.html

Rasmussen, J.B. ; Gunn, J.M. ; Sherwood, G.D. ; Iles, A.; Gagnon, A. ; Campbell, P.G.C.; Hontela, A. (2008) Direct and indirect (foodweb mediated) effects of metal exposure on the growth of yellow perch (*Perca flavescens*): implications for ecological risk assessment. *Human Ecol. Risk Assess.*, 14. 317-350.

Reynoldson, T.B.; Norris, R.H.; Resh, V.H.; Day, K.E.; Rosenberg, D.M. (1997) The Reference Condition: a comparison of multimetric and multivariate approaches to assess water-quality impairment using benthic macroinvertebrates. *J.North Amer. Benthol. Soc.*, 16(4), 833-852.

Santore, R.C.; Di Toro, D.M.; Paquin, P.R.; Allen, H.E.; Meyer, J.S. (2001) Biotic ligand model of the acute toxicity of metals. 2. Application to acute copper toxicity in freshwater fish and Daphnia. *Environ.Toxicol.Chem.*, 20(10), 2397-2402.

Sherwood GD, Pazzia I, Moeser A, Hontela A, Rasmussen JB (2002) Shifting gears: enzymatic evidence for the energetic advantage of switching diet in wild-living fish. Can. J. Fish. Aquat. Sci., 59(2): 229-241

Smith, R.M. and Martell, A.E. (2004) Critical constants for metal complexes. Standard Reference database 46. Gaithersburg (MD): US Department of Commerce, National Institute of Standards and Technology.

Suter, G.W., II. 1996. Overview of the ecological risk assessment framework. *In* Ecological risk assessment of contaminated sediments. C.G. Ingersoll, T. Dillon, and G.R. Biddinger (Edition). SETAC Press, Pensacola, Fla. pp. 1–6.

Schwartz, M.L.; Vigneault, B. (2007) Development and validation of a chronic copper biotic ligand model for *Ceriodaphnia dubia*. *Aquat.Toxicol.*, 84(2), 247-254.

Tipping E. (2002) Cation binding by humic substances. Cambridge environmental chemistry series. 434 p.

Warren, L.A., Tessier, A., and Hare, L., (1998) Modelling cadmium accumulation by benthic invertebrates in situ: The relative contributions of sediment and overlying water reservoirs to organisms cadmium concentrations. *Limnol. Oceanogr.* 43(7): 1442-1454.

INDEX

A

B

C

D

E

F

G

H

I

J

K

L

M

N

O

P

Q

R

S

T

U

V

W

Y

Z